Genetic Conservation
of Salmonid Fishes

NATO ASI Series

Advanced Science Institutes Series

A series presenting the results of activities sponsored by the NATO Science Committee, which aims at the dissemination of advanced scientific and technological knowledge, with a view to strengthening links between scientific communities.

The series is published by an international board of publishers in conjunction with the NATO Scientific Affairs Division

A	**Life Sciences**	Plenum Publishing Corporation
B	**Physics**	New York and London
C	**Mathematical and Physical Sciences**	Kluwer Academic Publishers
D	**Behavioral and Social Sciences**	Dordrecht, Boston, and London
E	**Applied Sciences**	
F	**Computer and Systems Sciences**	Springer-Verlag
G	**Ecological Sciences**	Berlin, Heidelberg, New York, London,
H	**Cell Biology**	Paris, Tokyo, Hong Kong, and Barcelona
I	**Global Environmental Change**	

Recent Volumes in this Series

Volume 244—Forest Development in Cold Climates
edited by John Alden, J. Louise Mastrantonio, and Søren Ødum

Volume 245—Biology of *Salmonella*
edited by Felipe Cabello, Carlos Hormaeche, Pasquale Mastroeni, and Letterio Bonina

Volume 246—New Developments in Lipid–Protein Interactions and Receptor Function
edited by K. W. A. Wirtz, L. Packer, J. Å. Gustafsson, A. E. Evangelopoulos, and J. P. Changeux

Volume 247—Bone Circulation and Vascularization in Normal and Pathological Conditions
edited by A. Schoutens, J. Arlet, J. W. M. Gardeniers, and S. P. F. Hughes

Volume 248—Genetic Conservation of Salmonid Fishes
edited by Joseph G. Cloud and Gary H. Thorgaard

Volume 249—Progress in Electrodermal Research
edited by Jean-Claude Roy, Wolfram Boucsein, Don C. Fowles, and John H. Gruzelier

Series A: Life Sciences

Genetic Conservation of Salmonid Fishes

Edited by

Joseph G. Cloud
University of Idaho
Moscow, Idaho

and

Gary H. Thorgaard
Washington State University
Pullman, Washington

Plenum Press
New York and London
Published in cooperation with NATO Scientific Affairs Division

Proceedings of a NATO Advanced Study Institute on
Genetic Conservation of Salmonid Fishes,
held June 23–July 5, 1991,
in Moscow, Idaho, and Pullman, Washington

NATO-PCO-DATA BASE

The electronic index to the NATO ASI Series provides full bibliographical references (with keywords and/or abstracts) to more than 30,000 contributions from international scientists published in all sections of the NATO ASI Series. Access to the NATO-PCO-DATA BASE is possible in two ways:

—via online FILE 128 (NATO-PCO-DATA BASE) hosted by ESRIN, Via Galileo Galilei, I-00044 Frascati, Italy

—via CD-ROM "NATO-PCO-DATA BASE" with user-friendly retrieval software in English, French, and German (©WTV GmbH and DATAWARE Technologies, Inc. 1989)

The CD-ROM can be ordered through any member of the Board of Publishers or through NATO-PCO, Overijse, Belgium.

Library of Congress Cataloging in Publication Data

Genetic conservation of salmonid fishes / edited by Joseph G. Cloud and Gary H. Thorgaard.
p. cm. -- (NATO ASI series. Series A, Life sciences ; v. 248)
"Proceedings of a NATO Advanced Study Institute on Genetic Conservation of Salmonid Fishes, held June 23-July 5, 1991, in Moscow, Idaho, and Pullman, Washington"--T.p. verso.
"Published in cooperation with NATO Scientific Affairs Division."
Includes bibliographical references and index.
ISBN 0-306-44532-8
1. Salmonidae--Germplasm resources--Congresses. 2. Salmonidae--Genetics--Congresses. 3. Fish populations--Congresses. I. Cloud, Joseph G. II. Thorgaard, Gary H. III. North Atlantic Treaty Organization. Scientific Affairs Division. IV. NATO Advanced Study Institute on Genetic Conservation of Salmonid Fishes (1991 : Moscow, Idaho, and Pullman, Wash.) V. Series.
QL638.S2G46 1993
639.3'755--dc20 93-23149

ISBN 0-306-44532-8

A Division of Plenum Publishing Corporation
233 Spring Street, New York, N.Y. 10013

Printed in the United States of America

Preface

As the human population increases and nations become more industrialized, the habitat and water quality required for the survival of fish continues to decline. In addition to these environmental factors, fish populations are directly or potentially affected by harvesting, enhancement programs and introgression with hatchery-propagated or transgenic fish. To our knowledge no other scientific meeting has been assembled to consider the breadth of the problem, to review the technology that is presently available for the preservation of the germ plasm of salmonid stocks and to identify the scientific advances that are required to overcome the problems.

Because many salmonids have spawning grounds within the confines of a specific region or county but will spend a large portion of their life cycle within the territorial waters of other countries or in the open ocean, the preservation of unique genes or gene pools in these animals requires international cooperation. This scientific meeting has provided a forum in which to discuss the problems, evaluate the present methods or technology for addressing the problems and suggest new directions or innovations that need to be implemented.

During this meeting we limited our discussion to salmonid fishes. However, the general conclusions about the factors that affect the population dynamics of fish stocks and the technical aspects concerning the preservation of germ plasm will be applicable to other fish species.

We are nearing the time in which the decline of specific fish stocks is irreversible. As members of the scientific community, we need to define which fish stocks to preserve and collectively decide on the most plausible way to reach that goal. Many of the topics in these proceedings have been presented by themselves at other meetings. However, because they are timely and are of great interest, this conference has attempted to provide a more complete appraisal of the problem and a presentation of the science and research methodology to correct or alleviate the problem.

The preservation of genetic materials is an international concern and has many possible ramifications for the future. As salmonids are increasingly used as food, maintaining the genes from many salmonid stocks is an important consideration for selective breeding programs as aquaculture expands as well as the intrinsic need to preserve the genetic legacy of these populations.

J.G. Cloud

October 1992

Contents

Methods to Describe Fish Stocks

RENÉ GUYOMARD

1. Introduction

The description of the genetic diversity of species is a prerequisite step for both basic (understanding of mechanisms involved in the adaptation and evolution of species) and applied (management of the genetic resources of species) purposes. This description includes two major phases. One is a proper identification of the species under consideration. From a theoretical point of view, the species is clearly defined as a community of individuals which can interbreed in the wild and produce fertile progenies. However, this biological concept of species is difficult or impossible to apply and, in practice, many species are still recognized on a morphological basis. In the case of the fishes which exhibit a large phenotypic plasticity (Allendorf et al., 1987), this typological approach does not lead to adequate identification of species in many situations. The second phase is the description of the intraspecific structure of the genetic variation and the identification of elementary breeding units. These units are defined as communities of individuals of opposite sex which have, *a priori*, the same probability to interbreed and to produce a fertile progeny. This mating system is called panmixy and each panmictic breeding unit a population. It must be emphasized that the population is the major evolutionary unit of the species and, therefore, should be the major management unit.

In this paper, we review different methods which can lead to a better identification of species and knowledge of the breeding structure of the species. We shall successively consider the methods to describe genetic variation, the methods of analysis of data and some sampling problems related to the description and interpretation of genetic variation.

2. Methods to Describe Genetic Variation

In general, the genetic diversity of a species cannot be estimated from phenotypic data directly collected in the wild because of the possible occurrence of environmental effects which preclude accurate interpretation of the observed variation. It is necessary to use methods which are based on experimental designs or laboratory methods. It is customary to distinguish quantitative, chromosomal and molecular methods.

2.1. Analysis of Quantitative Characters

Quantitative characters are characters which, in most cases, show continuous individual variations (e.g., size) or can display a large range of discrete values (e.g., number of piloric

Laboratoire de génétique des poissons, INRA-CRJ, 78352, Jouy-en-Josas, France

Genetic Conservation of Salmonid Fishes, Edited by J.G. Cloud
and G.H. Thorgaard, Plenum Press, New York, 1993

coeca). Their genetic control is unknown and is assumed polygenic. These characters are sensitive to environmental variations and should be analyzed in individuals reared in identical conditions. The analysis of quantitative characters in fish requires heavy experimental designs which are usually carried out for the study of important zootechnical traits.

2.1.1. Variation within Population

The analysis of quantitative characters is based on the following model: PHENOTYPE (P) = GENOTYPE (G)+ ENVIRONMENTAL EFFECT (E). G can be in turn partitioned into additive, dominance and epistatic components (see Falconer, 1989). The objective of the analysis of quantitative characters is to estimate the proportions of the phenotypic variability which can be explained by each of these different causes and, primarily, by additive variation. The ratio of the additive to the phenotypic variation is named heritability of the character (h^2) and is an important parameter for evaluating the response of a character to selection. These different components can be estimated from the covariance of phenotypic values between relatives (usually parents and offspring, halfsib or fullsib families). Table 1 (Falconer, 1989) gives the expression of these covariances in terms of different variance components assuming that the epistatic effects are restricted to two factor interactions.

Table 1. Contribution of the additive genetic variance (Va), dominance variance (Vd) and epistatic variance restricted to two factor interactions (Vaa, additive x additive variance, Vad and Vdd) to different covariances between relatives (from Falconer, 1989).

COVARIANCE	Va	Vd	Vaa	Vad	Vdd
Parent-offspring	1/2	0	1/4	0	0
Full-sib families	1/2	1/4	1/4	1/8	1/16
Half-sib families	1/4	0	1/16	0	0

None of these covariances gives a pure estimate of the additive genetic variance. In addition, fullsib and offspring-mother correlations will be contaminated by any maternal effect (genetic and non-genetic). In the case of characters which are fixed early in development, the influence of the maternal effects could be very strong and substantially inflate the heritability values. This could explain why the heritabilities obtained from fullsib or offspring-parents covariances are frequently high for meristic characters (Table 2). An experimental design which enables correct estimation of the additive and dominance components as well as the maternal effects should involve both fullsib and halfsib matings of parent pairs (Fig. 1).

Finally, it should be realized that the estimates of variance components are dependent not only on the characters, but also on the population and the experimental conditions during the experiment. For example, the heritability for growth in rainbow trout was found to be 0.09 (± 0.16) in families reared separately and 0.46 (± 0.18) when the same families were reared in competition (Chevassus, unpublished data).

2.1.2. Variation between Populations

Differences between populations are properly estimated from diallele crosses. A two-strain diallele cross provides estimates of the additive genetic differences between the two strains, the maternal effects and the heterosis. In theory, the comparison of the two strains should be achieved by the analysis of F2 factorial crosses which can also reveal genetic differences between these strains if chromosome mispairing or disruption of coadapted gene complexes occurs in F1 meiosis.

Table 2. Heritability value for meristic characters in salmonids. Heritabilities of the eight first characters have been estimated from the correlations between the mid-parents and their offspring values and could include strong non-genetic maternal effects. Such maternal effects have been demonstrated for the number of pyloric coeca in brown trout (Blanc et al., 1979) and could explain the differences between brown trout and rainbow trout for the heritability of the number of vertebrae.

Trait	Heritability (±se)	Correlation
Anal rays (RT)[1]	0.94 ± 0.50	Midparent offspring
Dorsal rays (RT)[1]	0.90 ± 0.27	"
Pelvic rays (RT)[1]	0.84 ± 0.11	"
Pectoral rays (RT)[1]	0.52 ± 0.09	"
Lower gillrakers (RT)[1]	0.37 ± 0.21	"
Upper gillrakers (RT)[1]	0.67 ± 0.11	"
Mandibular pores (RT)[1]	0.18 ± 0.22	"
Vertebrae (RT)[1]	0.84 ± 0.23	"
Pyloric coeca (BT)[2]	0.36 ± 0.15	Diallel crosses
Vertebrae (BT)[3]	0.34 ± 0.13	Half-sib families
Black punctuation (BT)[4]	0.40 ± 0.26	"

RT: rainbrow trout.
BT: brown trout.
1. Leary et al. (1985).
2. Blanc et al. (1979).
3. Blanc et al. (1982).
4. Krieg (1984).

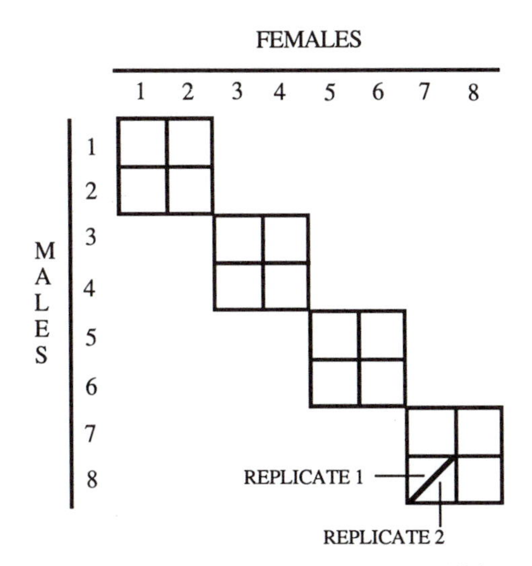

Figure 1. Proper estimates of paternal and maternal additive variance, dominance variance and maternal effects can be obtained from a series of diallele crosses involving halfsib and fullsib families.

2.2. Chromosomal Variation

Karyological methods can be used to detect two types of chromosomal variation: chromosome number and banding polymorphism. Intra and interpopulation variation in chromosome numbers have been observed in several species of salmonids. These variations have been interpreted as the result of centric fusions, but their genetic determinism has not yet been proven. Figure 2 shows this kind of variation in rainbow trout. The most extensive study in salmonids has been done in this species and has shown some degree of substructuring (Thorgaard, 1983). Differences in chromosome number have also been found between Canadian and European Atlantic salmon (2n = 56 and 58 respectively; Hartley and Horne, 1984). One interesting question related to this chromosomal polymorphism concerns the behaviour of chromosomes at meiosis when different centric fusions are involved in the same population and their effects on the individual viability. Banding methods do not work efficiently in fish and have not given valuable information on the genetic structure of salmonids.

In summary, it is generally admitted that karyological methods are not powerful tools for this purpose.

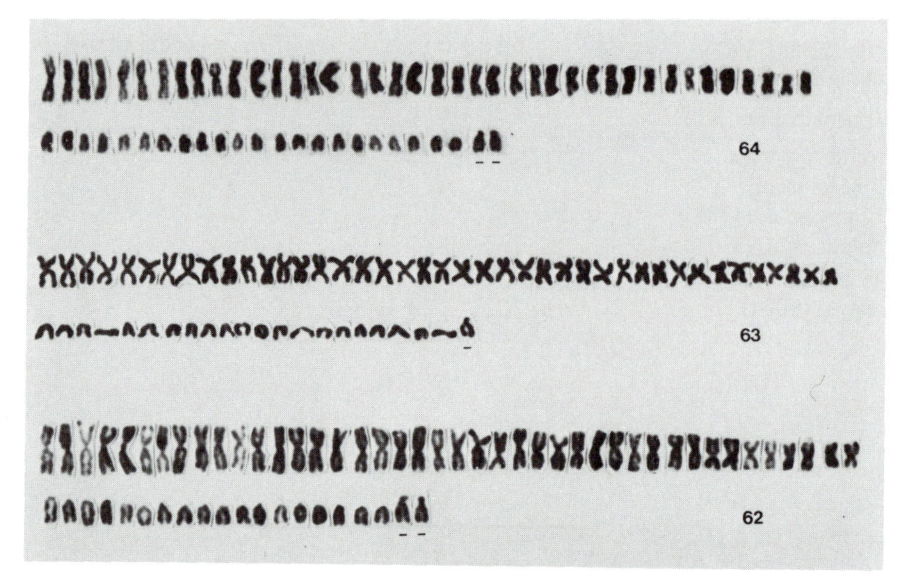

Figure 2. Chromosomes of three rainbow trouts differing in their chromosomes number (2n = 62 to 64), but having the same fundamental number (NF 104). These variations can be explained by centric fusions or fissions of chromosomes. Putative sexual chromosomes are underlined (from Chourrout and Happe, 1986).

2.3. Molecular Polymorphism

There is a large variety of methods which will detect molecular polymorphism, but the most widely used method involves electrophoresis. This method offers the following advantages over variation of quantitative characters and chromosome number in detecting polymorphism: (1) it is weakly influenced by environmental factors and can be used to compare directly samples collected in different locations; (2) it is under control of few loci and can be easily interpreted in terms of genotypes and allele frequencies. From these frequencies, heterozygosity level (diversity within population) and genetic distance (diversity between populations) can be calculated.

2.3.1. Protein Polymorphism

Protein electrophoresis was the first electrophoretic method developed for describing genetic variations. This method mainly detects mutations which lead to a change in the charge of proteins, that is approximately 25% of the mutations which occur at the nucleotide level (Fig. 3; Marshall and Brown, 1975). The principles of protein electrophoresis are briefly described in Figures 4 and 5, which show an electrophoregram of rainbow trout. Several papers and books are available for more details (e.g., see Pasteur et al., 1987; Utter et al., 1987).

Protein electrophoresis has already provided a considerable amount of original information on the genetic diversity and taxonomy of fish species. Figure 6 shows an example of a genetic structuration of French brown trout populations which was not detected before electrophoretic studies (Guyomard, 1989).

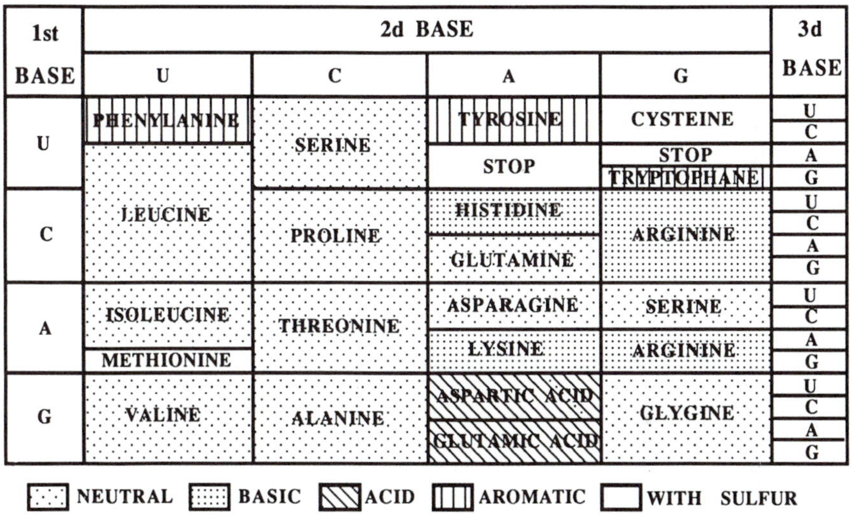

1st BASE	2d BASE: U	2d BASE: C	2d BASE: A	2d BASE: G	3d BASE
U	PHENYLANINE LEUCINE	SERINE	TYROSINE STOP	CYSTEINE STOP TRYPTOPHANE	U C A G
C	LEUCINE	PROLINE	HISTIDINE GLUTAMINE	ARGININE	U C A G
A	ISOLEUCINE METHIONINE	THREONINE	ASPARAGINE LYSINE	SERINE ARGININE	U C A G
G	VALINE	ALANINE	ASPARTIC ACID GLUTAMIC ACID	GLYGINE	U C A G

NEUTRAL BASIC ACID AROMATIC WITH SULFUR

Figure 3. Each of the 20 amino acids which can be found in the sequence of a protein are encoded at the DNA level by one or several particular sequences of three nucleotids (codon). The correspondance between the codons and the amino acids, the genetic code, is well known. It is possible to predict which amino acid change will occur for a given nucleotide substitution. Approximately 25% of them replace an amino acid by an other one with a different electric charge. These modifications are detected at the protein level by electrophoresis.

2.3.2. DNA Polymorphism

During the past decade, the studies on DNA polymorphism have been limited to mtDNA because this molecule can be purified in rather large amounts in individuals and generate easily interpretable electrophoretic patterns after restriction enzyme treatment. However, the simplification of DNA cloning procedures now provides the means to study the polymorphism of nuclear DNA. Two types of methods can be distinguished.

Analysis of restriction fragment length polymorphism. This approach uses the ability of a restriction enzyme to cut DNA at particular sites of its sequence. If a mutation occurs within the restriction site, this site is suppressed and the enzyme will generate a new pattern of restriction fragments. If we can visualize the sequence in which a restriction site is present or absent according to the individuals studied, we shall be able to reveal a restriction

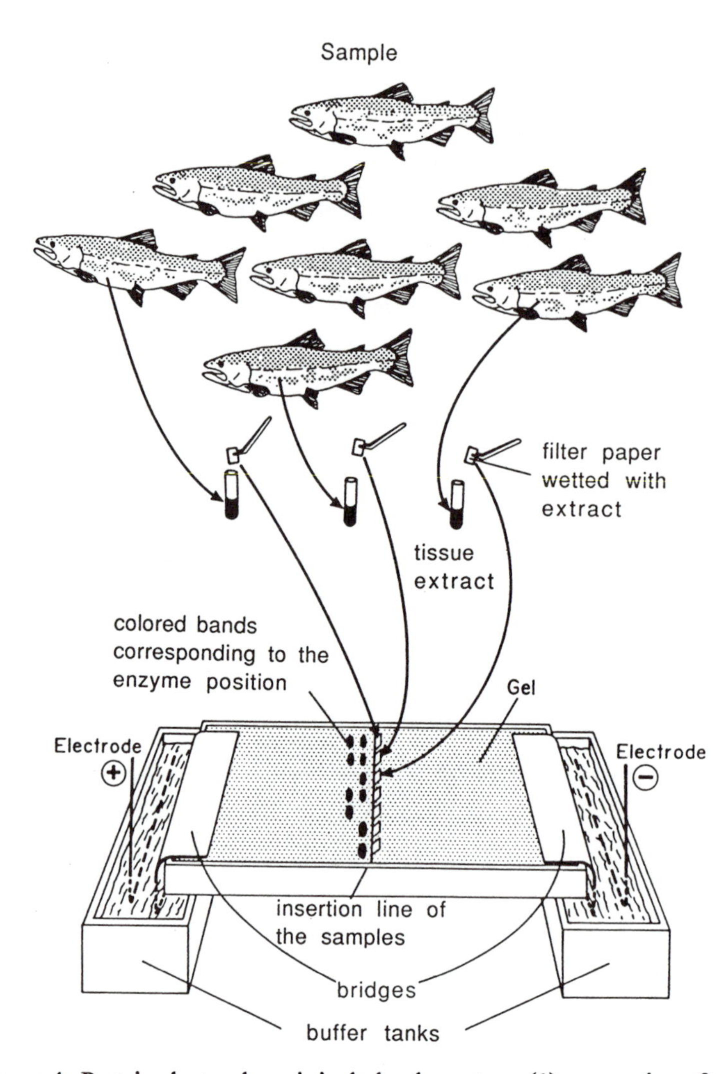

Figure 4. Protein electrophoresis includes three steps: (1) preparation of crude protein extracts from several individuals, (2) migration and separation of the proteins of these extracts under an electric field, (3) after a migration of two or three hours, visualization of a particular enzyme with a specific staining solution poured on the gel. The figure shows a monomeric enzyme encoded by one locus possessing two alleles.

fragment length polymorphism (RFLP) between them after separation of the fragments by DNA electrophoresis (Fig. 7).

This method has been first applied to mtDNA RFLP analysis for two main reasons: (1) this DNA is very easy to purify in quite large quantities, and (2) the DNA is present in a clonal state in each individual and its sequence is short so that discrete and interpretable bands are generally generated by restriction enzyme treatment of the whole molecule. Thus, no cloning step is required and mtDNA RFLP can be directly visualized with ethidium bromide (Fig. 7).

More sensitive methods involve endlabeling mtDNA fragments after digestion by restriction enzymes and autoradiography of the gel (alternatively, southern blotting on a membrane, probing with radiolabeled mtDNA molecules and autoradiography of the mem-

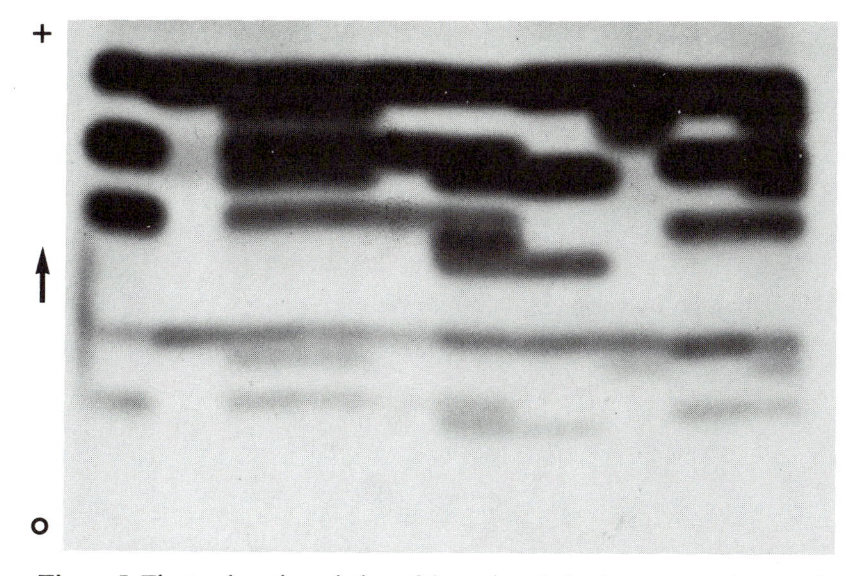

Figure 5. Electrophoretic variation of the malate deshydrogenase in the muscle of ten rainbow trout. A genetic model involving two loci and four different alleles allows to explain the electrophoretic variations shown on the figure. The most frequent allele is common to the two loci. O = origin.

brane). The methods to describe RFLP of mtDNA are described in more details by Ferris and Berg (1987) and Gyllensten and Wilson (1987). The choice between methods depends on the purpose of the study and the species under consideration. Gyllensten and Wilson (1987) recommended using 6-base enzymes and endlabeling for intraspecific studies in salmonids since mtDNA divergence within these species seems to be low.

For RFLP analysis of nuclear DNA, probes which can recognize a specific sequence are required (Fig. 7). A variety of probes, including heterologous ones, are available. Since 1984, the attention has been focused on the use of minisatellite probes (Jeffreys et al., 1985; Vassart et al., 1987). These minisatellite sequences are formed of tandem repeats of 10-20 base pair motives and are interspersed in the genome. The patterns of restriction fragments generated by these probes generally show an extensive polymorphism due to allelic variation in the number of tandem repeats at many different loci and are termed "fingerprints." Figure 8 shows an example of a fingerprint obtained in brown trout with M13. Similar results have been obtained with other minisatellite probes.

The complexity of the fingerprints precludes this approach for many population genetic studies. However, it is possible to visualize the variation which occurs at one minisatellite locus only. This requires cloning a probe which contains the flanking regions of the minisatellite tandem repeat at a single locus. Under very stringent washing conditions, the only visualized bands will be generated by the homologous hybridization of the single locus probe with the restriction fragments containing the flanking regions. Then, the polymorphism observed will result from variation in the number of tandem repeats at the locus characterized by the flanking regions (Fig. 9). Minisatellite single locus probes have already developed cloned in many different organisms, including salmonids where they have also revealed a high polymorphism (Taggart and Ferguson, 1990).

Detection of polymorphism by PCR. The polymerase chain reaction provides the means to amplify one or several particular fragments of the genome up to approximately two kilobases. The reaction requires primers for the initiation of the replication (Fig. 10). Such

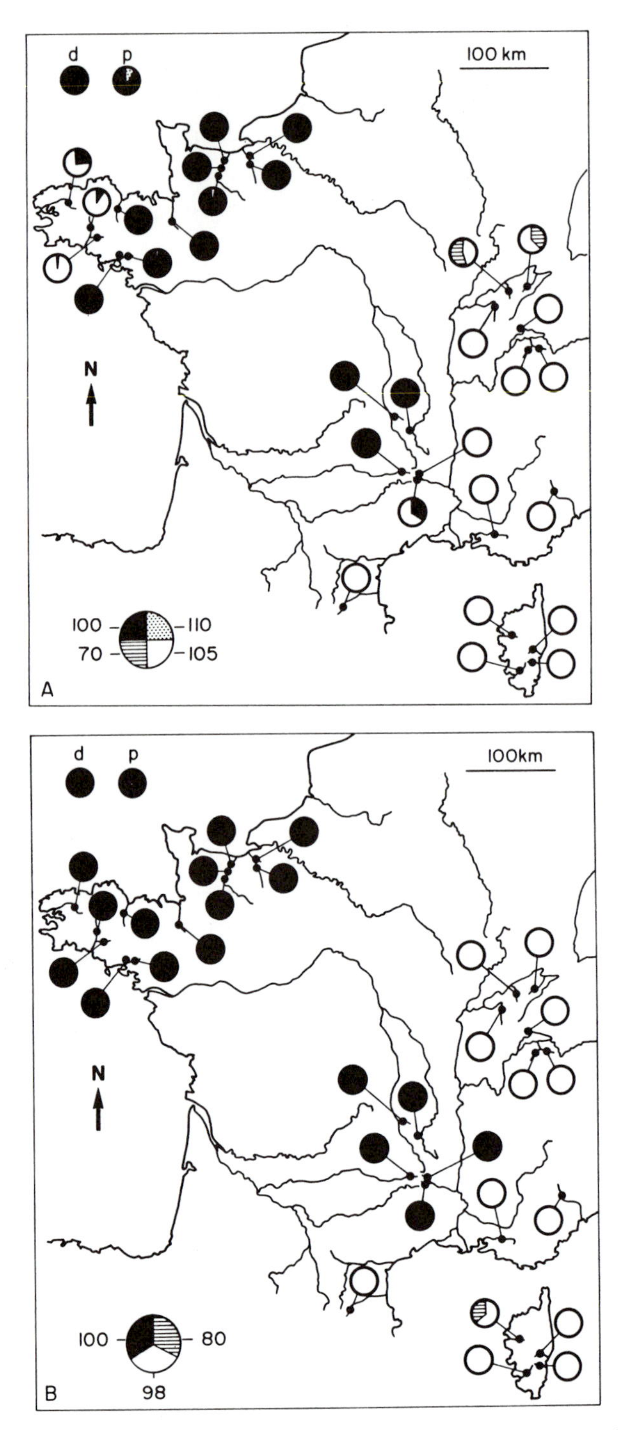

Figure 6. Geographic variation of allele frequencies at two loci in French domesticated (d) and natural populations of brown trout. A: LDH-5* (lactate deshydrogenase); B: TFN* (transferrine).

primers can be synthesized randomly and the PCR is then aspecific. This procedure has been successfully applied to gene mapping or strain identification (Williams et al., 1990). However, the most usual way to obtain primers is to define them from a sequence of interest. In this case, the PCR preferentially amplifies the fragment included between the two primers. PCR can be used to visualize small length variation of the amplified fragment in acrylamide gels. This polymorphism can be generated by variations in the number of short tandem repeats of 24 bp (Fig. 10). These sequences are called microsatellites by analogy with minisatellite sequences.

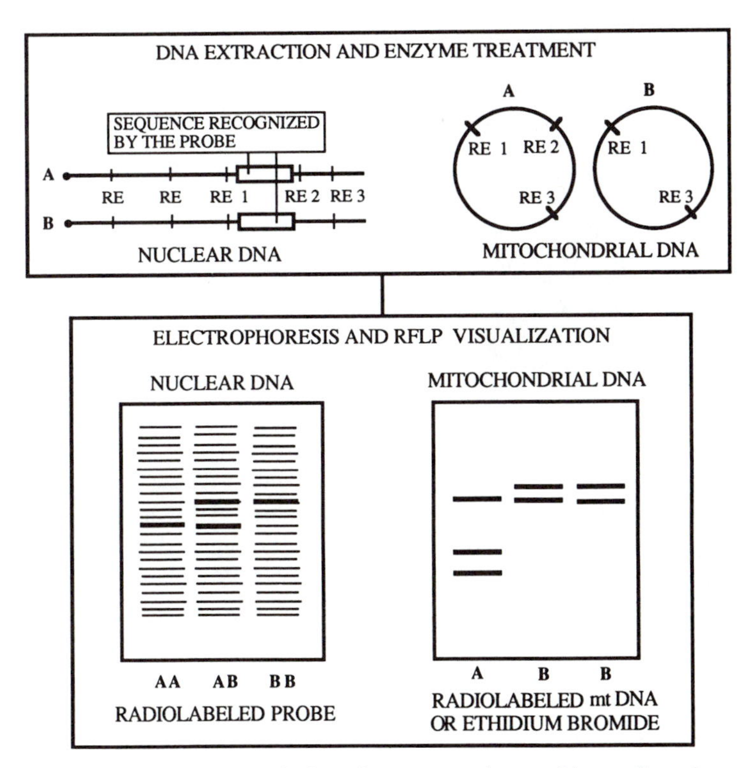

Figure 7. Analysis of restriction fragment polymorphism of nuclear and mitochondrial DNA (mtDNA). After extraction, DNA is digested by a restriction enzyme and submitted to electrophoresis. Restriction site variation of nuclear DNA is visualized with radiolabelled probe which recognizes a specific sequence. MtDNA restriction fragments can be visualized by ethidium bromide or radiolabelled mtDNA probe depending on the purity and quantity of the mtDNA analyzed. RE = restriction site.

Such sequences have been already characterized and found polymorphic in humans, whales, and drosophila (Tautz, 1989).

PCR products can be also sequenced. This method is now currently used for the study of mtDNA regions (the most frequently examined region is the D loop which contains the origin of replication of the mtDNA; Figure 11) again because mtDNA is a "naturally cloned" sequence.

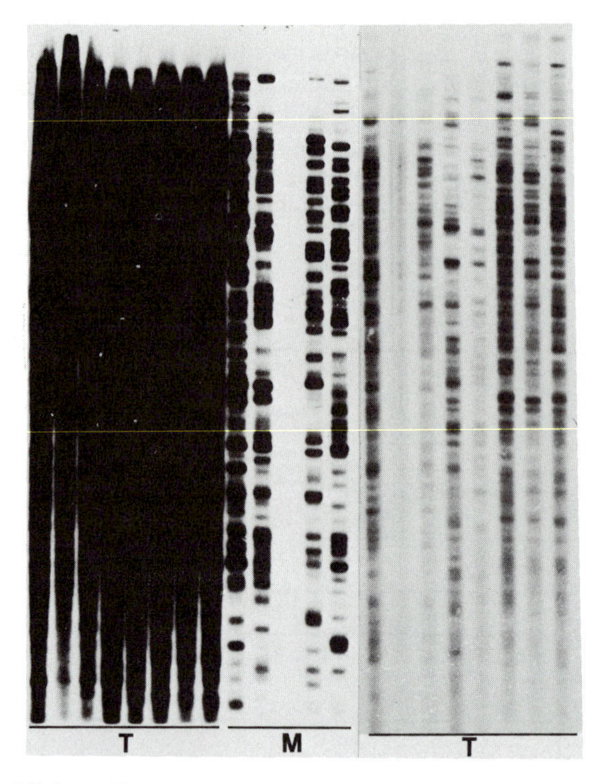

Figure 8. Mini-satellite sequences recognize tandem repeats of about 10-20 nucleotids which are interspersed in the genome. They lead to complex electrophoretic patterns called "fingerprints". The figure shows fingerprints obtained in brown trout with the M13 phage DNA as probe. Trout (T) and mammal (M) DNA are digested by Hinf I. The same eight samples of trout are shown for two different exposure times.

3. Comparative Advantages and Limitations of the Different Methods of Description

3.1. Quantitative versus Molecular Data

One major difference between these two categories of data is that the mean and variance of quantitative characters depend on the frequencies and also average effects of alleles occurring at the locus. In theory, quantitative characters should provide the information to identify components of the fitness or economical value of wild stocks. However, due to the possible effect of environmental factors, this identification requires complex experiments which are difficult or impossible to do in field studies. Most of them, like transplantation experiments or long-term collections of population parameters (such as the evolution of the size at first maturation) do not provide a means to distinguish genetic from environmental effects and hardly support any valid genetic conclusion.

In contrast, molecular data are easily interpreted in terms of genetic variation, but their biological significance remains unknown. Namely, it is usually impossible to determine which mechanisms (selection, genetic drift, migration, etc.) and characters are involved in the maintenance of the molecular polymorphisms or differentiation patterns observed.

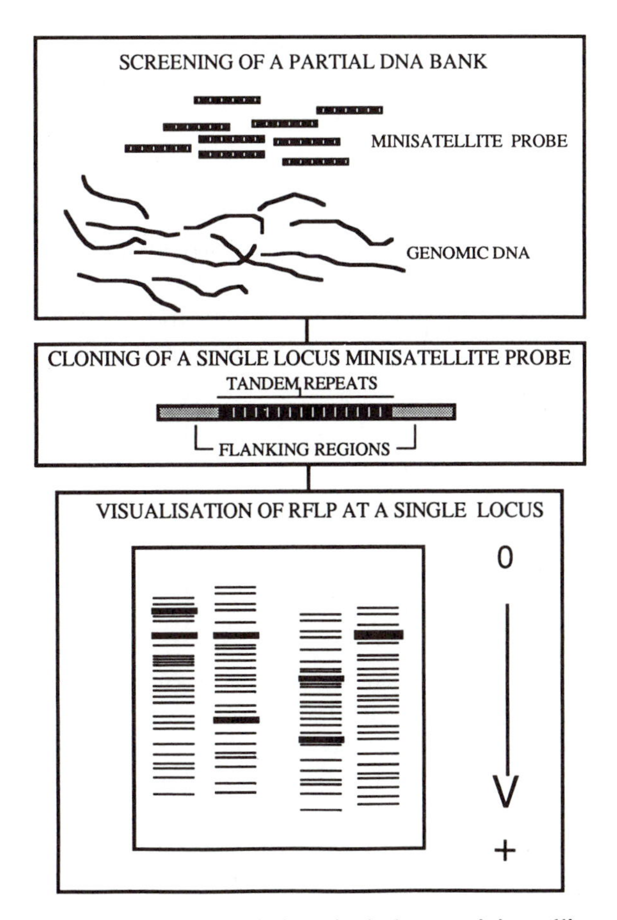

Figure 9. The procedure for isolating single-locus mini-satellite probes includes: (l) screening of a partial genomic bank usually made from 5 to 15 kb nuclear DNA fragments with a mini-satellite probe and (2) cloning of fragments which contain the tandemly repeated mini-satellite motives and the flanking regions. After hybridization of southern blots with such probes and high stringency washes, only the signals due to hybridization with the flanking regions (homologous hybridization) are clearly visible. The RFLP observed is still generated by variation of tandem repeat numbers, but is specific of the locus identified by the flanking regions.

3.2. Mitochondrial versus Molecular Data

The mitochondrial DNA is characterized by a maternal and clonal inheritance, the absence of recombination and, in some organisms a higher polymorphism than nuclear genes. The absence of recombination prevents the redistribution of mutations between clones. When these mutations are accumulated on the same molecule, mtDNA becomes a powerful marker for detecting traces of ancient allopatric differentiations, even in actually panmictic populations. The higher polymorphism of the mtDNA has often provided more detailed information on the genetic structure of species than enzymes (Avise et al., 1979). However, in some

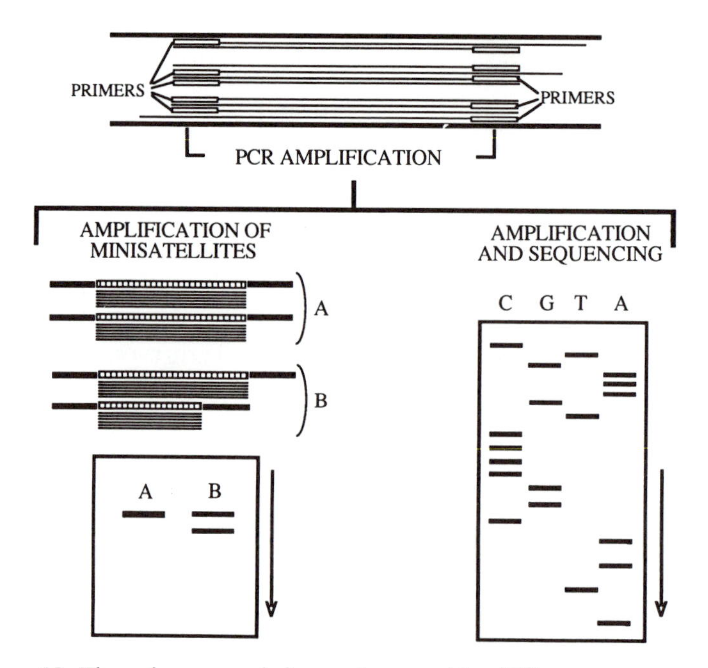

Figure 10. The polymerase chain reaction provides different ways to analyze the genetic variation at the DNA level. One possibility is to amplify a particular fragment and to examine if a polymorphism of size of the amplified fragment occurs between the individuals. This method is applied to the detection of single locus polymorphisms due to variation of the number of short tandem repeats (2-4 bp) called microsatellites. An other possibility is to sequence the amplified fragment and to compare the sequences between individuals.

organisms like drosophila (Solignac et al., 1986), there is no evidence that mtDNA is more polymorphic than nuclear DNA. It cannot be excluded that the relative evolution rate of mtDNA, with respect to nuclear evolution rate, varies substantially between organisms. In addition, the ratio of nuclear to mitochondrial nucleotide diversity also depends on populational parameters such as sex ratio, migrant male-to-migrant-female ratio and female population size (Birky et al., 1989). These parameters could account by themselves for the differences which are observed in some cases between mitochondrial and nuclear diversity.

The most important specific feature of mtDNA is its maternal inheritance. This implies that mtDNA study by itself does not provide adequate information on the present breeding structure of a species and should be systematically associated with nuclear DNA analysis.

3.3. Protein versus DNA Data

One limitation of protein electrophoresis is the high proportion of polymorphism (3/4) which cannot be detected by conventional studies. In most of them, a large majority of loci are found monomorphic and in few cases, all of them. This leads to an underestimation of variability between and within populations and to inaccurate estimates of genetic drift or gene flow between local populations. Depending on the geographic distribution of gene frequencies, the genetic diversity between stocks could be proportionally more or less underestimated than

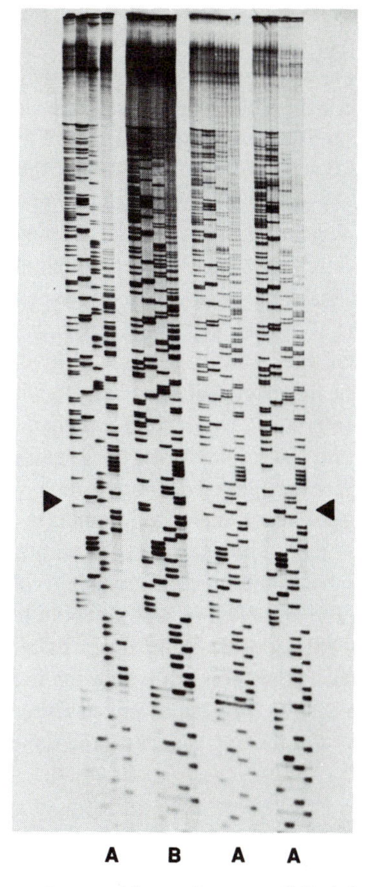

Figure 11. Sequence polymorphism of an amplified fragment of the control region (D loop) of brown trout mtDNA. Two genotypes (A and B) can be distinguished on the figure (the position of the mutation is given by the triangles).

diversity within populations. For example, if most of the diversity of the species is between populations, the genetic distances will be proportionally more underestimated than the heterozygosity levels.

Moreover, the fact that two different mutations can result in an identical electromorph increases the probability of parallelism and convergence in protein data and, consequently, the risk of misclassification when cladistic methods are used (see next section).

These problems can be overcome or substantially minimized in the case of DNA studies since variation can be described at the nucleotide level if necessary. DNA studies also offer the possibility to compare variation in coding and noncoding regions. However, due to its low cost and simplicity, protein electrophoresis remains an essential method in populations genetics and stock identification.

4. Data Analysis

This section is devoted to the treatment of the data which, after the descriptive phase, appear as a table of genotypes for each individual and each locus or character. Most of these methods can be applied to both morphormetric and molecular data.

4.1. Multidimensional Analysis

Individuals can be represented as points in a space which has as many dimensions as characters examined. The coordinates of an individual are defined by its values for each character. In the case of molecular data, the genotypes must be first transformed into numerical values (for example, AA = 1, AB = 2, BB = 3). The goal of the multidimensional analysis is to simplify the description of these clouds of observations by projecting them on planes. These planes are determined by axes which have particular properties. These properties depend on the type of multidimensional analysis which is performed, namely principal component analysis (PCA) or discriminant analysis. These methods are described in detail in specialized books (for example, see Lebart et al., 1984). Briefly, the PCA will determine the axis (principal axis) on which the projection of the cloud of points will give the largest dispersion of the observations. Figure 12 shows the first principal axis in a particular case where the observations form two clearly distinguishable groups. The second principal axis will be the axis showing the largest dispersion after axis 1. The PCA does not require regrouping individuals into populations and all the samples are considered as a single unit. When molecular data are used, it is recommended to use correspondence analysis. In contrast, discriminant analysis applies to *a priori* identified groups and determines the most discriminant axis and planes. The first discriminant analysis will be a linear function of the variables (characters) studied which shows the largest differences between the identified groups. It could be very different from the first component axis determined by PCA on the same set of data (Fig.12). Discriminant linear functions can be used to determine the group to which the individuals of an undetermined sample belong. They can be applied to the case of mixed fisheries or populations in stocked sites (see Pella and Milner, 1987, for the analysis of mixed fisheries).

4.2. Tree Construction

The major objective of tree construction is to clarify the phylogenetic relationships between species or populations. The methods to reconstruct trees can be classified in two types: phenetic and cladistic methods. A prerequisite step is to verify that the groups of individuals which will be treated with these methods really form true evolutionary units. In practice, this means that it is necessary to test if the samples collected in the wild may be considered as a single reproductive unit. In most cases, this is achieved by testing if there is no significant deviation from the HardyWeinberg proportions (test of panmixy) and no linkage disequilibrium. Multidimensional analysis can also detect deviation from panmictic or linkage equilibrium. In this section, we consider only some examples of methods of tree reconstruction and we recommend reading more complete reviews on these methods (e.g., Felsenstein, 1988; Weir, 1990). Figure 13 shows the two main types of tree reconstructions which can be done: phenetic and cladistic methods.

4.2.1. Phenetic Methods

These methods require the calculation of genetic distances which are a measure of the variability between populations. The simplest genetic distance is calculated like a geometric distance where the frequencies replace the usual coordinates, but many other distances can be used. Most of them are highly correlated at low values of divergences. One of the most-used distances is the standard genetic distance defined by Nei (1975). This distance is proportional to the divergence time in the case of constant evolution rate. When the genetic distances have been calculated, two different procedures can be applied: pairwise and clustering methods.

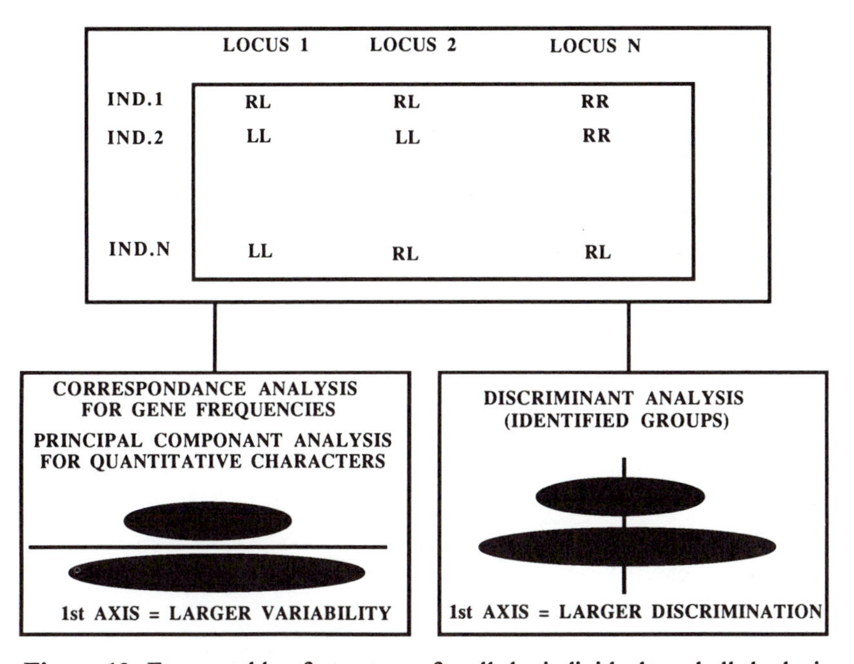

Figure 12. From a table of genotypes for all the individuals and all the loci analyzed, it is possible to perform different types of multidimensional analysis. When *a priori* identification of groups is not justified, all the individuals form a single sample and correspondence analysis or principal component analysis should be used depending on the category of data analysed. In the other cases, discriminant analysis should be preferred. The significance of the first axis is not the same for the two types of analyses.

Pairwise methods. This method first involves the construction of trees with an algorithm. The same algorithm allows for the construction of different trees assuming different initial regroupings. New distances can be derived from the branch lengths for each tree and compared with the initial distances. The best tree will be the tree which minimizes the total differences between these distances and the branch lengths. This type of method was first proposed by Fitch and Margoliash (1967).

Clustering methods. The basic procedure of these methods involves: (1) clustering of the two populations which show the smallest genetic distances; (2) calculation of a new genetic distance matrix assuming that these two populations now form a single unit; (3) clustering of the two closest populations in the new matrix and so on. The clustering methods differ from each other by the algorithm of recalculation of genetic distances after each clustering step. The most widely used is the UPGMA (unweighted pairgroup method using arithmetic average; Sneath and Sokal, 1973). This method gives good phylogenies when the mutation rates are the same along all branches of the trees (in Weir, 1990). Figure 14 represents the phenetic tree generated by UPGMA with Nei's standard distance between French domesticated and natural populations of brown trout. The highly divergent corsican lineage (pop. c2 and c3) could result from an acceleration of the evolution of population c3 by genetic drift (heterozygosity level = O) and not from a long time of divergence.

4.2.2. Cladistic Methods

The central idea of cladistic (Hennig, 1966) is to establish a phylogeny on the basis of presence or absence of new character states (derived forms or synapomorphism) which have arisen in the populations since they have started to diverge. Cladistic assumes that these derived forms have arisen only once and that no reversion can occur. The application of cladistic methods generally requires the ability to distinguish the derived states from the ancestral ones which were present in the ancestral population. The ancestral forms can be determined by comparing the species studied with the most closely related one (outgroup method). If there is

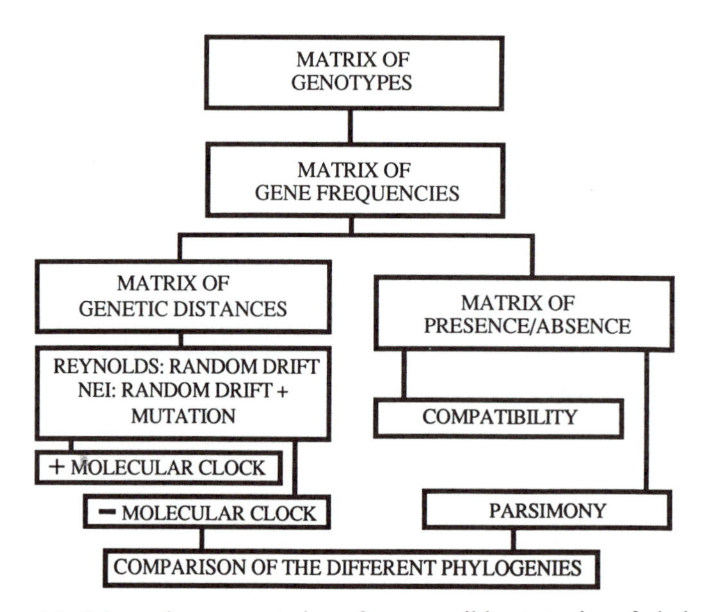

Figure 13. Schematic representation of two possible strategies of phylogeny construction from polymorphism data: (l) phenetic methods using matrices of genetic distances and (2) cladistic methods based on matrices of presence/absence of alleles (see the text).

no common form, the ancestral form remains undetermined. Figure 15 shows the results of a cladistic analysis performed on the same data as in figure 14 and using Atlantic salmon as the outgroup. On the cladogram, all the Corsican populations form a unique clade with respect to the other populations. Cladistic methods can be easily applied when all the characters are compatible. Figure 16 illustrates a case of incompatibility between two characters. Incompatibility can arise when the same mutation appears independently in two different lineages or when a reversion occurs. Such events occur frequently with electromorphs. Two different approachs are available when the data are incompatible: the compatibility and the parsimony methods.

Compatibility methods still assume that each derived state arises no more than once and that there is no reversion. The tree which shows the largest number of compatible characters is considered as the best phylogeny. The method can be readily applied as it is proven that pairwise compatible characters are compatible all together. However, the best trees generated frequently contain too few compatible characters.

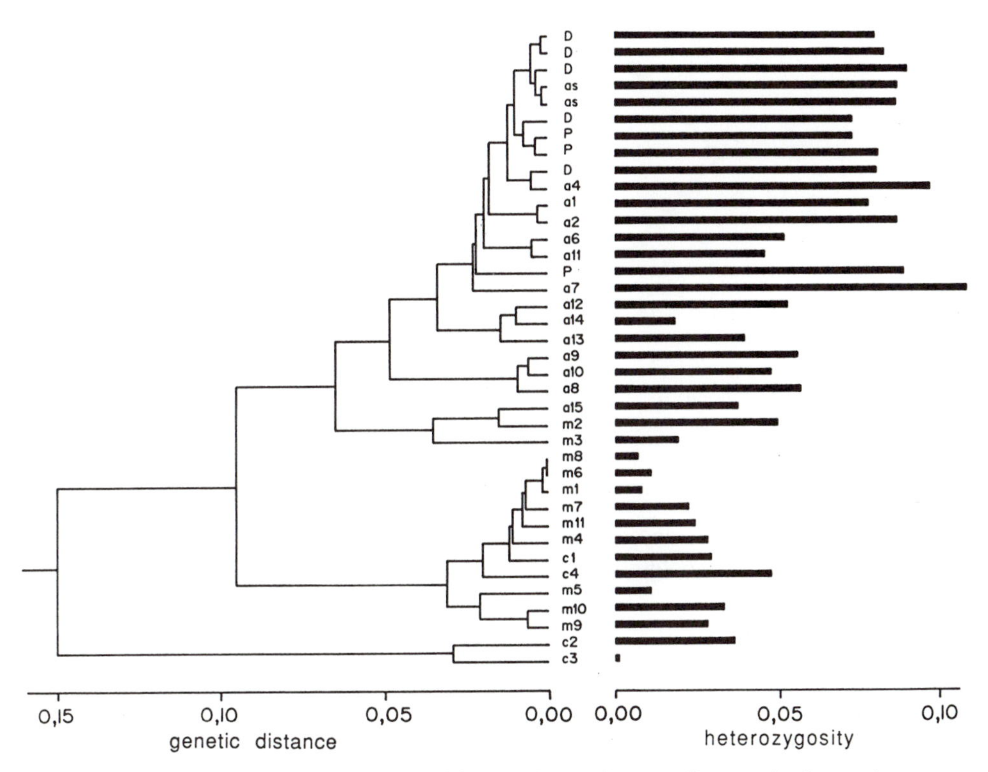

Figure 14. Phenogram generated by UPGMA from Nei's standard genetic distances and heterozygosity levels. D: domesticated stocks; P: Polish strains; a: Atlantic populations; m: continental French Mediterranean populations; c: Corsican populations. This tree represent true phylogenies when evolution rates are similar along all the branches. The occurrence of a highly divergent Corsican lineage could result from an increase of the evolution rate by bottleneck effect in c3.

Parsimony methods assume that the same derived form can have arisen independently more than once and/or that reversions occur. Here, the best phylogeny is given by the tree which minimizes the multiple origin and/or the reversion events. Table 3 shows different parsimony methods which have been proposed. In the case of electromorphs, the Wagner parsimony method seems to be justified. One limit to the application of parsimony methods is that the number of trees to compare is often too large.

Finally, it is intuitive that compatibility and parsimony methods perform well when the evolution rates or the total number of changes are not too high, a situation which is expected when conspecific populations or closely related species are considered.

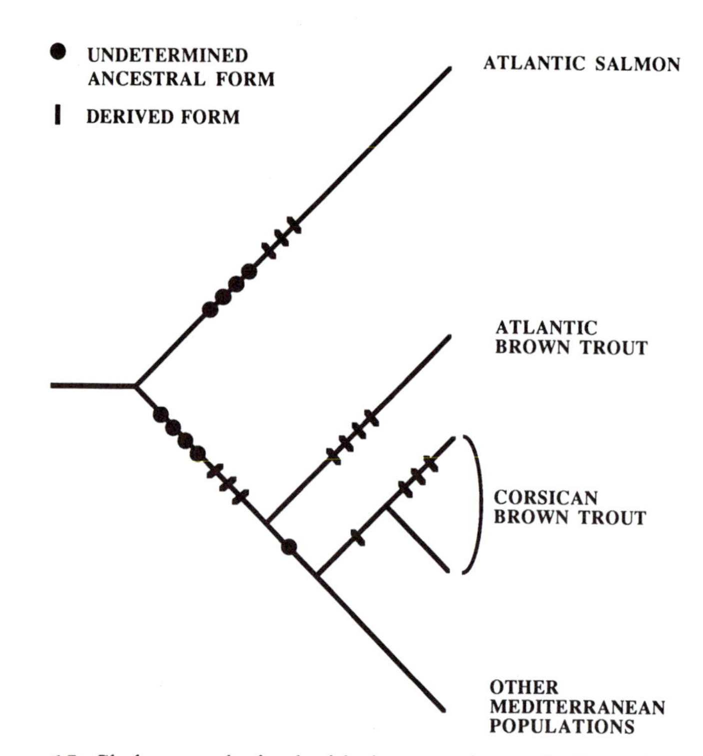

Figure 15. Cladogram obtained with the same data as in figure 12 and with Atlantic salmon as outgroup. Here, all the Corsican populations form an unique clade.

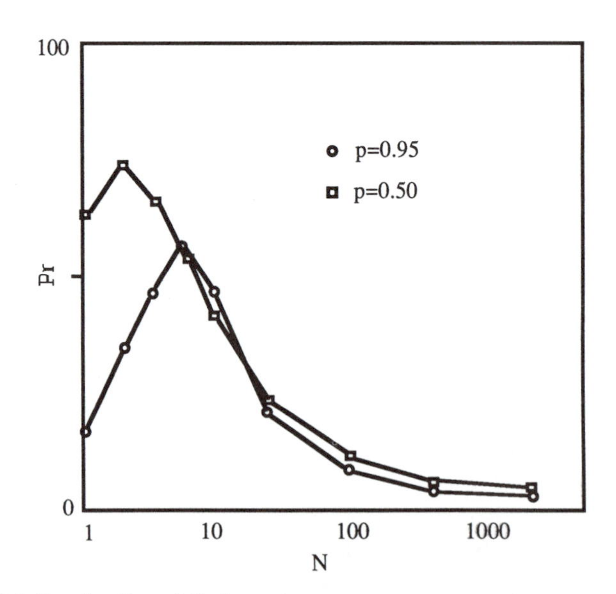

Figure 16. Species B and C share the same synapomorphism for character 1, while it is the case of A and B for character 2. The two cladograms are incompatible. 0: ancestral state; 1: derived state.

Table 3. Methods of parsimony for cladistic tree construction.

Method	Reversion From Derived Ancestral Form (0)	Multiple Origin of Derived Form (1)
Camin-Sokal (Camin and Sokal, 1965)	no	yes*
Dollo (Farris, 1977)	yes*	no
Wagner (Kluge and Farris, 1969)	yes*	yes*
Polymorphism (Farris, 1978)	no	no
	polymorphism*(0, 1 or 0 and 1)	

*events the number of which has to be minimized.

5. Sampling Problems

Sampling efforts have to be allocated between three main levels in stock description: the number of sampling sites, the number of individuals sampled per site and the number of characters examined per individual. The sampling strategy will depend on the genetic structure of the species and the aims of the study. In a survey of the genetic diversity of a species over its entire range of distribution, it is probably better to start with a geographically loose sampling and to adjust progressively on the basis of the results. If the samples are issued from panmictic populations, the accuracy of the genetic distance and heterozygosity estimates mainly depend on the number of loci and the number of individuals analyzed can be very low (Nei, 1978).

When the objectives are to estimate gene flow between adjacent populations or to verify panmictic or linkage equilibria, analyzing a large number of individuals is often required. Table 4 shows that the sample size necessary to detect a linkage disequilibrium is very large in most cases.

Finally, it is important to realize which biological conclusion can be drawn from a significant test. For example, let us consider a single population. At each generation, N spawners randomly drawn from this population reproduce in one tributary and N others in a second one. Figure 17 (from Allendorf et al., 1981) shows that the probability to detect significant allele differences between two samples (in Figure 17, for a sample size = 50) of the progenies issued from the two groups of spawners at the level 5% rapidly exceeds this value when N becomes lower than 50. Then, significant allele differences between samples do not necessarily signify that they belong to truly distinct breeding units. This possibility has to be kept in mind for salmonids which can frequently form small reproductive communities.

Table 4. Sample sizes required to detect a linkage disequilibrium (D') between two loci for different value of D' at the level 5%. pl and p2 are respectively the frequencies of one allele at locus 1 and one allele at locus 2 in the population. D is the observed value of linkage disequilibrium and Dmax the maximum value of linkage disequilibrium for the allele frequencies in the population (from Chakroborty, 1984)

D' = D/Dmax	p1=p2=0.2	p1=p2=0.5
-1.0	102	4
-0.4	974	57
-0.2	4081	251
0.2	272	251
0.4	68	57
1.0	12	11

6. Conclusion

The development of molecular methods and computerized methods of data treatment now provide very efficient tools to describe and analyze the genetic diversity of species. They have already been intensively applied to salmonids. Up to now, population genetic studies of salmonids have resulted in a static picture of the genetic structure of these species (in some cases, this elementary goal has not yet been achieved yet). This knowledge has not led to a

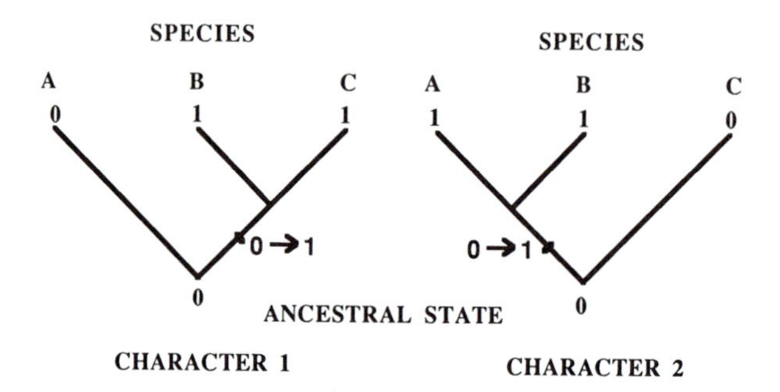

Figure 17. Probability of detection (Pr) of allele differences in two progenies each issued from N adults that are sampled at random from a single population. p = allele frequency in the adult population. Here, the size of the sample analyzed for each progeny is assumed to be 50 (from Allendorf and Phelps, 1981).

significant change of the management policies in most of the countries and, in few cases, has resulted in perverse effects (for example, construction of hatcheries instead of restoration of habitats). This stagnation could be partly due to our inability to clearly demonstrate by field studies the negative effects of the disappearance or genetic contamination of original gene pools on the adaptive value of the species and to identify clear priorities in the preservation of stocks. These goals are certainly more difficult to reach than a simple description of diversity and future studies should focus on them.

References

Allendorf, F.W., and S.R. Phelps. 1981. Use of allelic frequencies to describe population structure. *Canadian Journal of Fisheries and Aquatic Sciences*. 38:1507-1514.

Allendorf, F.W., N. Ryman, and F.M Utter. 1987. Genetics and fishery management. Past, present and future. In: "Population genetics and fishery management," N. Ryman and F. Utter, eds., pp. 1-19, University of Washington, Seattle and London.

Avise, J.C., C. Giblin-Davidson, J. Laerm, J.C. Patton, and R.A. Lansman. 1979. Mitochondrial DNA clones and matriarchal phylogeny within and among geographic populations of the pocket gopher, *Geomis pinetis*. *Proceedings of the National Academy of Sciences U.S.A.* 76:6694-6698.

Birky, C.W., P. Fuerst, and T. Maruyama. 1989. Organelle gene diversity under migration, mutation, and drift: equilibrium expectations, approach to equilibrium, effects of heteroplasmic cells, and comparison to nuclear genes. *Genetics* 121:613-627.

Blanc, J.M., B. Chevassus, and P. Bergot. 1979. Détermimisme génétique du nombre de coeca piloriques chez la truite commune (*Salmo trutta* L.) et la truite arc-en-ciel (*Salmo gairdneri* R.). III: effet du génotype et de la taille des oeufs sur la réalisation du caractère chez la truite fario. *Annales de génétique et de sélection animales* 11:79-92.

Blanc, J.M., H. Poisson, and R. Vibert. 1982. Variabilité génétique de la ponctuation noire sur la truitelle fario (*Salmo trutta*). *Annales de génétique et de sélection animales* 14:225-236.

Camin, J.H. and R.R. Sokal. 1965. A method from deducing branching sequences in phylogeny. *Evolution* 19:311-326.

Chakraborty, R.. 1984. Detection of nonrandom association of alleles from the distribution of the number of heterozygous loci in a sample. *Genetics* 108:719-731.

Falconer, D.S. 1989. "Introduction to quantitative genetics," Longman Scientific and Technical, Harlon, England.

Farris, S.D. 1977. Phylogenetic analysis under Dollo's law. *Systematic Zoology* 26:77-88.

Farris, S.D. 1978. Inferring phylogenic trees from chromosome inversion data. *Systematic Zoology* 27:275-284.

Farris, S.D. and W.J. Berg. 1987. The utility of mitochondrial DNA in fish genetics and fishery management in: "Population genetics and fishery management," N. Ryman and F. Utter, eds., pp. 277-299, University of Washington, Seattle and London.

Felsenstein, J. 1988. Phylogenies from molecular sequences: Inference and reliability. *Annual Review of Genetics* 22:521-565.

Fitch, W.M. and M. Margoliash. 1967. Construction of phylogenetic trees. *Science* 155:279-284.

Guyomard, R. 1989. Diversité génétique de la truite commune. *Bulletin français de pêche et de pisciculture* 314:118-135.

Gyllensten, U. and A.C. Wilson. 1987. Mitochondrial DNA of salmonids: inter and intraspecific variability detected with restriction enzymes in: "Population genetics and fishery management," N. Ryman and F. Utter, eds., pp. 301-317, University of Washington, Seattle and London.

Hartley, S.E. and M.T. Horne. 1984. Chromosome polymorphism and constitutive heterochromatin in Atlantic salmon, *Salmo salar*. *Chromosoma* 89:377-380.

Hennig, W. 1966. "Phylogenetic systematics," University of Illinois Press, Urbana.

Jeffreys, A.J., V. Wilson, and S.L. Thein. 1985. Hypervariable "minisatellite" regions in human DNA. *Nature* 314:67-73.

Kluge, A.G. and J.S. Farris. 1969. Quantitative phyletics and the evolution of anurians. *Systematic Zoology* 18:1-32.

Krieg, F. 1984. *Recherche d'une differenciation genetique entre populations de* Salmo trutta. Ph.D. thesis, universite de Paris-Sud, Orsay.

Leary, R.F., F.W. Allendorf, and K.L. Knudsen. 1985. Inheritance of meristic variation and the evolution of developmental stability in rainbow trout. *Evolution* 39:1318-1326.

Lebart, L., A. Morineau, and K.M. Worwick. 1984. "Multivariate descriptive analysis," John Wiley and Sons, New York.

Marshall, D.R. and A.H.D. Brown. 1975. The chargestate model of protein polymorphism in natural populations. *Journal of Molecular Evolution* 6:149-163.

Nei, M. 1975. "Population genetics and molecular evolution." North Holland, Amsterdam and New York.

Nei, M. 1978. Estimation of average heterozygosity and genetic distance from a small number of individuals. *Genetics* 89:583-590.

Pasteur, N., G. Pasteur, F. Bonhomme, J. Catalan, and J. BrittonDavidian. 1987. *Manuel technique pour l'électrophorese de protéines. Technique et documentation*, Lavoisier, Paris.

Sneath, P.H.A. and R.R. Sokal. 1973. *Numerical taxonomy*. Freeman, San Francisco.

Solignac, M., M. Monnerot, and J.C. Mounolou. 1986. Mitochondrial evolution in the *melanogaster* species subgroup of *Drosophila*. *Journal of Molecular Evolution* 23:31-40.

Taggart, J.B. and A. Ferguson. 1990. Hypervariable minisatellite DNA single probes for Atlantic salmon, *Salmo salar* L. *Journal of Fish Biology* 37:991-993.

Tautz, D. 1989. Hypervariability of simple sequences as a source for polymorphic DNA markers. *Nucleic Acids Research* 19:3756-3763.

Thorgaard, G.H. 1983. Chromosomal differences among rainbow trout populations. *Copeia*:650-662.

Vassart, G., M. Georges, R. Monsieur, H. Brocas, A.S. Lequarre, and D. Cristophe. 1987. A sequence of the M13 phage detects hypervariable minisatellites in human and animal DNA. *Science* 235:683-684.

Weir, B.S. 1990. "Genetic data analysis", Sinauer Associates, Inc. Publishers, Sunderland, Massachusetts.

Williams, J.G.K., A.R. Kubelik, K.J. Livak, J.A. Rafalski, and S.V. Tingey. 1990. DNA polymorphisms amplified by arbitrary primers are useful as genetic markers. *Nucleic Acids Research* 18:22-29.

Spatial Organization of Pacific Salmon: What To Conserve?

B.E. RIDDELL

1. Introduction

The rich biological diversity in salmonids has been recognized for centuries and has been a central premise in managing salmon fisheries in this century (the "Stock Concept"). But recently, as in many other biological resources (FAO, 1981; Oldfield, 1989), increased concern has been expressed about the loss of biological diversity and the impact of harvest management on Pacific salmon. Management of Pacific salmon (*Oncorhynchus* sp.) is probably as rich in social, economic, and political issues as its resource base is biologically, and the scope of this issue continues to expand. Multiple resource management principles, such as sustainable economic development (WCED, 1987), will increase harvest and environmental issues involved in salmon management decisions. Evidence for global climate changes increases uncertainty about future salmon production. Litigation is increasingly used to protect specific interest groups. Unfortunately, in many salmon management decisions, the non-biological interests have taken precedence over the biological resource (Wright, 1981; Fraidenburg and Lincoln, 1985; Walters and Riddell, 1986). Each of these may have been a responsible decision, but in aggregate they create a serious biological problem through the gradual but steady erosion of biological diversity. An adage for similar problems in other fields of resource management is The Tyranny of Small Decisions. However, decisions favoring biological conservation are becoming more frequent, even though their effects on resource use, other resources, and communities are becoming more controversial.

In the Pacific Northwestern USA, Nehlsen et al. (1991) have identified 214 salmonid stocks of concern, 159 of which are considered to be at moderate to high risk of extinction. Since 1990, the Sacramento winter Chinook and Snake River sockeye (Redfish Lake), and spring and summer Chinook stocks have been listed as threatened or endangered under the Endangered Species Act of the United States. In Canada, a comparable inventory of Pacific salmon has not been prepared. In southwestern British Columbia, however, one-third of the spawning populations known since the early 1950s have now been lost or decreased to such low numbers that spawners are not consistently monitored (Fig. 1). This area of British Columbia has been the center of urbanization and development and is not representative of the

Department of Fisheries and Oceans, Biological Sciences Branch, Pacific Biological Station, Nanaimo, B.C. V9R 5K6

Genetic Conservation of Salmonid Fishes, Edited by J.G. Cloud
and G.H. Thorgaard, Plenum Press, New York, 1993

salmon resource in the province generally, but northwestern USA is not unique in their concern for conservation of Pacific salmon populations.

In the immediate future, resource managers can anticipate being asked more frequently "what to conserve" and policy makers will have to consider "at what cost". The latter issue will be an essential one but will not be considered in this paper (see Norton, 1986). The essential biological issue is what to conserve, but there is no single answer. The answer will vary between situations depending on the species and the remaining distribution of breeding populations, population dynamics and integration of the species with its ecological system, the status of the

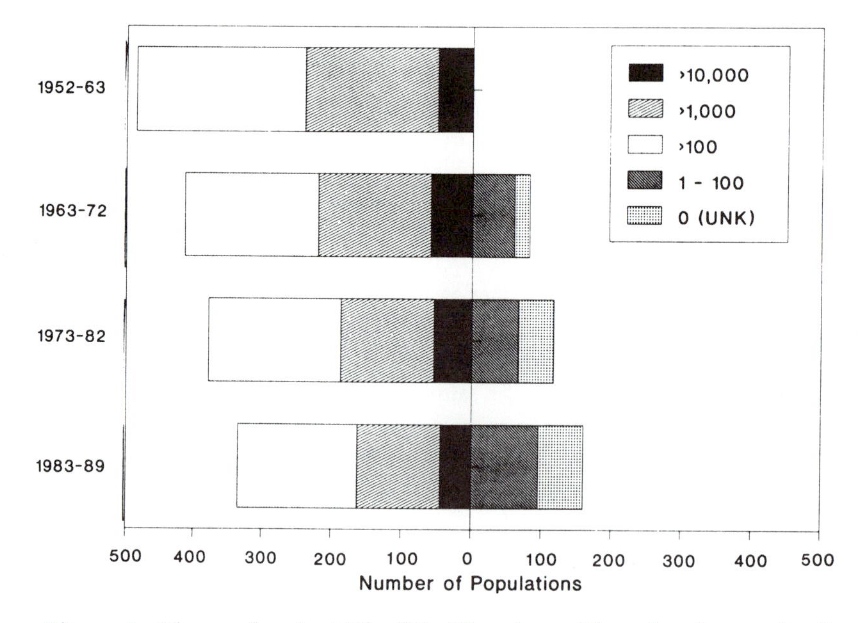

Figure 1. The number (n=495) of Pacific salmon (*Oncorhynchus nerka, O. keta, O. gorbuscha, O. kisutch, O. tshawytscha*) spawning populations by size classes (numbers of spawners by species) and time periods in southwestern British Columbia (east coast Vancouver Island and adjacent mainland areas including the lower Fraser River). The number of populations with zero spawners or unknown numbers (UNK) were combined because once a population consistently shows none or very few spawners the field staff may not monitor the stream.

resource, and the causes of the conservation problem. Each answer will be information intensive, expensive, and likely controversial. The answer to such complex situations are seldom unique or unanimously agreed upon. Further, many of the future debates will be results of past mistakes.

This paper presents a series of principles for management of Pacific salmon to conserve genetic diversity. The guidelines are admittedly pragmatic and their applicability may vary. However, they are a step towards recognizing and integrating a conservation genetic objective in Pacific salmon management while also recognizing the limitations of our knowledge in the population genetics of Pacific salmon. More scientific research is needed to improve this knowledge but the incorporation of genetic advice in fishery management decisions should not wait for the development of more rigorous quantitative guidelines.

2. Genetic Diversity in Pacific Salmon

The life history and biology of Pacific salmon have recently been reviewed in Groot and Margolis (1991). Reproduction and early juvenile rearing (of varying periods) are in

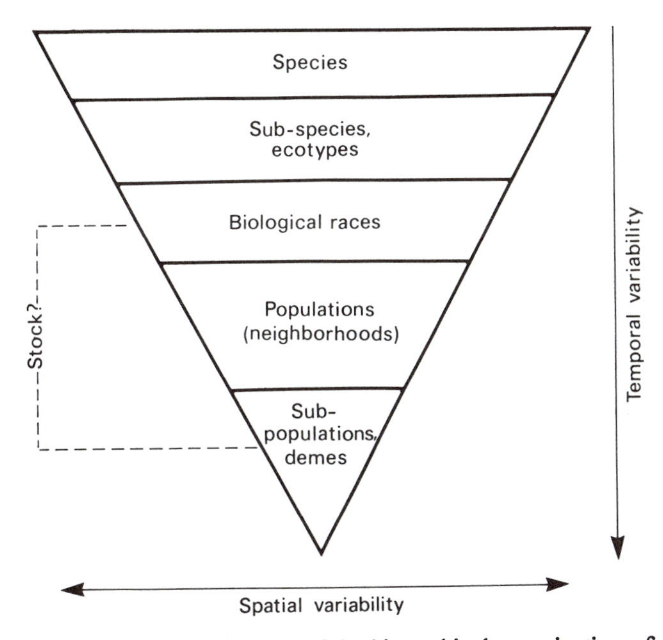

Figure 2. Schematic representation of the hierarchical organization of genetic diversity in Pacific salmon. The inverted triangle emphasizes that locally adapted, and largely reproductively isolated sub-populations or demes are the basic unit of diversity in these species. Varying definitions of "stock" would place this term in the range identified by the dashed lines.

freshwater but juveniles migrate to the sea for feeding until maturity and then return to their natal stream to spawn and die. Patterns of historical colonization following glaciation, adaptation to local spawning and rearing environments, and recent anthropogenic impacts have resulted in fragmented spatial distributions of locally adapted spawning populations (reviewed recently in Altukhov and Salmenkova, 1991; Taylor, 1991). Biological diversity within the Pacific salmon species naturally forms a hierarchical organization (Fig. 2).

Higher levels of the organization share more common life history traits, for example races of salmon defined by adult migration timing (spring and fall Chinook salmon, winter and summer steelhead trout). Lower levels within each higher division consist of more localized spawning groups adapted to finer scale environmental differences, and being largely reproductively isolated from other local groups. In population genetic theory, these isolated local groups are commonly referred to as demes (Gilmour and Gregor, 1939; Wright, 1969). The differentiation between a deme and larger population units becomes more arbitrary depending on the exchange of genes between population units. In Wright's treatise (1969), he describes several models for spatially sub-divided populations up to and including the neighborhood, defined as the population of a region in a spatial continuum. In conservation biology, the concept of the metapopulation (see Simberloff, 1988) is analogous to Wright's view of populations as being a loosely but connected group of demes.

The total genetic diversity in a species is the sum of variability between each hierarchical level and within each level, the latter including the cumulative variability from each lower level. Over time, the between population variation should increase as the species invades new territory or continues to adapt to environments within its range (temporal variability, Fig. 2). However, total diversity may not change and may even decrease over time depending on the spatial range maintained, the balance of selective versus disruptive genetic factors within the breeding populations, and the population dynamics of the species. The latter being of particular concern in Pacific salmon because of their long history of exploitation and anthropocentric impacts on freshwater habitats. At any one time though, the greatest diversity in such a hierarchically organized species exists over all hierarchical levels (populations within races, races within ecotypes, etc.) within the spatial range of the species (spatial variability, Fig. 2). Unfortunately, such a truism does not assist us in determining how much diversity is conserved over a specified area or group of populations; and the degree of differentiation between hierarchical levels differs between salmon species (Ryman, 1983; Altukhov and Salmenkova, 1991). What we can deduce from this organization is that maintaining maximum biological diversity requires maximizing the number of demes conserved over time and space.

The term "stock" has been used extensively in referring to a population of salmon, but its definition and application has been inconsistent and, at times, confusing. Consequently, stock may refer to various organizational levels in Figure 2 and may not be a useful term in conservation discussions. The word "stock" is from old English use meaning descent or lines of ancestor. Its use in fisheries literature follows from the 1938 Conference on Salmon Problems (Moulton, 1939) when stock was selected as the most desirable term to indicate that there was not necessarily a hereditary basis for differences observed between groups of salmon. No evidence was presented at this conference for hereditary factors in differences between "river stocks". Hereditary factors have, of course, now been demonstrated and, for Pacific salmon, summarized in Ricker's (1972) famous paper on hereditary and environmental factors affecting salmonid populations. Confusion in the use of stock results from its initial application in identifying intra-specific groups for fishery management, and numerous subsequent efforts to define stock in terms of population structure. A major conclusion of the 1980 Stock Concept symposium, however, was that the latter is not as important as acceptance of the stock concept to provide a genetic perspective in management decisions (MacLean and Evans, 1981). The stock concept is a conceptual summary of the genetic basis to the spatial hierarchy in Figure 2. The term "stock" though should probably not be used to describe population units in conservation since the population level it refers to is uncertain and likely varies between individuals (see Dizon et al., 1992). A common terminology to describe the spatial organization of Pacific salmon should be accepted to promote understanding of biological diversity. Terms in Figure 2 are commonly used in population genetics literature .

The term "stock" remains useful in managing fisheries and may be defined as a manageable or recognizable group of population units (Larkin, 1972). Larkin also acknowledges that "what we define as a stock is partly an arbitrary decision taken for human convenience. Any practical management definition of a stock requires some degree of treating groups collectively rather than separately." Pacific salmon demes are unlikely to be harvested discretely or to be identifiable uniquely. Consequently, managing stocks and conserving demes involves two issues: identifying the stocks in a fishery, and limiting harvest rates to maintain inter-demic diversity within stocks.

3. What Is It We Are Preserving?

Conserving biological diversity will involve trade-offs with other management objectives and will incur costs. It is appropriate then to briefly consider the values of conserving this diversity, particularly since the necessity for maintaining diversity will continue to be questioned. A notable example of this is Larkin's (1981) perspective on population genetics and fisheries management. Although Larkin had been more conservative in previous papers, in this one, he questions the emphasis on between "stock" differences and suggests that "Insofar as genetics is concerned, we should not become too hysterical about population declines to low levels. " His latter suggestion should now be rejected (see Soule, 1987), although hysterical seems to over state the point, but information on the relative fitness of different populations is still limited.

3.1. Economic Values

The most obvious values are the financial returns from fishing and tourism and the potential for future development. If the only objective in salmon management was maximizing the sustainable catch for commercial fisheries, then maintaining broad population diversity may not be necessary or consistent with this objective. Production could be sustained by the most productive natural and enhanced populations but less productive ones would be over exploited (Ricker, 1958; Kope, 1992), possibly resulting in extinction. Management objectives are not so simple, however, and over the past 15 to 20 years three major changes favouring diversity have occurred. The most significant change has likely been the development of specific catch sharing agreements (allocation) between an increasing number of user groups. These agreements have limited harvest in mixed stock and species fisheries to allow certain species and individual stocks to escape to fisheries closer to their natal streams. The obvious example of this is the sharing of Pacific salmon catches between native and non-native fishers in the United States (Clark, 1985; Blumm, 1990). The rapid expansion of recreational fishing and investment in tourism has also contributed to change. Along the west coast of Canada, recreational fishing and related tourism have been estimated to now equal the value from the Pacific salmon commercial fishery. The third change has been the substantial increase in world salmon production through mariculture, and from hatchery and wild populations. Recent high catches of salmon have reduced prices and are changing industry concerns from the supply of salmon to optimizing economic benefits from a catch limited by available markets. Distributing fishing effort and catch over time and space and stabilizing salmon production over more populations would benefit the industry and favour population diversity.

There are also indirect economic benefits. Maintaining a diverse population basis will increase habitat utilization and the stability of their natural ecosystem. Protecting and utilizing productive fish habitats may become important in negotiations with other resource developers. Efforts to preserve these habitats for salmon would be weakened if it was not utilized, particularly since natural re-colonization will be slow and salmon transfers have a very low success rate in re-establishing self-sustaining populations (Withler, 1982).

3.2. Social Values

Pacific salmon are an integral part of the social heritage of the Pacific coast of North America. The cultural significance of Pacific salmon to native peoples is without question, and salmon fishing has also been fundamental in the development of non-native communities (Lyons, 1969; Netboy, 1980; Roos, 1991). But even in the general public, Pacific salmon are an integral part of the aesthetics in this region. Large numbers of people volunteer to participate in habitat improvement and small scale salmon enhancement projects. Educational programs to foster a conservation ethic are an essential part of salmon enhancement programs. Access to salmon fishing is a valued recreational opportunity and its economic importance was noted above. It is probably not an exaggeration to suggest that a healthy Pacific salmon resource is part of the moral values in these west coast communities. Healthy salmon populations are equated to a healthy environment and good resource management practices.

Comparing these societal values with other values which can be determined quantitatively (for example, the economic value of another commodity) will be a problem, but the importance of social values in maintaining biological diversity should not be under-estimated. People live throughout most of the range of Pacific salmon and will naturally value maintaining populations in their local area. E.O. Wilson (1984) has argued that human nature has developed "in a good part because of the particular way we affiliate with other organisms." Along the Pacific coast, responsible stewardship of localized salmon populations probably has strong societal support, but it must be expressed and heard. Procedures likely need to be established to ensure that these values are considered in management decisions (Scarnecchia, 1988).

3.3. Biological Values

In a practical sense, the biological diversity presently observed is a non-renewable resource, and only an instant in the dynamic evolutionary process. The diversity has resulted from colonization events, innumerable events which changed genetic variation, and the differential fitness of individuals over past environments. Once a spawning population is lost, any unique traits it may have possessed are realistically gone forever. Consequently, the principal biological values are adaptedness in the existing populations, maintenance of population structures and the evolutionary process, and, very simply, maintaining the spatial and temporal basis for salmon production. The latter would be true even if the diversity between spawning populations did not have a genetic component since salmon accurately return to their natal habitat (summarized in Table 1 of Quinn, 1990; Gharrett and Smoker, 1993). This point was first made by W.M. Rich (1939) at the 1938 Conference on Salmon Problems. But the value of maintaining diversity for production is still frequently confused with debates about genetic variation between spawning populations and how much to conserve. The distribution of existing spawning populations and protection of the rearing habitat provides the basis for production. The existing production may not be maximal due to depressed population sizes or unutilized habitat but maximizing the number of demes maintained will likely increase juvenile production (and presumably total production unless the marine environment is limiting). Maintaining most of the demes would limit exploitation rates to those sustainable by demes with lower productivity. Spawning numbers in the demes with higher productivity would increase and juvenile production more fully utilize the habitat. Further, large spawning escapements may stimulate colonization of unused spawning and rearing environments.

Another value, but one which will probably receive little weight compared to the above, is the knowledge created by studying the present population structure and biological diversity. Our understanding of the population genetics of Pacific salmon is still very limited and concerns have been expressed about the genetic impact of harvest practices (Ricker, 1981; Healey, 1986; Riddell, 1986), and hatchery practices and production (Goodman, 1990; Hilborn, 1992). Studies of the remaining natural populations are essential to develop genetic guidelines for restoration programs and conservation of the species.

4. What Are the Major Sources of Impacts?

It is impractical in this essay to consider the variety and combinations of specific factors which have reduced diversity and population sizes. The factors involved and their relative importance will vary and will have to be identified and managed within each situation. However, to summarize sources of past impacts and concerns for the future, five categories of impacts have been identified, but their impacts are not independent. For example, the catch sustainable from a population size varies with the productivity of the population and the habitat carrying capacity. If productivity is reduced through habitat loss or environmental change then fishing pressures must be reduced to sustain that population size.

4.1. Fishing

Pacific salmon are heavily exploited species. Historically, fishing was very near or in rivers, but with the development of refrigeration in the early 1900's, fishing began to move off-shore becoming more distant from the rivers. This led to the development of ocean fisheries harvesting mixtures of many salmon populations, sequential exploitation by several fisheries on an individual population, and intense competition both internationally and domestically. The biological results were decreased spawning population sizes, changes in the biological characteristics of the spawning populations, and the creation of complex salmon management problems. The Canadian Commission on Pacific Fisheries Policy concluded that "the immediate cause of continuing declines and low levels of abundance (of salmon) is overfishing" (p.14, Pearse, 1982). Fishing has contributed to decreased diversity in Pacific salmon, but fishing is also the first impact targeted for conservation actions since fishing can be immediately controlled when necessary.

While resource managers are now acutely aware of many problems, resolving these problems and controlling fishing impacts will remain an unenviable task. Efforts to maintain a diverse resource base and increase production from depressed populations will be increasingly disruptive to fisheries, local communities, and other activities potentially affecting fish habitat. But denial and inaction will only exacerbate the eventual impact, whether it is extinction of a population or increased costs to maintain it. Frequently, the controversy about a conservation action involves denial of responsibility for the problem...some other group was the principal cause. Individual fishing groups do not believe that their impact is enough to cause the problem or lead to extinction, and they should not, therefore, be heavily impacted by the conservation plan. Management actions also seem disproportionate to the contribution of the population to a fishery. Controversy naturally arises from this situation since any one group is likely correct, but it is the aggregate impact of all groups that must be managed. Further, these debates have to be addressed even when the fishery may not be the principal cause. Future salmon production can only be developed from what is present and conserved today. Consequently, if the causes require time to correct then the population can be conserved by reducing the harvest impacts, or possibly, by more artificial actions such as gene banks (Bergan et al., 1991).

4.2. Point Source Problems

A point source problem refers to a wide variety of localized anthropocentric impacts which reduce the productive capacity of salmon habitat. The impact of 28 dams in the Columbia River hydropower system is an obvious example (Raymond, 1988). Water regulation for power, irrigation, and industry is a wide spread concern along the Pacific coast (Dorcey, 1991; Mundie, 1991) but is also only one of many habitat impacts. In a recent review of habitat impacts in the Fraser River, twelve impacts associated with human activities were identified (Table 1). Henderson (1991) associates habitat impacts (temperature, flow, etc.) with each of these activity types by habitat types (lakes, tributaries, mainstem, estuary). It is not unexpected though that a large variety of impacts were identified when 86% of the province's population lives in British Columbia's largest watershed.

Table 1. Human activities affecting Pacific salmon habitat in the Fraser River, B.C. Canada (summarized from Henderson, 1991).

Dykes and stream channelization	Argiculture and water abstraction
Municipal and industrial landfills	Logging
Port development	Silviculture
Dredging and log storage	Roads, railways, and transport of dangerous goods
Urbanization and municipal effluents	Pulp mill effluent and wood preservatives
Mining	Dams

Early development in the Fraser severely depressed salmon production following the 1913 rock slide at Hell's Gate in the Fraser canyon, and logging dams at the outlets of Quesnel and Adams lakes (Ricker, 1987; Roos, 1991). Sockeye runs to these lakes were amongst the largest sockeye runs in the Fraser. Sockeye production in the Quesnel has now recovered but returns to the Upper Adams river remain very small and the original Upper Adams sockeye are likely extinct. The logging dam on the Adams River was built in 1908 (Fig. 3) and in 1911 no sockeye were reported above the dam. Small numbers of sockeye now return to Upper Adams spawning streams but 16 transplants of non-local populations were released between 1949 and 1975 (Williams, 1987). If the original runs were not extinct following the dam, then the genetic race has likely now been lost through genetic drift and introgression. Williams reports that the Upper Adams River has 1.2 million m^2 of spawning area. Consequently, the

Figure 3. The Adams River logging dam 1908-1921. This dam blocked passage of summer run sockeye to the Upper Adams River, populations in this river have not recovered.

loss of this one population has cost the fishery, on average, six million sockeye per cycle year based on the productivity of other summer sockeye populations in the area (Cass, 1989).

Point source problems will have to be identified on a case-by-case basis and may not be easily addressed. Frequently, their impact on salmon production is unknown and difficult to partition from the annual environmental variation observed in salmon populations. Further, these impacts result from large industrial economic and urbanization bases. Conservation discussions will therefore contrast values associated with salmon diversity with values and ethics of other industries and interest groups (Norton, 1986; Callicott, 1991).

4.3. Urbanization and Population Growth

Population growth and urban development are obviously associated with Fishing and the Point Source Problems but are an increasingly important concern for the future. The population in British Columbia has increased by approximately 50% since the 1971 census and is expected to double by about year 2010 (Ministry of Finance and Corporate Relations, Province of BC). The direct impacts of this in salmon management are likely to be loss of small stream habitat, increased water regulation, and increased recreational fishing effort. The indirect effect may be increasingly polarized debates between conservation values and economic and health necessities of an expanding population. Resource managers would be well advised to begin planning how to meet these demands.

4.4. Biological Limitations and Climate Change

As in the Point Source Problems, numerous biological factors may influence the productivity and existence of salmon populations but each situation will likely differ. The types of biological factors involved include: intra and interspecific competition, predation, exotic introductions, and disease. Loss of biological diversity is not commonly attributed to biological interactions except for the notable exception of exotic introductions: sea lamprey (*Petromyzon marinus*) in the Great Lakes (Smith, 1968; Smith and Tibbles, 1980), the parasite *Gyrodactylus salaris* in Norway (Johnsen and Jensen, 1991), and the opossum shrimp (*Mysis relicta*) in Idaho lakes (Spencer et al., 1991). The ecological and genetic effects of fish introductions were the topic of a recent international symposium (FIN, May 17-19, 1990, Windsor, Ontario). In his synthesis paper Allendorf (1991) concludes: "Purposeful introduction rarely have achieved their objectives. Moreover, both intentional and unintentional introductions usually have been harmful to native fishes and other taxa through predation, competition, hybridization, and the introduction of disease." Several papers in that symposium showed that genetic effects of introductions constitute a threat to the long-term existence of wild populations and species.

The primary intention of identifying this type of impact was that biological factors may become more important when attempting to restore population sizes and if climate changes. If a population has been in low abundance for many years, its community may have adjusted such that the species can not expand its share of the resources or its abundance. Projections about climate change in Canada (Hengeveld, 1991; McBean et al., 1991) and the Pacific northwestern United States (Leovy and Sarachik, 1991; Neitzel et al., 1991) suggest it could have a significant effect on biological diversity in Pacific salmon, particularly in the southern portions of species' ranges. Climate change will reduce the freshwater productivity of southern salmon populations, and may threaten the survival of populations if they can not adapt. The latter will be of particular concern in small, spatially isolated demes that may not have the genetic variation remaining to adjust to a rapidly changing environment.

4.5. Hatchery Impacts

The use of hatcheries has a long history in Pacific salmon management. Hatcheries and spawning channels are used to augment catches, mitigate environmental impacts, and to supplement numbers of natural spawners. The numbers of Pacific salmon juveniles released from facilities are staggering (over 5 billion per year in the late 1980's) and McNeil (1991) has estimated that 55% of the world salmon harvest in 1990 was cultured salmon (mariculture plus ocean ranched). The economic importance of this contribution is obvious but there is clearly growing concern about the long term impact of cultured production on the genetics of natural populations (Helle, 1981; White, 1989; Goodman, 1990; Hindar et al., 1991; Nehlsen et al., 1991; Waples, 1991; Hilborn, 1992; Meffe, 1992). Waples (1991) summarizes the concerns as three issues: direct genetic effects (caused by hybridization and introgression), indirect genetic effects (due to altered selection regimes or reductions in population sizes caused by competition, predation, disease, or other factors), and genetic changes in hatchery populations which magnify consequences of hybridization with wild fish. Waples (1991) and Hindar et al. (1991) both document that hatchery production can have substantial direct and indirect genetic effects on wild fish.

Managing these concerns will again be controversial. Hatchery production is seen by user groups as a technical solution to difficult management problems or the loss of productive habitat. Basically, the controversy will contrast production objectives of users (to sustain catch) and longer term management objectives to conserve genetic diversity. However, as Waples (1991) also states, production and conservation objectives are inseparable in the stewardship of the Pacific salmon resource. The issues are how to utilize hatchery production while controlling harvest to protect wild populations, and to develop genetic guidelines for culture programs so that direct and indirect genetic effects are minimized.

5. What to Conserve?

The rhetorical response is simple: "Everything". In practice though, the response seems to have been "as much as is practical". The latter response reflects past emphasis on salmon harvest objectives and, I suggest, management confusion in applying the stock concept. In British Columbia, if a population is not large enough or identifiable in a fishery, then fishery managers refer to these populations as being passively managed. The designation resulted from practical limitations but signifies acknowledgment of a risk of population losses. Under the broader set of values discussed above, more balance between short (present harvest) and long (genetic diversity and sustained production) term objectives would be anticipated. To achieve this, managers will require clear policy statements about fishery management goals, and advice on the population dynamics, biological characteristics of the species, and how to conserve genetic diversity. The following principles are proposed for the latter, but are likely more general than a manager would desire. However, more specific advice would quickly lose its general applicability and could be applied incorrectly without a manager's appreciation. Given our limited knowledge of population genetics in Pacific salmon and their fine-scale spatial organization, it seems appropriate to advise from a conservative perspective.

These principles are listed in decreasing order of importance and assume the hierarchical model of biological diversity, and a broad spatial scale of populations impacted by harvest, habitat alterations, and the presence of hatchery populations. The priority of these principles may vary between situations in more localized areas.

(i) In the absence of other proof, manage Pacific salmon from the premise that localized spawning populations are genetically different, and valuable to the long term production of this resource.

To borrow from Waples (1991) and the conclusion of the FIN symposium, "First, do no harm." This is obviously an idealistic principle but aptly emphasizes the importance of maintaining genetic diversity. By managing from this premise and protecting habitat, resource managers are, first, stewards of the resource for long term production and, secondly, managers of short term utilization and impacts. Managing from the opposite premise will continue to result in lost diversity and production. **It is simply untenable to expect managers to prove value in each localized population before it will be conserved.**

(ii) Identify higher levels of organization which are threatened.

Genetic variation between the higher organizational units (sub-species, races, etc.) resulted from largely independent evolutionary lines. Differences observed between these units are important components of diversity but their independence also implies importance as reserves of rarer genes.

(iii) Maintain a broad perspective of the spatial and temporal impacts.

Too narrow a focus on production has led to conservation problems but too narrow a focus on one conservation problem may also be counterproductive. A narrow focus could mask sources of the conservation problem, generate new problems by neglecting other local populations or ecological issues, and could be detrimental to credibility in the broader resource management community.

(iv) Maintain genetic variation within populations by maximizing the spatial and temporal distribution of demes.

The likelihood of maintaining genetic variation, and therefore adaptability, increases as the number of salmon reproducing per deme increases and the number of demes increase. Further, maintaining demes in marginal environments is possibly more important than currently appreciated in salmon management (Scudder, 1989). Scudder suggests that conserving marginal populations and habitats is one of the "best" ways to conserve genetic diversity. By maximizing the spatial and temporal distribution of spawning populations, the numbers of spawners per deme would increase, exchange between demes would be facilitated, and new demes may develop as spawners disperse from more productive habitats

As for advice concerning a minimum spawning population size, more evaluation of this issue is required. The concept of a minimum viable population size (see Soule, 1987) for long term survival has not, to my knowledge, been applied to Pacific salmon. Further, the concept will not provide a single value to be applied to all populations (nor is this appropriate since each situation is likely unique). Rather, it considers the probability of a population surviving over a specified time, environmental variation, and the genetic effective population size (N_e, for Pacific salmon see Waples, 1990). Managers must recognize, however, that the genetic N_e can be substantially smaller than the census population observed on spawning grounds. The two values can not be confused.

(v) Maintain groupings of fragmented populations/races, and contiguous distributions between fragments to facilitate gene flow.

The loss of genetic material in small populations is dominated by random events (genetic, demographic, environmental). Consequently, the loss of genetic material in different populations should be independent. Over many such populations, a large proportion of the original genetic variation should exist.

Managers should not label these small populations as "biologically unviable" or "economically extinct" (phrases from author's experience). Their economic value is diversity and opportunities for future production, and the viability debate frequently confuses small numbers with low productivity. If viability refers to population continuance then it depends on why the population is small (habitat capacity, overfishing, etc.), productivity of the population, and stochastic variation. Small populations are at greater risk of random extinction but are not necessarily unproductive. If over-fishing is the cause, the maximum exploitation rate a population can sustain is a function of its productivity (the maximum rate of adult returns per spawner at low population size) only. For the reasons presented in principles (*iii*) and (*iv*), small populations should not be ignored and actually merit some cost to conserve them.

(vi) Identify remaining wild populations and/or areas of least disrupted habitats, and protect these over a broad spatial range.

Undisturbed habitat and populations are Nature's in situ "gene bank" and are increasingly important as biological controls and study sites. Unfortunately, over large portions of the Pacific salmons' ranges it is difficult to identify such sites. Further, for such refuges to be valuable in conserving diversity, a wide spatial distribution of them will be required and whole watersheds may be necessary to maintain ecological interactions.

(vii) Manage intensive culture programs to maintain genetic variation within the cultured populations and to minimize genetic effects on natural populations.

These topics have recently been reviewed by Allendorf and Ryman (1987) and Waples (1991). Maintaining variation in cultured populations may increase productivity and reduce concerns about direct genetic effects on natural populations. However, whether selection can be averted in intensive culture situations is uncertain. The use of non-local populations in brood stock has largely been stopped and is strongly advised against, unless in extreme situations and after thorough public review.

The development of mariculture is an additional threat of genetic effects on natural populations. The mixing of mariculture fish with wild fish will occur less frequently than from ocean ranching of hatchery fish, but the risk of genetic consequence is greater. Catastrophic mixing may occur following large scale escapes from sea pens, and the genetic composition of the mariculture population will differ from the wild because of selection and the use of non-local population in brood stocks. To protect natural diversity, the frequency of escapes must be minimized (an objective obviously shared by the industry) and guidelines developed to minimize their genetic impact.

(viii)Maintain populations with unique genetic traits or, at least, with genetic traits of important local value.

These populations are non-renewable resources and must be conserved to protect present or future opportunities. This principle seems self-evident but risks continue to be imposed on such populations. Once the gene complexes controlling these traits are altered or lost, reestablishing them is unlikely.

(ix) Maintain original source populations used in resource developments.

Maintaining large numbers of natural spawners in populations used for developments (for example, original brood sources for a hatchery or mariculture program, or a localized fishing opportunity) protects diversity and the development investment. The development of brood stocks for mariculture is an important example. If problems develop in the brood stock then the natural population provides a common source of genetic material to correct the problem. On the other hand, a large natural population may be needed to counteract potentially large numbers of escapees from the mariculture site.

(x) Maintain populations occupying atypical habitats or expressing unusual phenotypic traits.

Experience with Pacific salmon clearly indicates that phenotypic variation between populations is usually associated with some genetic variation. Further, the utilization of different habitats may have a genetic basis and these marginal environments may be important in maintaining genetic diversity (Scudder, 1989).

6. Discussion

Post-glacial re-colonization, habitat patterns, precise homing of Pacific salmon to natal streams, and evolutionary processes have resulted in a spatial hierarchy in the genetic organization of Pacific salmon. The hierarchy is based on locally adapted, largely isolated spawning groups or demes. Anthropocentric impacts during the past century have increasingly disrupted the genetic structure and habitat basis of salmon production. However, over the past 10-20 years, the focus of Pacific salmon management decisions has diversified from principally one of maximum production for commercial fisheries. Changes in resource allocation and broadening of economic benefits, coupled with rising environmentalism and expression of social values, provide increased opportunity to conserve and rehabilitate salmon populations and habitats. Genetic variation, within and between hierarchical levels, and productive habitat are the resource base of Pacific salmon, both for long-term sustainable production and continuing evolutionary processes. Opportunities to conserve this base should be vigorously pursued and tested, presumably by explicitly incorporating conservation objectives in salmon management planning and practice (for example, see Riggs, 1990; PMFC, 1992). Rehabilitation of populations (number, distribution, and size) and protection of habitat will also be essential for minimizing impacts of climate change.

The principles in this paper provide advice about how to conserve genetic diversity through Pacific salmon management. Unfortunately, a plan of how to conserve may be inadequate for successful conservation. Success will require commitment to genetic conservation in salmon management policy, processes to consider economic, social, and biological values in decision making, plus four activities also identified in Allendorf (1991): education, cooperation between management agencies, regulation, and research. Some of these will require more time to develop, particularly methods to compare different types of values and the development of decision processes, but others should proceed immediately.

An appeal for more research has almost become a cliche in scientific literature, but is appropriate in population genetics and conservation biology (including habitat requirements) of Pacific salmon. Nelson and Soule (1987) express this need very well:

> "In surveying the causes of loss of genetic diversity we are struck by how often the conspirators are not the expected Ignorance and Greed but, rather, the equally dangerous Partial Knowledge and Good Intentions."

Those authors emphasized the need for studies of population structure (determining N_e and numbers of migrants) and the effect of selective pressures generated by exploitation on life history traits. Additional requirements are for studies of genetic variation within hatchery populations (inadvertent selection, N_e, inbreeding, operational guidelines) and the genetic interaction of hatchery and wild fish. In response to concerns about climate change, genetic studies of thermal tolerance and correlated traits are advised for populations in areas expected to be affected. It is notable that increased emphasis on genetic conservation may also change public perspectives about research costs in salmon management. Under principle (i), limited knowledge should limit utilization. Investments in research may have a more tangible benefit since utilization could become less restricted as we learn what and how to conserve.

This work should proceed immediately to provide a solid information base for management but results of genetic studies will not be available for several years. In the interim, conservation interests are best addressed through evaluation of the existing resource base, and education of the public and management agencies. Evaluation includes an inventory of existing genetic and habitat resources, and assessment of present versus potential production (stock assessment, see Gulland, 1983). If spawning population sizes are less than the management goals then an immediate conservation benefit can be achieved by increasing the number of spawners and/or increasing the productivity of the population. Conservation will also be benefited by maintaining the broadest spatial and temporal distribution of demes, including the small fragmented and/or marginal demes. Education now becomes paramount. Immediate increases in the number of spawners and populations can only be achieved by reallocating catch (user groups' benefits) to breeding populations. Beneficiaries from the salmon resource are usually supportive of conservation needs but will, understandably, argue to minimize disruption of their usage.

Geneticists advising on conservation of Pacific salmon have a responsibility to user groups and the public to explain, in understandable terms, what genetic diversity is, the importance of conserving it, and the basis for recommended management principles. The essential role of education in resource management was also emphasized by the FIN symposium (Allendorf, 1991):

> "Education is a key in dealing with these issues. Many of the past and current arguments in favor of introductions have been based upon perceived societal demands for food, recreation, or economic benefits. There are two central educational issues. First, the history of introductions tells us that such introductions rarely achieve their objectives. Second, society must realize that such introductions also involve a 'cost' and we usually do not understand natural systems sufficiently to know what the cost will be."

Similar information and education problems exist for Pacific salmon. Management debates frequently involve a small but recurring set of fallacies:

a. lowering harvest rates to restore production will result in continually lower catch
b. small populations are unproductive populations
c. hatcheries are solutions to management and habitat issues (referred to as the Technological Fix Syndrome, Hilborn 1992)
d. sustaining production and conserving genetic diversity are incompatible objectives
e. only a small number of salmon are required for the preservation of populations.

These fallacies must be explained and their impacts identified before managers should expect an increased commitment to an explicit conservation objective in salmon management. Each has been addressed to some extent in this paper.

Genetic conservation in Pacific salmon is fundamentally sustainable development (WCED, 1987) providing a longer term perspective of utilization for a broader set of beneficiaries. The resource bases for sustaining production are genetic diversity (genetic variation within and between all levels of the organizational hierarchy) and the habitat utilized by all life stages of the species. Genetic diversity provides for the continuing evolutionary processes and the biological basis of future production. The first step in sustaining production is improved stewardship of existing resources, but a greater challenge for conservation may be maintaining productive habitat. Human population growth and associated economic development plus climate change can be anticipated to increasingly threaten salmon habitat. Successful conservation will also require habitat protection, educational programs, increased research, and the establishment of accountable decision processes to consider multiple resources and all types of resource values. Realistically, salmon conservation can not proceed independently of other natural resources, particularly the increasing demands for freshwater.

References

Allendorf, F. W. 1991. Ecological and genetic effects of fish introductions: synthesis and recommendations. *Canadian Journal of Fisheries and Aquatic Sciences* 48 (Suppl. 1):178-181.

Allendorf, F.W. and N. Ryman. 1987. Genetic management of hatchery stocks, in: 'Population Genetics and Fishery Management," N. Ryman and F. Utter, eds., pp.141-159, University of Washington Press, Seattle.

Altukhov, Y.P. and E.A. Salmenkova. 1991. The genetics structure of salmon populations. *Aquaculture* 98:11-40.

Bergan, P.I., D. Gausen, and L. P. Hansen. 1991. Attempts to reduce the impact of reared Atlantic salmon on wild in Norway. *Aquaculture* 98:319-324.

Blumm, M. 1990. Anadromous fish law—1979-90. Anadromous Fish Law Memo 50. Extension/Sea Grant Program. Oregon State University, Corvallis, Oregon, USA.

Callicott, J.B. 1991. Conservation ethics and fishery management. *Fisheries* (Bethesda) 16:22-28.

Cass, A. 1989. Stock status of Fraser River sockeye. Canadian Technical Report of Fisheries and Aquatic Sciences 1674.

Clark, W.G. 1985. Fishing a sea of court orders: Puget Sound salmon management 10 years after the Boldt decision. *North American Journal of Fisheries Management* 5:417-434.

Dorcey, A.H.J. ed. 1991. Perspectives on sustainable development in water management: Towards agreement in the Fraser River basin. Vl. Westwater Research Centre, University of British Columbia, Vancouver, BC.

Dizon, A.E., C. Lockyer, W.F. Perrin, D.P. Demaster, and J. Sisson. 1992. Rethinking the stock concept: a phylogeographic approach. *Conservation Biology* 6:24-36.

FAO (Food and Agriculture Organization of the United Nations). 1981. Conservation of the genetic resources of fish: problems and recommendations. Report of the Expert Consultation on the Genetic Resources of Fish, Rome, 9-13 June 1980. Technical Report Paper 217. Food and Agricultural Organization of the United Nations, Rome.

Fraidenburg, M. and R. Lincoln. 1985. Wild chinook salmon management: an international conservation challenge. *North American Journal of Fisheries Management* 5: 311-329.

Gharrett, A. and W.W. Smoker. 1993. Genetic components in life history traits contribute to population structure, in : "Genetic conservation of Salmonid Fishes," J.G. Cloud, ed., pp. 192-198, these proceedings.

Gilmour, J.S.L. and J. W. Gregor. 1939. Demes: a suggested new terminology. *Nature* 144: 333 .

Goodman, M.L. 1990. Preserving the genetic diversity of salmonid stocks: a call for federal regulation of hatchery programs. *Environmental Law* 20:111-166.

Groot, C. and L. Margolis, eds. 1991. "Pacific salmon life histories," University of British Columbia Press, University of British Columbia, Vancouver.

Gulland, J. 1983. Stock assessment: Why? Fisheries Circular 759. Food and Agriculture Organization of the United Nations, Rome.

Healey, M.C. 1986. Optimal size and age at maturity in Pacific salmon and effects of size-selelective fisheries, in: "Salmonid age at maturity," D.J. Meerburg ed., pp.39-52, Canadian Special Publication of Fisheries and Aquatic Sciences 89.

Helle, J.H. 1981. Significance of the stock concept in artificial propagation of salmonids in Alaska. *Canadian Journal of Fisheries and Aquatic Sciences* 38:1665-1671.

Henderson, M A. 1991. Sustainable development of the Pacific salmon resources in the Fraser River basin, in: 'Perspectives on sustainable development in water management: Towards agreement in the Fraser River basin, "A.H.J. Dorcey, ed., Vol 1, pp. 133-144. Westwater Research Centre, University of British Columbia, Vancouver.

Hengeveld, H. 1991. Understanding atmospheric change: a survey of the background science and implications of climate change and ozone depletion. Environment Canada, State of the Environment Report 91-2, Ottawa.

Hilborn, R. 1992. Hatcheries and the future of salmon in the northwest. *Fisheries* (Bethesda) 17:5-8.

Hindar, K., N. Ryman and F. Utter. 1991. Genetic effects of cultured fish on natural fish populations. *Canadian Journal of Fisheries and Aquatic Sciences* 48(5):945-957.

Johnsen, B.0. and A.J. Jensen. 1991. The *Gyrodactylus* story in Norway. *Aquaculture* 98:289-302.

Kope, R.G. 1992. Optimal harvest rates for mixed stocks of natural and hatchery fish. *Canadian Journal of Fisheries and Aquatic Sciences* 49(5) :931-938.

Larkin, P.A. 1972. The stock concept and management of Pacific salmon, in: 'The stock concept in Pacific salmon," R.C. Simon and P.A. Larkin, eds., pp. 11-15, H.R. MacMillan Lectures in Fisheries, University of British Columbia, Vancouver.

Larkin, P.A. 1981. A perspective on population genetics and salmon management. *Canadian Journal of Fisheries and Aquatic Sciences* 38(12):1469-1475.

Leovy, C. and E.S. Sarachik. 1991. Predicting climate change for the Pacific northwest. *Northwest Environmental Journal* 7:169-201.

Lyons, C. 1969. "Salmon our heritage, the story of a province and an industry," Mitchell Press Ltd., Vancouver.

MacLean, J.A. and D.O. Evans. 1981. The stock concept, discreteness of fish stocks, and fisheries management. *Canadian Journal of Fisheries and Aquatic Sciences* 38(12):1889-1898.

McBean, G.A., O. Slaymaker, T. Northcote, Pl LeBlond, and T.S. Parsons. 1991. Review of models for climate change and impacts on hydrology, coastal currents, and fisheries in British Columbia. Canadian Climate Centre, Downsview, Ont. CS Report 91-11.

McNeil, W.J. 1991. Expansion of cultured Pacific salmon into marine ecosystems. *Aquaculture* 98:173-183.

Meffe, G.K. 1992. Techno-arrogance and halfway technologies: salmon hatcheries on the Pacific coast of North America. *Conservation Biology* 6:350-354.

Moulton, F.R., editor. 1939. The migration and conservation of salmon. Publication of the American Association for the Advancement of Science 8.

Mundie, J.H. 1991. Overview of effects of Pacific coast river regulation on salmonids and the opportunities for mitigation. American Fisheries Society Symposium 10:1-11.

Nehlsen, W., J.E. Williams, and J.A. Lichatowich. 1991. Pacific salmon at the crossroads: stocks at risk from California, Oregon, Idaho, and Washington. *Fisheries* (Bethesda) 16:4-21.

Neitzel, D.A., M. J. Scott, S. A. Shankle, and J. C. Chatters. 1991. The effect of climate change on stream environments: the salmonid resource of the Columbia River basin. *Northwest Environmental Journal* 7:271-293.

Nelson, K., and M. Soule. 1987. Genetical conservation and exploited fishes, in: 'Population Genetics and Fishery Management," N. Ryman and F. Utter, eds., pp 345-368, University of Washington Press, Seattle, Washington.

Netboy, A. 1980. "The Columbia River Salmon and Steelhead Trout, Their Fight for Survival," University of Washington Press. Seattle, Washington.

Norton, B.G., ed. 1986. "The Preservation of Species, the Value of Biological Diversity," Princeton University Press, Princeton, New Jersey.

Oldfield, M. L. 1989. "The Value of Conserving Genetic Resources," Sinauer Association Incorporated. Sunderland, MA.

Pearse, P. H. 1982. Turning the Tide: A new policy for Canada's Pacific fisheries. Department of Fisheries and Oceans, Communications Branch, 555 West Hastings Street, Vancouver, British Columbia.

PMFC (Pacific Marine Fisheries Commission). 1992. Preseason report III. Final environmental assessment of proposed emergency management measures adopted by the Council for the 1992 ocean salmon season. Pacific Marine Fisheries Council, Portland, Oregon, USA.

Quinn, T.P. 1990. Current controversies in the study of salmon homing. *Ethology Ecology & Evolution* 2:49-63.

Raymond, H.L. 1988. Effects of hydroelectric development and fisheries enhancement on spring and summer chinook salmon and steelhead in the Columbia River basin. *North American Journal of Fisheries Management* 8:1-24.

Rich, W.H. 1939. Local populations and migration in relation to the conservation of Pacific salmon in the western states and Alaska, in: 'The Migration and Conservation of

Salmon," F.R. Moulton, ed., pp. 45-50, Publication of the American Association for the Advancement of Science 8.

Ricker, W.E. 1958. Maximum sustainable yields from fluctuating environments and mixed stocks. *Journal of the Fisheries Research Board of Canada* 15 (5):991-1006.

Ricker, W.E. 1972. Heredity and environmental factors affecting certain salmonid populations, in: 'The Stock Concept in Pacific Salmon," R.C. Simon and P.A. Larkin eds., pp. 19-160, H.R. MacMillan Lecture Fisheries. University of British Columbia, Vancouver.

Ricker, W.E. 1981. Changes in the average size and average age of Pacific salmon. *Canadian Journal of Fisheries and Aquatic Sciences* 38(12):1636-1656 .

Ricker, W.E. 1987. Effects of the fishery and of obstacles to migration on the abundance of Fraser River sockeye salmon (*Oncorhynchus nerka*). Canadian Technical Report of Fisheries and Aquatic Sciences 1522.

Riddell, B .E. 1986. Assessment of selective fishing on age at maturity in Atlantic salmon (*Salmo salar*): a genetic perspective, in: "Salmonid age at maturity," D.J. Meerburg, ed., pp. 102-109, Canadian Special Publication of Fisheries and Aquatic Sciences 89.

Riggs, L.A. 1990. Principles for genetic conservation and production quality. Report of GENREC/Genetic Resource Consulting to Northwest Power Planning Council, Portland, Oregon.

Roos, J. E. 1991. Restoring Fraser River salmon. The Pacific Salmon Commission, Vancouver, British Columbia.

Ryman, N. 1983. Patterns of distribution of biochemical genetic variation in salmonids: differences between species. *Aquaculture* 33:1-21.

Scarnecchia, D.L. 1988. Salmon management and the search for values. *Canadian Journal of Fisheries and Aquatic Sciences* 45(11):2042-2050.

Scudder, G.G.E. 1989. The adaptive significance of marginal population: a general perspective, in: 'Proceedings of the National Workshop on Effects of Habitat Alteration on Salmonid Stocks," C.D. Levings, L.B. Holtby, and M.A. Henderson, eds., pp. 180-185, Canadian Special Publication of Fisheries and Aquatic Sciences 105.

Simberloff, D. 1988. The contribution of population and community biology to conservation science. *Annual Review of Ecology and Systematics* 19:473-511.

Smith, S.H. 1968. Species succession and fishery exploitation in the Great Lakes. *Journal of the Fisheries Research Board of Canada* 25(4):667-693.

Smith, S.H. and J.J. Tibbles. 1980. Sea Lamprey (*Petromyzon marinus*) in lakes Huron, Michigan, and Superior: history of invasion and control, 1936-78. *Canadian Journal of Fisheries and Aquatic Sciences* 37(11):1780-1801.

Soule, M.E., ed. 1987. "Viable Populations for Conservation," Cambridge University Press, Cambridge.

Spencer, C.N., B.R. McClelland, and J.A. Stanford. 1991. Shrimp stocking, salmon collapse, and eagle displacement. *Bioscience* 41:14-21.

Taylor, E.B. 1991. A review of local adaptation in Salmonidae, with particular reference to Pacific and Atlantic salmon. *Aquaculture* 98:185-207.

Walters, C. and B.E. Riddell. 1986. Multiple objectives in salmon management: the chinook sport fishery in the Strait of Georgia, Britsh Columbia. *Northwest Environmental Journal* 2:1-15 .

Waples, R.S. 1990. Conservation genetics of Pacific salmon. II. Effective population size and the rate of loss of genetic variability. *Journal of Heredity* 81:267-276.

Waples, R.S. 1991 Genetic interactions between hatchery and wild salmonids: lessons from the Pacific Northwest. *Canadian Journal of Fisheries and Aquatic Sciences* 48(Suppl. 1):124-133.

WCED (World Commission on Environment and Development). 1987. Our common future. Oxford University Press, Oxford.

White, R. 1989. We're going wild: a 30-year transition from hatcheries to habitat. *Trout* (Summer 1989):14-49.

Williams, I.V. 1987. Attempts to re-establish sockeye salmon (*Oncorhynchus nerka*) populations in the Upper Adams River, British Columbia 1949-84, in: "Sockeye Salmon (*Oncorhynchus nerka*) Population Biology and Future Managenent," H.D. Smith, L. Margolis, and C.C. Wood, eds., pp. 235-242, Canadian Special Publication of Fisheries and Aquatic Sciencs 96.

Wilson, E. 0 1984. Biophilia. Harvard University Press, Cambridge, MA.

Withler, F.C. 1982. Transplanting Pacific salmon. Canadian Technical Report of Fisheries and Aquatic Sciences 1079.

Wright, S. 1981. Contemporary Pacific salmon fisheries management. *North American Journal of Fisheries Management* 1:29-40.

Wright, S. 1969. "Evolution and the Genetics of Populations, Volume 2, The Theory of Gene Frequencies," University of Chicago Press, Chicago.

Status of Biodiversity of Taxa and Nontaxa of Salmonid Fishes: Contemporary Problems of Classification and Conservation

ROBERT J. BEHNKE

1. Introduction

The recent publication by Nehlsen et al. (1991) makes it clear that the greatest challenge in preserving the genetic diversity of salmonid fishes concerns the protection of nontaxa. These authors list 214 "stocks" of Pacific salmon and steelhead at risk of extinction in California, Oregon, Washington, and Idaho. Of these, 101 are judged to be at high risk and 18 may already be extinct. These stocks of salmon and trout do not meet any of the generally accepted criteria for taxa recognition. They are not geographically isolated and although they maintain their integrity by reproductive isolation as a result of homing, they exhibit virtually no distinguishing morphological nor quantifiable genetic characteristics. Their intraspecific differentiation concerns life history and ecological adaptations to specific environments. Intraspecific fracturing of a species into numerous discrete stocks or populations is driven by natural selection to expand future evolutionary options for genetic continuity and acts to maximize the abundance of a species. Obviously, any conservation program to preserve biodiversity, to be truly effective, must begin at the lowest nontaxon level.

The social, political, and economic ramifications of protection of nontaxa under the Endangered Species Act are enormous. In this regard, I foresee considerable turmoil surrounding the identification and classification of significant evolutionary units of nontaxa for protection. The same problems resulting from diverse concepts, theories, and methods used to classify taxa (especially species and subspecies) that create current taxonomic controversies also apply to the classification of nontaxa. An understanding of the basis of taxa classification controversy is useful for developing a priority system for arranging nontaxa based on significance of life history adaptations—a concept of significant evolutionary units.

2. What Is a Species?

Are species real? Are they the only natural units of classification or only a concept of the human mind? Are species individuals or members of a class? Reflection on these controversial questions should make it clear that no definition of a species will find broad, universal acceptance. If we briefly examine historical concepts and definitions and seek

Department of Fishery and Wildlife Biology, Colorado State University, Fort Collins, CO 80523, U.S.A.

Genetic Conservation of Salmonid Fishes, Edited by J.G. Cloud and G.H. Thorgaard, Plenum Press, New York, 1993

elements of commonality, it is possible to find generally accepted and agreed upon aspects and proceed to phylogenetic branching points where controversy begins.

If we begin with the 10th edition of Linnaeus' *Systema Naturae* of 1758, we find that the validity of species and their classification was a simple matter unclouded by theoretical and philosophical concepts. Linnaeus' species was monophyletic, it maintained cohesion and integrity through time—all members of a species would be more closely-related to each other than to any member of another species. With these attributes, we have a beginning toward the development of a concept and definition of a species. The problem, of course, concerns the imposition of evolutionary theory on a taxonomy based on immutable unchanging species.

Just about 100 years after Linnaeus initiated formal taxonomy with his system of binomial nomenclature, Charles Darwin published *On the Origin of Species* (1859). Darwin's thesis on the origin of species can be summarized in four basic steps. (1) Variation exists within each interbreeding population; (2) more offspring are produced in each generation than can survive to maturity; (3) differential survival to reproduction occurs under the influence of natural selection so that the surviving members possess those variations most favorable to survival and pass these "survival adaptations" on to their offspring (in a modern sense, this adaptiveness of natural selection is basic for a rationale for the preservation of intraspecific diversity); (4) generation by generation, the slow, gradual accumulation of many small differences leads to the origin of new species.

With evolution creating continual gradual change, the conceptual gap between species became blurred. Where does intraspecific diversity end and interspecific diversity begin in phylogenetic branching? Darwin himself believed a species to be undefinable. He wrote: "In determining whether a form should be ranked as a species or a variety, the opinion of naturalists having sound judgment and wide experience seems the only guide to follow" (Darwin 1859: 47). This appeal to the authority of naturalists of sound judgment regarding species recognition is still apparent today. The American Fisheries Society's committee on names of fishes decides on an official list of fish species recognized as valid. I and others of "sound judgment" may find disagreement among the species officially recognized or not recognized by the American Fisheries Society. This illustrates the point that in cases where the magnitude of divergence among the terminal branches of a monophyletic phylogeny is largely unknown (for example, the phylogeny that leads to diverse forms of Arctic char of the *Salvelinus alpinus* complex) the recognition of species becomes a matter of professional judgment. Different professionals using "sound judgment" will likely disagree on species validity because their perceptions are colored by different theoretical and philosophical concepts and/or by different interpretations of evidence derived from different taxonomic methods. That is, we are only human and human subjectivity cannot be eliminated.

Darwin's work reflects his amazing ability to correctly interpret the patterns of evolution although the processes of evolution (the genetic basis) were unknown at the time. The lack of a genetic basis for Darwin's evolutionary theory and his Lamarckian interpretation of variation, directly induced by environmental factors, resulted in a period when the Darwinian view of evolution by natural selection was largely out of favor in academia.

In the early 1900s the discovery of Mendel's papers stimulated the founding of genetics as a discipline. The early geneticists, however, developed a theory on the origin of species by saltation (macromutation). In the strict sense, the species of the early geneticists was quite clear-cut. It arose by a macromutation, suddenly in a single generation. Natural selection played no role in the origin of such a species. Natural selection could only accept or reject the new species. Intraspecific diversity was thought to be caused by micromutations and of a random and nonadaptive nature. The implications for rejecting natural selection as the guiding force in the origin of species and the associated rejection of the adaptiveness of intraspecific diversity were highly negative in regards to the protection and preservation of intraspecific (nontaxa) genetic diversity. When a series of dams began to be constructed on the Columbia River in the 1930s, under the prevailing geneticists' concept of evolution, it was generally believed that the abundance of anadromous species such as chinook salmon could be maintained and increased by large-scale hatchery propagation of a generic stock. If the multitude of diverse chinook salmon stocks in the Columbia basin were not adaptive — that is, there was no survival or abundance advantage for a species to fraction into numerous local populations to utilize different sections of the basin — then there need be no concern for preservation of intraspecific genetic diversity and abundance could be maintained by a common hatchery stock. We now know how wrong this evolutionary concept was, but it is too late to rectify all wrongs committed. Most of the 214 salmonid stocks listed as at risk of extinction by Nehlsen et al. (1991) have been placed at risk or have had their extinction risk exacerbated because of the stocking of nonnative hatchery stocks. A bad species concept leading to the erroneous assumption that the abundance of a species (whose abundance depended on the maintenance of a wealth of intraspecific diversity) could be increased by the artificial propagation of a few generic hatchery stocks, has been a major cause in the depletion of diversity. Goodman (1990) pointed out that hatchery programs may violate the Endangered Species Act by further jeopardizing the continued existence of endangered wild stocks.

In the 1930s the work of such systematists and geneticists as Ernst Mayr, G. G. Simpson, T. Dobzhansky, and R. A. Fisher effectively reconciled the Darwinian view of evolution with contemporary genetics by interpreting evolution, essentially as conceived by Darwin, but explaining selection within a genetic context. Julian Huxley's synthesis of the mainstream evolutionary concepts of the time in two major works: *The New Systematics* (1940) and *Evolution: The Modern Synthesis.* (1942), ushered in a new era of unanimity on the species concept and an answer to the question: What is a species?

For about the next 30 years Mayr's biological species concept became the dominant guideline for species definition. The species attributes of internal cohesion, integrity, and stability could be defined by reproductive isolation — "actual," in the case of sympatry with a most closely-related species or "potential," in the case of allopatric forms. Species became more real, more "individualized." Of course there was a problem for making a decision in the case of "potential" reproductive isolation with allopatric distribution of distinct forms. How could a rational decision be made regarding species or subspecies recognition for distinct allopatric groups? The 1953 text, *Methods and Principles of Systematic Zoology*, by Mayr, Lindsey, and Usinger, gives advice for applying the biological species concept to decide if species or subspecies recognition should be given. Table 3 of chapter 5 is a "discrimination grid" in which the first column choices are given between morphologically identical or morphologically different, each with a choice of sympatric or allopatric. The second column is "not reproductively isolated," the third column, "reproductively isolated." If, for example, I have two distinct forms of trout in allopatry, I would have to make a judgment if they would have potential reproductive isolation (recognize as species) or that they would lack reproductive barriers to hybridization if they occurred in sympatry (recognize as subspecies). I might favor recognition as subspecies but another taxonomist using the same evidence and "sound judgment" might recognize them as species. The Mayr, Lindsey, and Usinger text also addressed this problem with a quantitative method for taxa recognition. A coefficient of difference formula denoted the percent of non-overlap among diagnostic characters between two samples. A CD value of 1.28 equalled 90% joint non-overlap and this was recommended as a minimum value for subspecies recognition. A 95% or 100% joint non-overlap might be interpreted as species recognition.

During this period many student theses and publications occurred naming many new subspecies, simplistically defined by their CD values. Systematics and taxonomy may have enjoyed an era of good will and general agreement that, finally after 200 years, the riddle of the species concept was resolved and the question, what is a species, adequately answered by the biological species concept, but these disciplines probably became dull and uninteresting to many brighter students.

A basic problem with the biological species concept for defining species on the basis of reproductive isolation in sympatry with a closely-related form is the erroneous assumption that reproductive isolation must be the results of considerable genetic differentiation accumulated in allopatry for a long time (perhaps a million years or more). What we now know about genetic relatedness of sympatric reproductively isolated intraspecific forms of salmon, trout, char, and whitefishes such as anadromous and resident life history types, several distinct intraspecific runs of salmon all sympatric in the same river basin, noninterbreeding populations coexisting in a single lake, etc. is that reproductive isolation may come about with very slight and virtually imperceptible genetic change. This is true in species with specific reproductive behavior such as homing to natal sites, that can preserve the integrity and cohesion of life history differences of populations in sympatry.

About 20 years ago, systematics and taxonomy were invigorated (some might say devastated) by the impacts of philosophy, new concepts of classification such as numerical taxonomy and cladistics, and widespread application of new methods of quantitative genetic analysis.

Although systematics and taxonomy have become more exciting and intellectually challenging in recent years, concepts of species and species definitions are now so diverse and controversial that it is every man for himself—anything goes. Unless one is a zealous member of a subcult of a cult professing the true knowledge of evolution and taxonomy, the term species has become more vague and undefinable for contemporary taxonomists—essentially a return to Darwin's vague species concept; a species is what an authority (or a committee), using "sound judgment," says it is.

In a larger sense, what, if anything, do the new theories, philosophies, and methods hold for the improvement of programs for the conservation of biodiversity—for defining evolutionary units (especially nontaxa) for a conservation classification scheme? The opportunities are great but I foresee a period of misuse and misunderstanding by biologists, administrators, and politicians concerning their interpretation of subject matter with which they have no more comprehension than they do of atomic physics.

3. Implications and Examples

To illustrate the influence of modern methods and species concepts on contemporary ichthyology I cite a paper by Markle et al. (1991). The cyprinid genus *Oregonichthys* has been considered monotypic with a single species distributed in the Willamette and Umpqua river drainages. Electrophoretic and morphological analyses showed great similarity and overlap between the Umpqua and Willamette forms except that the percent nonoverlap of the breast scalation character exceeded 90%. During the dominance of the biological species concept, such a study might be a typical graduate student thesis research project whereby the student would apply the discriminant grid and calculate a CD value exceeding 1.28 resulting in the recognition of a new subspecies. The present paper, however, emphasizes that the Umpqua form spawns on rocks and the Willamette form spawns on vegetation, which the authors interpret as endowing the two allopatric forms with "potential" reproductive isolation. This potential reproductive isolation justifies species recognition, according to the authors, under the phylogenetic, biological, or cohesion species concepts. The authors stated that they were guided by the "Simpson-Wiley evolutionary species" concept which includes the idea of ancestor-descendant relationships, identity maintenance, and the definition that species have their own evolutionary tendencies and historical fate (but wouldn't subspecies, stocks and populations also fit this concept?). Although the description of a new species of *Oregonichthys* is a taxonomic paper, the pertinent message concerns the preservation of a unique cyprinid phylogeny represented in the genus *Oregonichthys*. Both the Willammette and Umpqua forms (whether species or subspecies) are at risk of extinction. The data on life history, habitat characteristics, and identification of factors threatening their continued existence represent a modern taxonomic work describing a doubtful species, but serving a higher purpose—a data base for conservation.

Modern methods and concepts are currently being applied to salmon species, especially chinook salmon, in the Columbia River basin proposed for listing and protection under the Endangered Species Act. The bureaucratic mindset typically finds the quantitativeness and repeatability of simplistic notions such as the discrimination grid or CD values to be irresistibly seductive. Such minds are readily deluded by the illusion of technique. The danger that I foresee in applying modern methods of quantitative genetics to identity and define "significant" evolutionary units for conservation purposes concerns the lack of association between degrees of genetic differentiation as interpreted from various types of DNA analysis or electrophoresis and degrees of life history and ecological differentiation determined in the regulatory genome—the "qualitative" genetic basis for niche filling as site-specific coevolved evolutionary units.

Papers by Meyer et al. (1990) and Avise (1990) on the species flock of cichlid fishes in Lake Victoria, Africa, makes this point perfectly clear in regards to the limitations of quantitative genetic data to accurately diagnose "significant" evolutionary units. About 200 species of 20 genera of cichlid fishes are recognized as endemic to Lake Victoria. These species exhibit considerable morphological and ecological divergence to fill a great range of niches to utilize all levels of trophic resources, yet all of the multitude of species appear to have been evolved from one common ancestor within the past 200,000 years. Detailed analysis of the Cytochrome b gene, transfer RNA genes, and mtDNA of 14 species of 9 genera, in total found less quantifiable genetic differentiation than is found within *Homo sapiens* (which is not a highly variable species).

I believe Charles Darwin would not be surprised to find that the sum of our present knowledge serves to verify the basic elements of his theory on the origin of species by natural selection. He would likely be surprised in regards to how powerful a force natural selection can be to fraction a species into locally adapted populations, often maintaining reproductive isolation in contact with other populations of the species. He would not be surprised to find that in 1991, the answer to the question, what is a species, is about the same as in 1859. We still use his appeal to authority method, only now we have many diverse sources of authority to which we may appeal in order to justify preconceived notions.

References

Avise, J. C. 1990. Flocks of African Fishes. *Nature* 347:512-513.

Darwin, C. 1859. "On the origin of species by means of natural selection of the preservation of favored races in the struggle for life." John Murray, London.

Goodman, M. L. 1990. Preserving the genetic diversity of salmonid stocks: a call for federal regulation of hatchery programs, *Environmental Law* 20:11-166.

Huxley, J.S. 1940. "The new systematics." Clarendon Press, Oxford.

Huxley, J.S. 1942. "Evolution: The modern synthesis." Allen and Unwin, Ltd.

Linnaeus, C. 1758. "Systema naturae per regna tria naturae, secundum classes, ordines, genera, species cum charactenibus, differentiis, synonymis, locis." Editiodecima, reformata, Tom. I. Laurentii Salvii, Holmiae.

Markle, D. F., T.N. Pearsons, and D.T. Bills. 1991. Natural history of *Oregonichthys* (Pisces: Cypninidae), with a description of a new species from the Umpqua River of Oregon. *Copeia* 1991(2):277-293.

Mayr, E., E.G. Linsley, and R.L. Usinger. 1953. "Methods and principles of systematic zoology." McGraw-Hill.

Meyer, A., T.D. Kochner, P. Basasibwaki, and A. Wilson. 1990. Monophyletic origin of Lake Victoria cichlid fishes suggested by mitochondrial DNA sequences. *Nature* 347:550-553.

Nehlsen, W., J.E. Williams, and J.A. Lichatowich. 1991. Pacific salmon at the crossroads: stocks at risk from California, Oregon, Idaho, and Washington. *Fisheries* 16:4-21.

Requirements for Genetic Data on Adaptations to Environment and Habitats of Salmonids

C. D. LEVINGS

1. Introduction

In this paper I review some of the literature on how salmonids, especially the genera *Oncorhynchus* spp. and *Salmo* spp., are adapted to specific environments and habitats. The versatility and colonizing ability of salmon populations may arise from genetic material in marginal populations (Scudder, 1989); for example those at the boundaries of ranges or in unique habitats that can be disrupted by man's activities. Historically, genetic material must have been present to allow salmon populations to cope with natural environmental variation and to use specific habitats to their advantage for survival. If the original genetic structure is no longer present, if environments vary outside the natural range, or if restored habitats are different than those originally present, restoration of salmonid populations may be difficult.

Knowledge of specific adaptations to environments and habitats is therefore crucial to the success of salmon management initiatives. Recently, environmental and management policies have led to the development of plans which seek to protect, restore and manage salmon habitat. For example, a stated goal of the Green Plan for Canada, an environmental initiative (Anon, 1990), is to double the size of salmon populations (*Oncorhynchus* spp) in the Fraser River basin in western Canada. In this river, the productivity of some species of salmonids have been increasing in recent years (e.g., sockeye salmon, *O. nerka*; Fig. 1a), but further production from salmon populations is to be achieved by restoration of habitat and cleanup of pollution, together with fishery management actions to allow additional escapement to the spawning grounds. In the Columbia River basin in the United States, a stated goal of management agencies is to double the size of existing runs of salmon and steelhead trout (Northwest Power Planning Council, 1987). Recovery of salmon populations is also being attempted in some rivers in Europe and eastern North America where salmon have virtually disappeared (e.g., the Rhine; Fig. 1b). In the United Kingdom, the Atlantic salmon (*Salmo salar*) has now returned to tributaries of the River Clyde in Scotland, following cessation of pollution in the river system (Maitland, 1986). Recovery of the Atlantic salmon in the Thames and the Rhine Rivers has begun but doubts remain about the success of the initiatives (Maitland, 1986; deGroot, 1990).

Fisheries and Oceans, West Vancouver Laboratory, 4160 Marine Drive, West Vancouver, B.C., Canada V7V 1N6

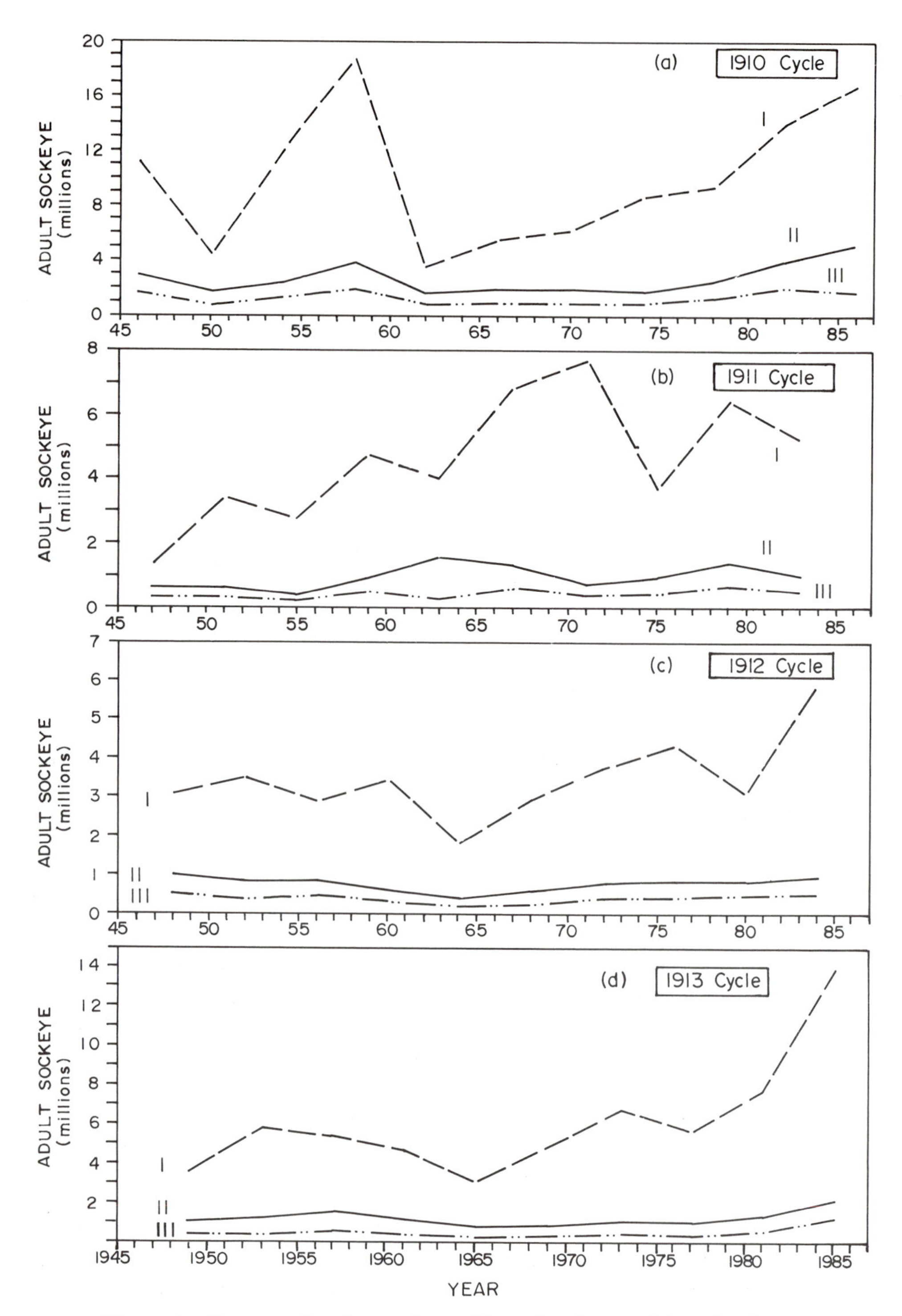

Figure 1a. Returns of sockeye salmon (*Oncorhynchus nerka*) to the Fraser River, 1945 - 1985. I - total run; II - total escapement; III - effective female spawners (from Servizi, 1989).

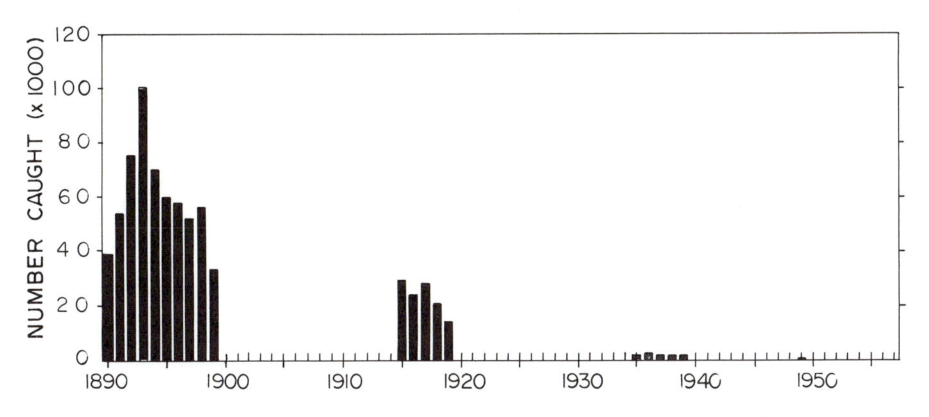

Figure 1b. Catches of Atlantic salmon (*Salmo salar*) by Dutch fishermen in the Rhine River (data from Cazemier, 1988).

2. Approach

This paper reviews the evidence of life history adaptations by environment, habitat and life history type, and I have made a distinction between environment and habitat. Ryder and Kerr (1989) described environment as the following: heat, light, dissolved oxygen, and nutrients. In their scheme, habitat acts as a structuring agent that shapes the way environment is provided to organisms in an ecosystem. For example, water flow within a drainage basin may be configured by geological features into stream and lake habitat, each with particular temperature regimes.

3. Fresh Water Environment

In the following I assume that differences between drainage basins reflect regional changes in key environmental factors such as temperature, which in turn are correlated with other significant variables such as river discharge. Genetic differences between regions and drainage basins were identified by Stahl (1986) for Atlantic salmon in North America and Europe, indicating that changes in populations due to regional environmental changes would be difficult to sort out from inter-basin variation. Broad scale geographic differences in the population structure of sockeye, chinook, pink, chum, and coho salmon and rainbow and cutthroat trout in the northeast Pacific were identified using electrophoretic techniques by Utter et al., (1980). More recent work has refined these patterns for chinook salmon (*O. tshawytscha*) in western North America (Waples, these Proceedings) so that differences between major basins can now be discriminated.

The influence of fresh water environmental factors on salmon survival varies between species, probably because of the differences in rearing periods in fresh water and the ocean, and because of variation in longevity. In pink salmon (*O. gorbusha*), for example, fresh water survival in the Fraser River has been shown to fluctuate between 9 and 19% (Servizi, 1989). Since this species moves to the sea very soon after hatching, variation in survival must arise

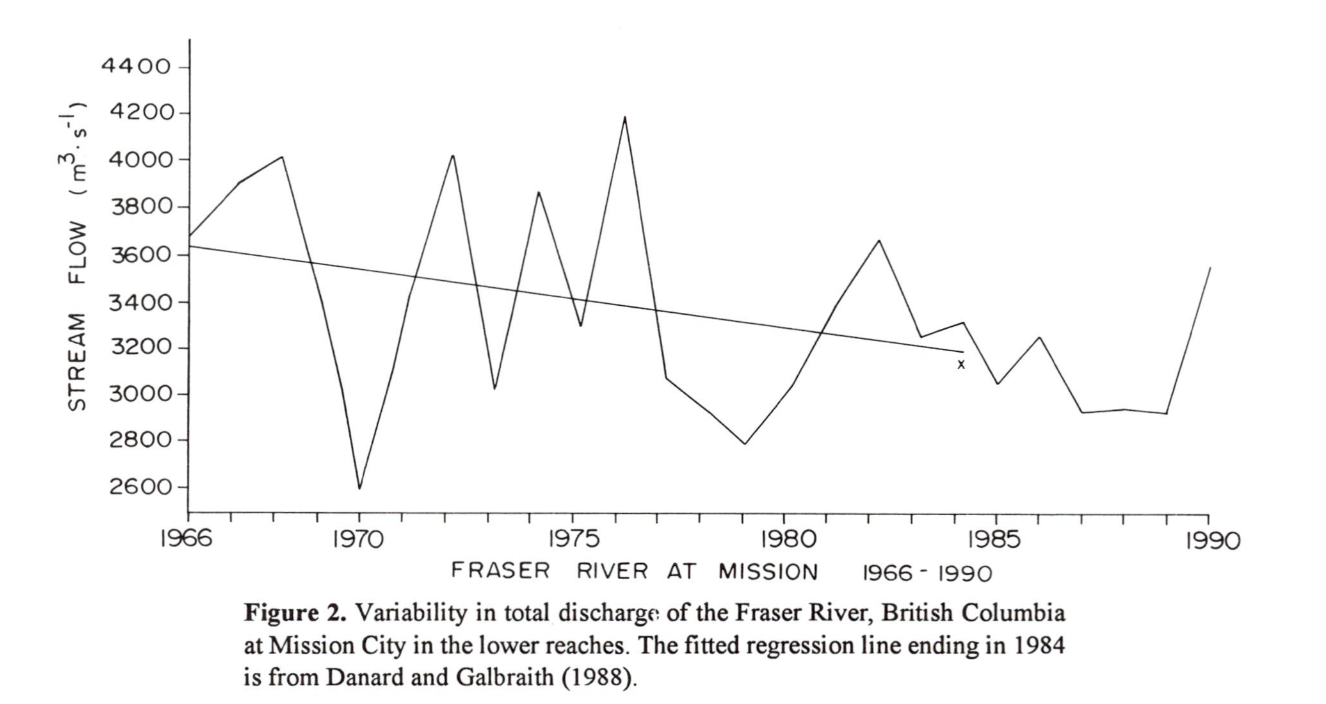

Figure 2. Variability in total discharge of the Fraser River, British Columbia at Mission City in the lower reaches. The fitted regression line ending in 1984 is from Danard and Galbraith (1988).

from environmental factors affecting incubation. Variability in discharge of the Fraser River (Fig. 2), with concomitant effects on scouring and egg survival, may be implicated. Fresh water survival of Atlantic salmon, in contrast, was fairly constant (coefficient of variation 13 to 15%) in two rivers in Newfoundland (Chadwick, 1988), perhaps because this species has multiple year classes that use rather stable rivers and lakes for up to three years.

Regional scale differences in climate over land are projected to occur with a doubling of atmospheric carbon dioxide and warming and this could affect the temperature regime of the fresh water environments that salmon use. In Canada, for example, regions at mid-latitude are expected to experience a general warming, with an increase in the number of degree-days and a possible change in the distribution of the ecoclimatic regions of the country (Zoltai, 1988). In northern British Columbia, the interior cordilleran region could shift to the northwest (Fig. 3), resulting in warmer conditions for sockeye salmon using the upper tributaries of the Fraser River, specifically Stuart and Takla Lakes. Since adult sockeye salmon in the Fraser basin are affected by prespawn mortality related to high temperatures (Gilhousen, 1990), climate warming could result in the elimination of populations that are intolerant of higher temperature levels. Hatching period or development time is also obviously related to temperature and was shown to vary between coastal and interior populations of chinook and sockeye salmon (Beacham and Murray, 1989).

Temperature can also have effects in the incubation and rearing phases, but data are lacking on the genetic basis of the effects relative to phenotypic changes. Winter temperature could be one of the limiting factors for coho salmon (*O. kisutch*), as this species spends one or two winters in fresh water. In the Fraser River basin in British Columbia, rearing fry of coho salmon are subjected to temperatures down to 0°C. (Swales et al., 1986) in mainstem rearing habitats. Overwintering fry survived better in warmer, off-channel ponds. For coho salmon, climate warming could therefore result in an expansion of their range inland, into habitats currently characterized by severe winters. The timing of emergence of coho salmon fry is affected by temperature (Holtby, 1988), and in Carnation Creek warmer temperatures increased the proportion of one year smolts relative to two year smolts.

The effects of global warming could influence the productivity of Atlantic salmon by shifts in smolt age over the geographic range of the species. The mean age of smolting for Atlantic salmon is strongly related to temperature and photoperiod, which vary with latitude over the range of the species. An index of growth opportunity, which combined temperature and photoperiod, was a good predictor of mean smolt age (Metcalfe and Thorpe, 1990). However, temperature may be a mediator influencing food supply for and acquisition by individual fish. Little is known about specific adaptations of Atlantic salmon populations to temperature, but the available evidence indicates much of the variation in rate processes such as growth are phenotypically controlled. Thorpe (1989) summarized data from laboratory and field experiments and showed that it was the rate of energy storage after midsummer, which could be a genetic trait, that determined whether or not an individual Atlantic salmon would smolt in the following year.

Data on the genetic basis of temperature tolerance for salmonids are not available. This area of research would be a particularly important topic for study. Redband trout (*Salmo* spp.) have been shown to be genetically distinct from other resident trout populations using electrophoresis (Wishard et al., 1984) and were thought to be specifically adapted to warm temperatures and arid environments. However, there are no published data to support this hypothesis.

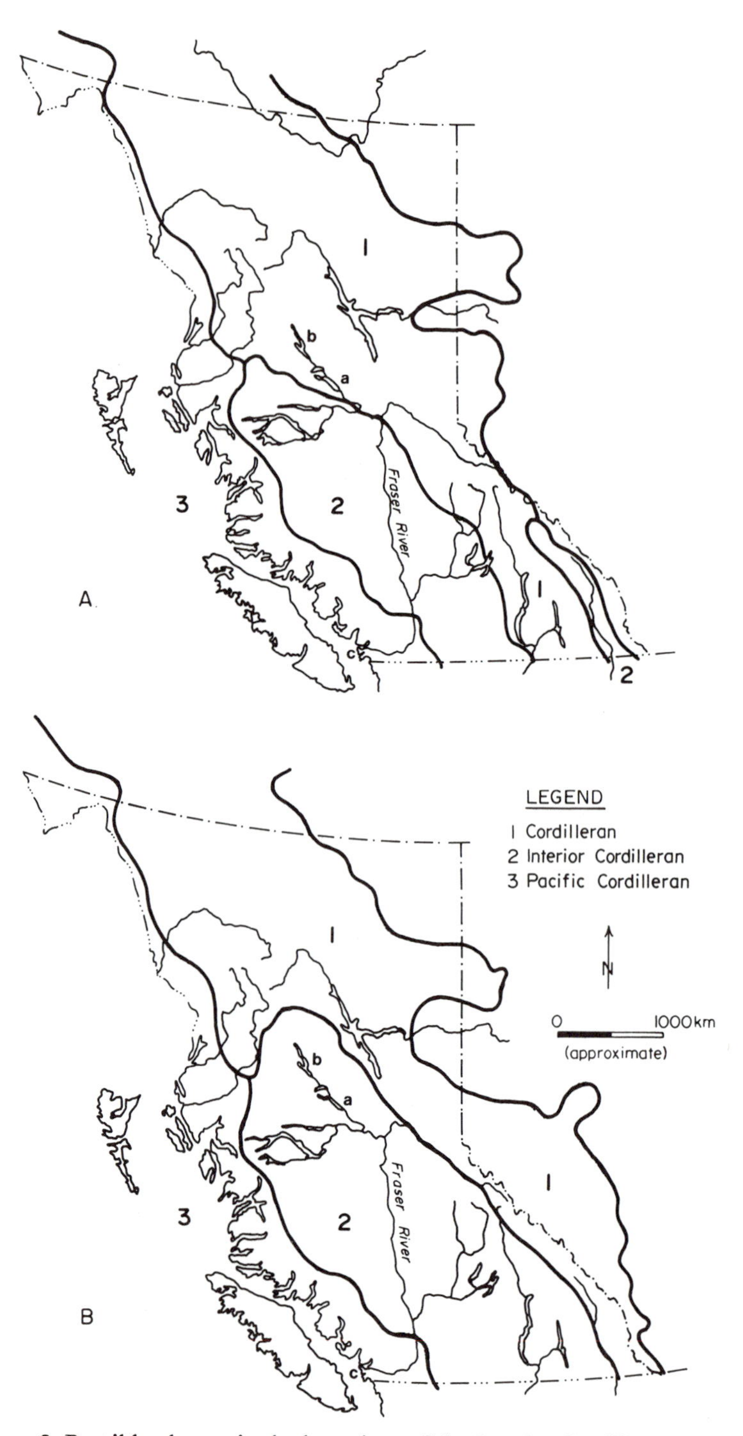

Figure 3. Possible change in the boundary of the Interior Cordilleran ecoclimatic region in British Columbia with global warming (from Zoltai, 1988). a - Stuart Lake; b - Takla Lake.

4. Fresh Water Habitats

4.1. Migratory Pathways

Many species of salmonids are adapted to move as fry or smolts from lotic habitats (rivers, streams) to lentic habitats (lakes, oceans) and then return to lotic habitats for reproduction. Migrations to and from spawning habitats can range from a few kilometres up to over 1000 km.

The migration of fry and smolts was one of the earliest behavioural traits of salmonids demonstrated to be under genetic control. Brannon (1967) showed that sockeye salmon fry from the Stellako River population moved upstream to rearing lakes, whereas Adams River fry moved downstream to rear. Populations of rainbow trout (*O. mykiss*) living above and below impassable waterfalls were found to show different migratory behaviour (Northcote et al., 1970) and electrophoretic data indicated the genetic basis for migration or residency in these populations. Stream type chinook salmon and Atlantic salmon have been documented to undergo lengthy rearing migrations or reside in fresh water for 1-5 years (Levings and Lauzier, 1991; Cuinat, 1988). Stream type chinook salmon are genetically distinct from other life history types within the species (Carl and Healey, 1984; see below) and therefore the prolonged migration must have adaptive significance. The migration of juvenile chinook during winter may be related to river discharge (Levings and Lauzier, 1991) which is correlated with temperature but "stimulation" of movement by temperature change was not demonstrated in laboratory tests by Bjornn (1971). In disrupted migratory habitats, such as the Columbia River, wild chinook migrating at low river flows showed a lower survival rate (Raymond, 1988). Survival may have been reduced because movement through impoundments was slow relative to the natural migration rate.

The upstream migration of adult salmon can be severely disrupted by habitat change, but in some instances, populations with the genetic potential for recolonization have persisted. For example, adult pink salmon movement through the Fraser River canyon in British Columbia was limited by a rock slide in 1913. Although a few fish could migrate upstream, most pink salmon spawned below the canyon until about 1945 when fishways were built (Ricker, 1989). Pink salmon colonized the spawning habitat in the upstream tributaries, indicating that sufficient genetic variation in the population remained to enable rebuilding of the upstream populations. More recently, fewer fish are again spawning upstream of the canyon relative to the lower river, perhaps because reduced discharge in the Fraser River has induced migration blocks or because adult pink salmon have been smaller in recent years, with reduced swimming ability. The reduced size of pink salmon may in turn be an effect of size selection by the fishing fleet (Ricker, 1989). The ability to move through migration blocks may be specific to certain populations as the pink salmon population using the Seton River has not declined at the same rate as that in the Thompson River and both are upstream of the canyon (Williams et al., 1986). Effects of dams on chinook salmon populations of the Columbia River appear to have been more pronounced. Winans (1989) suggested that the lack of genetic diversity currently found in chinook salmon populations of the Snake River basin in Oregon was due to bottleneck effects resulting from reductions in number of fish on the spawning grounds for interbreeding. Bottleneck effects due to habitat changes were also reported for chinook salmon from California river systems (Bartley and Gall, 1990).

4.2. Spawning Habitat

Destruction of spawning grounds can lead to elimination of specific populations, as homing to particular spawning beds has been well documented, indicating that homing is genetic controlled (Bams, 1976). Spawning habitats have been affected by a wide variety of factors and adult salmon will not necessarily avoid a disrupted spawning area to reproduce in an undisrupted site. Excessive silt due to hydraulic mining has affected the gravel quality of spawning beds of chinook salmon in California (Bartley and Gall, 1990) as have reductions in water flow for irrigation (e.g., California Advisory Committee on salmon and Steelhead Trout, 1988). In the mainstem Columbia River only one free-flowing reach (Hanford Reach at river km 613) remains as a viable chinook salmon spawning habitat - all other spawning habitats have been permanently submerged due to reservoir construction (Rickard and Watson, 1985; Fraidenburg and Lincoln, 1985). The populations of chinook salmon that were adapted to use these habitats are presumably lost. Efforts to restore the landlocked Atlantic salmon of Lake Vanern in Sweden have been impaired by the permanent loss of spawning areas from hydroelectric developments. 20 out of 22 tributary streams have suffered total loss of salmon from hydroelectric developments (Ros, 1981) and now restoration of the individual populations is proving to be extremely difficult (Lundholm, 1988).

4.3. Rearing Habitat

Some species of anadromous salmonids show forms that are genetically distinct and adapted to conduct their entire life history in fresh water and use lakes for rearing to the adult stage (e.g., kokanee form of sockeye salmon and ouananiche form of Atlantic salmon). Other species of anadromous salmonids (e.g., pink, coho, and chinook salmon) appear to have some physiological traits that enable them to survive to adulthood without going to sea, judging from their survival in the Great Lakes. Natural reproduction occurs in tributary streams to the lakes (Johnson and Ringler, 1981), but factors such as varying flows limit productivity of the Pacific salmon in these habitats.

Juveniles of anadromous and resident salmonids have evolved specific characteristics of habitat use which vary between and within drainage basins. For example, populations of juvenile Atlantic salmon in two tributaries of the Miramichi River in eastern Canada exhibited morphological adaptations to streamflow which were genetically based (Riddell et al., 1981). Fish rearing in the larger tributary were more robust and were adapted to faster streamflow. There may also be adaptations to specific types of rearing habitats such as river margins, wetlands, estuaries, and lakes within a drainage. Coho salmon rearing in lake habitat were morphologically different than those rearing in a river in the Cowichan River basin in British Columbia (Swain and Holtby, 1989). The two types of coho salmon may require separate population and habitat approaches if they are genetically distinct. Habitat compensation schemes, which sometimes "trade off" one type of rearing area for another, could be in error if populations are adapted to specific habitats. Seasonally-flooded wetlands, located off the main channel of the river, produced up to 25% of the coho salmon smolts from Carnation Creek in British Columbia (Brown and Hartman, 1988), while the remainder are produced in pool habitat on the mainstem creek. Chinook salmon in British Columbia show three different life history types (ocean type, 90 day type, and stream type), and these have been shown to be genetically distinct within a particular river system (Carl and Healey, 1984). Thus, ocean type juvenile chinook salmon migrate seaward soon after emergence to use estuarine habitat whereas the stream type fish are adapted for longer life in the river. Adaptations which have been shown to have a genetic basis in the fry stage of stream type fish include a higher level of aggressive and territorial behaviour (Fig. 4).

In some drainage basins exposure to pollutants may have led to the selection of salmonid populations that are tolerant to certain types of contaminants, possibly in the rearing phase. For example, the decrease in pH in western Europe and eastern North America may have led to the apparent selection of acid-resistant populations of brook trout (*Salvelinus fontinalis*) (Robinson et al., 1976). Stream habitats in agricultural and urban areas are

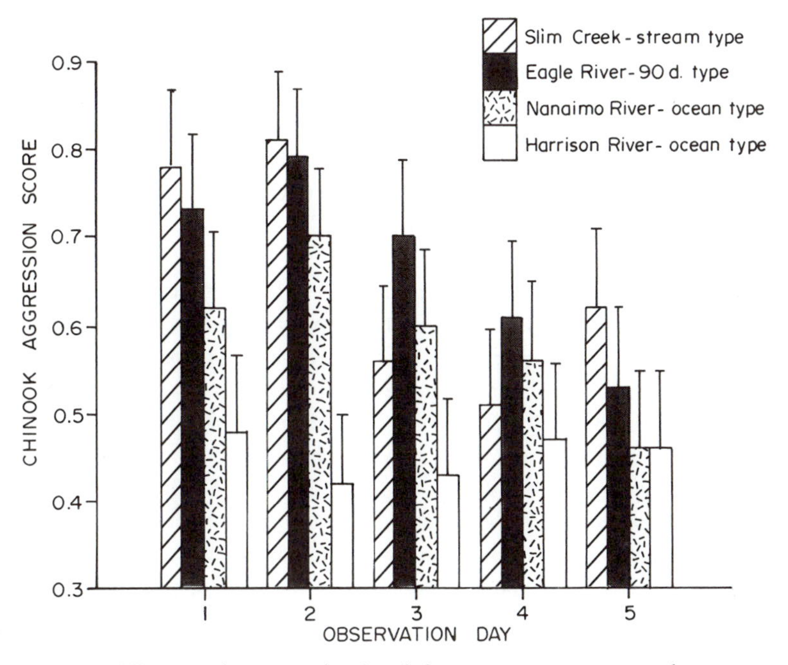

Figure 4. Differences in aggression levels between stream-type and ocean-type juvenile chinook salmon (*Oncorhynchus tshawytscha*) from the Fraser River system, British Columbia (from Taylor, 1988).

particularly vulnerable to fish kills and selection may therefore be occurring with other kinds of contaminants. O'Hara (1988) reported that fish kills in rivers outnumbered those in lakes (274 *vs* 107 over 1976 - 1984) in northwest England. In the lower Fraser Valley, British Columbia, juvenile coho salmon are the species most frequently affected by the accidental discharges of contaminants (Servizi, 1989). Usually 100-1000 fish are reported killed and, as the mortality occurs on very small populations, such incidents can lead to extinction of the total stock in urban streams.

5. Estuaries

5.1. Spawning Habitats

Pink and chum salmon in British Columbia and Alaska use estuaries for spawning habitat (e.g., Helle et al., 1964; Groot, 1989). The incubating eggs of certain populations are obviously tolerant of brackish conditions. Whether this attribute is specific to these populations, as suggested by Helle et al., (1964), or is inherent in all populations of pink and chum salmon, is not known. However, as sea level rises, as projected with global warming, brackish water spawning habitats may become more widespread. With higher sea levels, salt water will penetrate further into river mouths than at present, replacing fresh water spawning areas.

5.2. Migratory Habitats

Estuaries are migratory habitats where adult salmon sometimes delay their movement into rivers. The fish are particularly vulnerable to harvesting by gillnets and seines, and so interactions with fisheries management are particularly strong in these habitats. For example, the early segment of a population of coho salmon has been selected by gillnetting in the Columbia River estuary (Cramer, 1991). Most of the coho salmon from this population now arrive at the estuary over a rather narrow time interval, whereas in the past, run timing was extended.

The time that salmon arrive at the estuary from the open ocean is specific for each population, and generally the fish that are destined to spawning grounds furthest upriver arrive early. Timing must therefore be a genotypic trait. Sockeye salmon travelling to the Stuart River, which is 800 km upstream from the estuary, are the first to arrive at the mouth of the Fraser River each year (Gilhousen, 1990). The exact timing of the immigration of adult salmon into estuaries and movement to spawning areas may be stimulated by increasing discharge from the river, as shown by Chadwick (1988) for Atlantic salmon. The length of time that adult salmon spend in estuaries is therefore dependent on conditions in fresh water. Migration blocks in the estuary can have an obvious effect on success of spawning and hence could affect the reproductive success of specific salmon populations by causing prespawn mortality. For example low dissolved levels were cited by Alabaster et al (1991) as a limiting factor for migrating Atlantic salmon in the Thames estuary.

5.3. Rearing Habitats

Ocean type chinook salmon and chum salmon juveniles may be adapted to use specific types of habitats within estuaries, specifically wetlands such as eelgrass beds and marshes (Levings et al., 1989), but there is no information on the genetic bases for these habitat preferences. The majority of juvenile anadromous salmonids use estuaries as migratory habitats after smoltification, when they move from the river to the ocean. However, there are important exceptions, for example, ocean type chinook salmon and chum salmon. Ocean type chinook salmon rear in estuaries and are estuarine dependent, judging from analyses of adult scales (Reimers, 1973) and transfer experiments (Levings et al., 1989). There are also populations of coho salmon that are adapted to use brackish water as a rearing habitat, and coho fry have been found in certain estuaries in British Columbia (e.g., Carnation Creek; Hartman and Scrivener, 1990). However, earlier authors (e.g., Mason, 1975) concluded coho fry using estuaries did not survive to adulthood. In southern California, rearing in brackish lagoons has been recognized for some time as a life history strategy for steelhead trout (Shapovalov and Taft, 1954; Smith, 1990). If these coho salmon and steelhead trout populations are specifically adapted to rear in brackish waters, their habitats could expand if sea level rises with global warming.

The sea trout (*Salmo trutta*) found in the western Atlantic, considered to be an diadromous form of the brown trout, rears in sea lochs and fjords (e.g., Gibson and Ezzi, 1990) and may also be estuary-dependent. There is some evidence that the diadromous form is genetically distinct from the non-migratory form, judging from transfer experiments into lakes above and below a waterfall in Norway (Jonsson, 1982). However the situation is complex as smoltification can be affected by size which in turn is affected by food supplies (Jonsson, 1989). Tagging experiments with sea trout in fjords in Scotland (Gibson and Ezzi, 1990) and in Norway (Jonsson, 1985) showed that many trout stayed within about 100 km of the river mouth from which they emigrated. Diet analyses indicated the fish caught in fjords were using food organisms from brackish habitats (Pemberton, 1976).

6. Ocean Environments

6.1. Rearing

Preference for particular habitats or water masses by salmon in the North Pacific is probably adaptive and genetically determined (Burgner, 1980), as each of the 7 species of anadromous salmonids appears to occupy a specific portion of the ocean for rearing. Salmon adapted to a particular oceanic area apparently remain in the area even if conditions are not optimal for rearing. For example, the food production of coastal areas where chinook and coho salmon rear in the northeast Pacific is higher when coastal upwelling is more pronounced, and the strength of upwelling is influenced by wind fields closer to the Equator. When the "El Nino" condition is prevalent in the tropical ocean, upwelling, and hence survival, is reduced. The survival of coho salmon populations from Oregon coastal rivers, for example, decreased because of reductions in upwelling strength during 1982-1983 (Fig. 5; Fisher and Pearcy, 1988). Survival of Atlantic salmon that rear in the northwest Atlantic Ocean has also been shown to be highly variable relative to survival in fresh water, and populations of this species also use specific water masses (Chadwick, 1988).

As ocean temperatures are projected to increase due to global warming, salmonids at transitional areas in the oceans or at the extremes of their ranges may shift their distribution and survival patterns (Francis, 1990). The interactive effects of temperature on the fresh water life history of salmon would obviously have to be taken into account in any prognosis (Kope and Botsford, 1990). Ware and McFarlane (1989) described the portion of the North Pacific from approximately Vancouver Island to Northern California as a transitional zone in terms of fisheries production and upwelling. Salmonids rearing in this area may be particularly susceptible to fluctuations in food supply related to El Nino events. Warmer water can also bring predators such as Pacific mackerel into contact with rearing salmonids (Ashton et al., 1985).

6.2. Migration Pathways

Factors influencing the migration of adult salmon from mid-ocean rearing areas to the river mouths are poorly understood and the routes chosen by the fish to the coastline are not well documented. However, for most species the returns are highly regularized, as shown by the major runs of sockeye salmon from the Fraser River system (e.g., Gilhousen, 1990), which indicates time of return has a strong genetic component (Burgner, 1980). Ocean temperature in winter in the centre of the Gulf of Alaska is also thought to have a major influence on the time of return for Fraser River sockeye salmon populations (Blackbourn, 1987). The routes that Fraser River sockeye salmon use to migrate past Vancouver Island in British Columbia to reach the river mouth have changed in recent years. For the period 1974 to 1984 more fish moved inside Vancouver Island, through Johnstone Strait, compared to around the southern

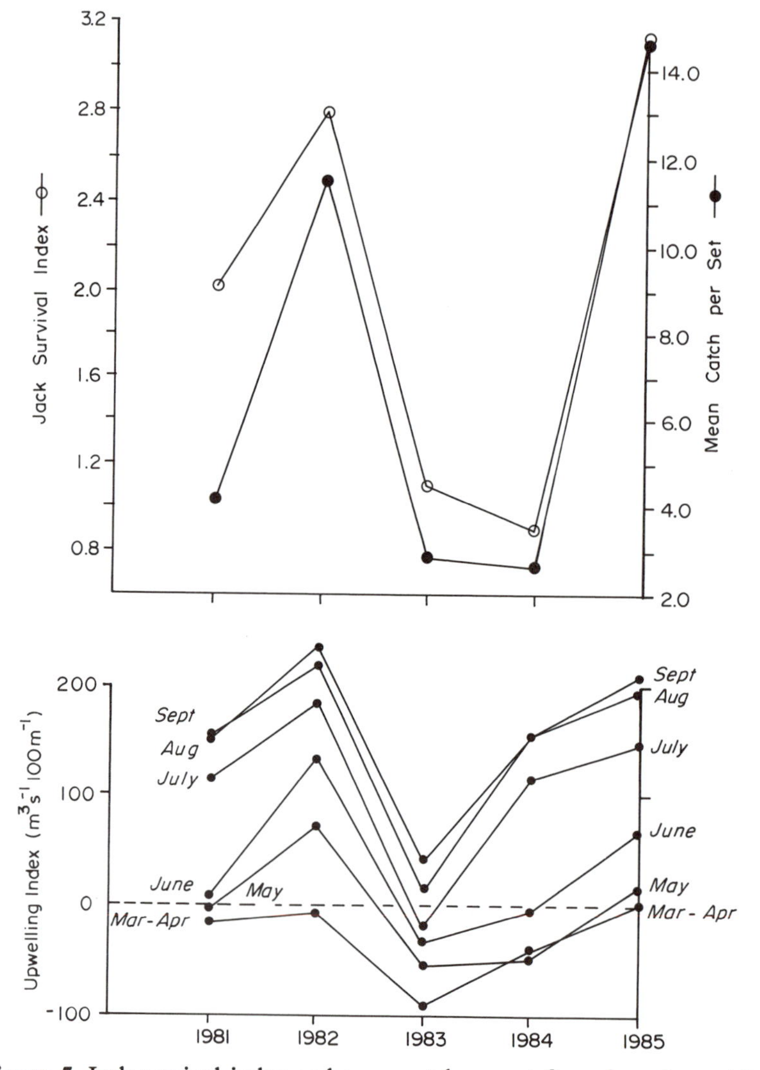

Figure 5. Jack survival index and mean catch per set for coho salmon (*Oncorhynchus kisutch*) rearing off the Oregon coast in relation to an upwelling index (from Fisher and Pearcy, 1988).

end of the Island, through Juan de Fuca Strait. Whether this is because of changes in the genetic or population composition or because the fish are responding to changes in environmental cues is not understood (Mysak et al., 1986). However, changes in climate could clearly change variables such as salinity and temperature.

7. Conclusions

Our understanding of the adaptations of wild salmonids to environmental and habitat factors is very limited, but there is currently strong public interest in conserving and rehabilitating salmon habitats and populations (e.g., Nehlsen et al., 1991). Better data on ecological genetics could help ensure the success of these initiatives. Improvement of the data bases may require some reorientation of the current research on fish genetics. Most of the emphasis by fish geneticists has been on cultured populations where the objective has been to select genotypes which maximize fish production in hatcheries or on fish farms by minimizing effects of environment and habitat. For example, aquaculturists working with Atlantic salmon have determined the heritabilities of 10 characters, all of which relate to survival on fish farms or are features which improve marketability of farmed fish (Gjedrem, 1983). In comparison, there are no data on the heritability of the tolerances of wild salmonids to key environmental factors such as temperature.

Data on the adaptations of salmon to specific environments and habitats will be vital for fisheries managers concerned with wild salmon production and conservation under a varying world climate. Management policies will be difficult to implement without this knowledge. Unfortunately, in parts of the world where hatcheries and fish farms have introduced cultured fish into ecosystems and interbreeding with wild fish has occurred, it may now be impossible to accurately research the genetic bases of environment/habitat adaptations. However, in relatively undeveloped parts of the world, undisrupted and protected populations of salmonids can still be found. The Noatak National Preserve near the Bering Strait in Alaska, for example, has been designated an International Biosphere Reserve under UNESCO and the Noatak River supports one of the most northerly chum salmon populations in the world. The Noatak River basin is a relatively isolated area, but even here, concerns have been raised about the operation of a salmon hatchery near the boundaries of the Preserve (Kelly et al., 1990). More effort is needed to establish and maintain wild salmon populations in intact habitat elsewhere in Europe, Asia, and North America. Maintenance of undisrupted salmonid populations will require considerable effort by agencies but there is no alternative if society is to implement the recommendation of the World Commission on Environment and Development (WCED, 1987) for preservation of biodiversity. Reserves with wild salmon populations will be important areas for salmonid geneticists to carry out long term research on adaptations, and conservation of wild salmonids in designated areas should be afforded international concern.

References

Alabaster, J.S., P.J. Gough, and W.J. Brooker. 1991. The environmental requirements of Atlantic salmon, *Salmo salar L.*, during their passage through the Thames estuary, 1982-1989. *J. Fish. Biol.* 38:741-762.

Anon. 1990. Canada's Green Plan, Government of Canada, Ottawa, Ontario, Cat. No. En21-94/1990E.

Ashton, H.J., V. Haist, and D.M. Ware. 1985. Observations on abundance and diet of Pacific mackerel (*Scomber japonicus*) caught off the west coast of Vancouver Island, September 1984. *Can. Tech. Rep. Fish. Aquat. Sci. 1394.*

Bams, R.A. 1976. Survival and propensity for homing as affected by presence or absence of locally adapted paternal genes in two transplanted populations of pink salmon. *J. Fish. Res. Bd. Can.* 33:2716 - 2725.

Bartley, D.M. and G.A.E. Gall. 1990. Genetic structure and gene flow in chinook salmon populations of California. *Trans. Amer. Fish. Soc.* 119:55-71.

Beacham, T.D. and C.B. Murray. 1989. Variation in developmental biology of sockeye salmon (*Oncorhynchus nerka*) and chinook salmon (*O. tshawytscha*) in British Columbia. *Can. J. Zool.* 67:2081-2089.

Bjornn, T.C. 1971. Trout and salmon movements in two Idaho streams as related to temperature, food, stream flow, cover, and population density. *Trans. Amer. Fish. Soc.* 100:423-438.

Blackbourn, D.J. 1987. Sea surface temperature and the pre-season prediction of return timing in Fraser River sockeye salmon (*Oncorhynchus nerka*), in: "Sockeye salmon (*Oncorhynchus nerka*) population biology and future management," H. D. Smith, L. Margolis, and C. C. Wood, eds., pp. 296-306, *Can. Spec. Publ. Fish. Aquat. Sci.* 96.

Brannon, E.L. 1967. Genetic control of migrating behavior of newly emerged sockeye salmon fry, International Pacific Salmon Fisheries Commission, *Progress Report No. 16.*

Brown, T.G. and G.F. Hartman. 1988. Contribution of seasonally flooded lands and minor tributaries to coho (*Oncorhynchus kisutch*) salmon smolt production in Carnation Creek, a small coastal stream in British Columbia. *Trans. Amer. Fish. Soc.* 117:546-551.

Burgner, R.L. 1980. Some features of ocean migrations and timing of Pacific salmon, in: "Salmonid Ecosystems of the North Pacific," W. J. McNeil and D. C. Himsworth, eds., pp. 153-164, Oregon State University Press and Oregon State University Sea Grant Program, Corvallis.

California Advisory Committee on Salmon and Steelhead Trout. 1988. Restoring the Balance, California Advisory Committee on Salmon and Steelhead Trout, 1988 Annual Report, 120 Schoonmaker Point, Foot of Spring Street, Sausalito, CA.

Carl, L.M. and M.C. Healey. 1984. Differences in enzyme frequency and body morphology among three juvenile life history types of chinook salmon (*Oncorhynchus tshawytscha*) in the Nanaimo River, British Columbia. *Can. J. Fish. Aquat. Sci.* 41:1070-1077.

Cazemier, W.G. 1988. Fish and their environment in large European river ecosystems - the Dutch part of the river Rhine. *Sciences de l'Eau* 7:95-114.

Chadwick, E.M.P. 1988. Relationship between Atlantic salmon smolts and adults in Canadian rivers, in: "Atlantic salmon: planning for the future," D. Mills and D. Piggins, eds., pp. 301-324, Proc. Third Inter. Atlantic Salmon Symp., Biarritz, France, 1986. Timber Press, Portland.

Cramer, S.P. 1991. Dynamics of the decline in wild coho (*Oncorhynchus kisutch*) populations in the lower Columbia River, Abstract, p. 38, Int. Symp. on Biological Interactions of Enhanced and Wild Salmonids, June 17-20, 1991, Nanaimo, B.C.

Cuinat, R. 1988. Atlantic salmon in an extensive French river system: the Loire-Allier, in: "Atlantic salmon: planning for the future," D. Mills and D. Piggins, eds., pp. 389-399, Proc. Third Inter. Atlantic Salmon Symp., Biarritz, France, 1986. Timber Press, Portland.

deGroot, S.J. 1990, Is the recovery of andromous fish species in the Rhine a reality? I. The Atlantic salmon (*Salmo salar*). *De Levende Natuur* (1990) 3:82-88 (in Dutch).

Danard, M., and J. Galbraith. 1988. Recent trends in runoff, precipitation, and temperature in the Fraser River watershed. Unpublished manuscript, Institute of Ocean Sciences, Sidney, British Columbia, Canada.

Fisher, J.P. and W.G. Pearcy. 1988. Growth of juvenile coho salmon (*Oncorhynchus kistuch*) off Oregon and Washington U.S.A. in years of differing coastal upwelling. *Can. J. Fish. Aquat. Sci.* 45:1036-1044.

Fraidenburg, M.E. and R.H. Lincoln. 1985. Wild chinook salmon management: an international conservation challenge. *North Amer. J. Fish. Man.* 5:311-329.

Francis, R.C. 1990. Climate change and marine fisheries. *Fisheries* (American Fisheries Society) 15:7-9.

Gibson, R.N. and I.A. Ezzi. 1990. A comparative study of sea trout (*Salmo trutta L*) in west Highland sea lochs with special reference to Loch Feochan, in: 'The Sea Trout in Scotland," M. J. Picken and W. M. Shearer, eds., pp. 61-70, Proc. Symposium held at the Dunstaffnage Marine Research Laboratory, 18-19 June 1987, Published by the Natural Environment Research Council.

Gilhousen, P. 1990. Prespawning mortalities of sockeye salmon in the Fraser River system and possible causal factors. International Pacific Salmon Fisheries Commission *Bulletin.* XXVI.

Gjedrem, T. 1983. Possibilities of genetic changes in the salmonids. *Roczniki Nauk Rolniczych* 100:65-78.

Groot, E.P. 1989. Intertidal spawning of chum salmon: saltwater tolerance of the early life stages to actual and simulated intertidal conditions, M.Sc. Thesis, University of British Columbia, Vancouver, B.C.

Hartman, G.F. and J.C. Scrivener. 1990. Impacts of forestry practices on a coastal stream ecosystem, Carnation Creek. *Can. Bull. Fish. Aquat. Sci.* 223.

Helle, J.H., R.S. Williamson, and J.E. Bailey. 1964. Intertidal ecology and life history of pink salmon at Olsen Creek, Prince William Sound, Alaska. U.S. Fish and Wildlife Service, Special Scientific Report No. 483.

Holtby, L.B. 1988. Effects of logging on stream temperatures in Carnation Creek, British Columbia, and resultant impacts on the coho salmon (*Oncorhynchus kisutch*). *Can. J. Fish. Aquat. Sci.* 45:502-515.

Johnson, J.H. and N.H. Ringler. 1981. Natural reproduction and juvenile ecology of Pacific salmon and rainbow trout in tributaries of the Salmon River, New York. *New York Fish and Game Journal* 28:49-60.

Jonsson, B. 1982. Diadromous and resident trout *Salmo trutta*: is their difference due to genetics?. *Oikos* 38:297-300.

Jonsson, B. 1985. Life history patterns of freshwater resident and sea-run migrant brown trout in Norway. *Trans. Amer. Fish. Soc.* 114:182-194.

Jonsson, B. 1989. Life history and habitat use of Norwegian brown trout (*Salmo trutta*) *Freshwater Biology*. 21:71-86.

Kelly, M.D., P.O. McMillan, and W.J. Wilson. 1990. North Pacific salmonid enhancement programs and genetic resources: issues and concerns. U.S. Dept. of the Interior, National Park Service, Tech. Rep. NPS/NRARo/NRTR-90/03, Washington, D.C.

Kope, R.G. and L.W. Botsford. 1990. Determination of factors affecting recruitment of chinook salmon *Oncorhynchus tshawytscha* in central California. *Fish. Bull.* (U.S.) 88:257-269.

Levings, C.D., C.D. McAllister, J.S. Macdonald, T.J. Brown, M.S. Kotyk, and B.A. Kask. 1989. Chinook salmon (*Oncorhynchus tshawytscha*) and estuarine habitat: a transfer experiment can help evaluate estuary dependency, in: "Proc. National Workshop on Effects of Habitat Alteration on Salmonid Stocks," C. D. Levings, B. Holtby, and M. A. Henderson, eds., pp. 116-122, *Can. Spec. Pub. Fish. Aquat. Sci.* 105.

Levings, C.D. and R.M. Lauzier. 1991. Extensive use of the Fraser River basin as winter habitat by juvenile chinook salmon (*Oncorhynchus tshawytscha*). *Can. J. Zool.* 69:1759-1767.

Lundholm, B. 1988. Fishes, fishing, and pollution in Lake Vanern (Sweden), in: 'Toxic Contamination in Large Lakes," N.W. Schmidtke, ed., vol. II, Impact of Toxic Contaminants on Fisheries Management, pp. 229-250, Lewis Publishers Inc., 121 South Main Street, Chelsea, Michigan.

Maitland, P.S. 1986. The potential impact of fish culture on wild stocks of Atlantic salmon in Scotland, in: 'The Status of the Atlantic Salmon in Scotland," D. Jenkins and W.M. Shearer, eds., pp. 72-78, Proceedings Institute of Terrestrial Ecology Symposium No. 15, Monks Wood Experimental Station, Abbots Ripton, Huntingdon, U.K.

Mason, J.C. 1975. Seaward movement of juvenile fishes, including lunar periodicity, in the movement of coho salmon (*Oncorhynchus kistuch*) fry. *J. Fish. Res. Bd. Can.* 32:2542-2546.

Metcalfe, N.B. and J.E. Thorpe. 1990. Determinants of geographical variation in the age of seaward-migrating salmon, *Salmo salar*. *J. Anim. Ecol.* 59:135-145.

Mysak, L.A., C. Groot, and K. Hamilton. 1986. A study of climate and fisheries: interannual variability of the northeast Pacific ocean and its influence on homing migration routes of sockeye salmon. *Climatological Bull.* 20:26-35.

Nehlsen, W., J.E. Williams, and J.A. Lichatowich. 1991. Pacific salmon at the crossroads: stocks at risk from California, Oregon, Idaho, and Washington. *Fisheries* (American Fisheries Society) 16:4-21.

Northcote, T.G., S.N. Williscroft, and H. Tsuyuki. 1970. Meristic and lactate dehydrogenase genotype differences in stream populations of rainbow trout below and above a waterfall. *J. Fish. Res. Bd. Can.* 27:1987-1995.

Northwest Power Planning Council. 1987. 1987 Columbia River Basin Fish and Wildlife Program, Northwest Power Planning Council, 850 S.W. Broadway, Suite 1100, Portland, Oregon 97205.

O'Hara, K. 1988. Freshwater fisheries management and pollution in Britain: an overview, in: 'Toxic Contamination in Large Lakes," N.W. Schmidtke, ed., vol. II, Impact of Toxic Contaminants on Fisheries Management, pp. 265-280, Lewis Publishers Inc., 121 South Main Street, Chelsea, Michigan.

Pemberton, R. 1976. Sea trout in North Argyll sea lochs: II. diet. *J. Fish. Biol.* 9:195-208.

Raymond, H.L. 1988. Effects of hydroelectric development and fisheries enhancement on spring and summer chinook salmon and steelhead in the Columbia River basin. *North American Journal of Fisheries Management* 8: 1-24.

Reimers, P.E. 1973. The length of residence of juvenile fall chinook salmon in the Sixes River, Oregon. *Fish. Comm. Oregon Res. Briefs* 4(2):1-43.

Rickard, W.H. and D.G. Watson. 1985. Four decades of environmental change and their influence upon native wildlife and fish on the mid-Columbia River, Washington, U.S.A. *Environmental Conservation* 12:241-248.

Ricker, W.E. 1989. History and present state of the odd-year pink salmon runs of the Fraser River region. *Can. Tech. Rep. Fish. Aquat. Sci.* 1702.

Riddell, B.E., W.C. Leggett, and R.L. Saunders. 1981. Evidence of polygenic variation between two populations of Atlantic salmon (*Salmo salar*) native to tributaries of the S.W. Miramichi River, N.B. *Can. J. Fish. Aquat. Sci.* 38:321-333.

Robinson, G.D., W.A. Dunson, J.E. Wright, and G.E. Mamolito. 1976. Differences in low pH tolerance among strains of brook trout (*Salvelinus fontinalis*). *J. Fish. Biol.* 8:5-17.

Ros, T. 1981. Salmonids in the Lake Vanern area, in: 'Fish Gene Pools," N. Ryman, ed., pp. 21-31, Ecol. Bull. (Stockholm) 34, FRN, Box 6710, S-11385, Stockholm.

Ryder, R.A. and S.R. Kerr. 1989. Environmental priorities: placing habitat in hierarchic perspective, in: 'Proc. National Workshop on Effects of Habitat Alteration on Salmonid Stocks," C.D. Levings, B. Holtby, and M. A. Henderson, eds., pp. 2-12, *Can. Spec. Pub. Fish. Aquat. Sci.* 105.

Scudder, G.G.E. 1989. The adaptive significance of marginal populations: a general perspective, in: "Proc. National Workshop on Effects of Habitat Alteration on Salmonid Stocks," C. D. Levings, L. B. Holtby, and M. A. Henderson, eds., pp. 180-185, *Can. Spec. Pub. Fish. Aquat. Sci.* 105.

Servizi, J.A. 1989. Protecting Fraser River salmon (*Oncorhynchus* spp) from wastewaters, in: 'Proc. National Workshop on Effects of Habitat Alteration on Salmonid Stocks," C.D. Levings, B. Holtby, and M. A. Henderson, eds., pp. 136-153, *Can. Spec. Pub. Fish. Aquat. Sci.* 105.

Shapovalov, L. and A.C. Taft. 1954. The life histories of steelhead trout (*Salmo gairdneri gairdneri*) and silver salmon (*Oncorhynchus kisutch*) with special reference to Waddell Creek, California, and recommendations regarding their management. *Cal. Dept. of Fish and Game, Fish. Bull.* 98.

Smith, J.J. 1990. The effects of sandbar formation and inflows on aquatic habitat and fish utilization in Pescadero, San Gregorio, Waddell and Pomponio Creek estuary/lagoon systems, 1985-1989. Department of Biological Sciences, San Jose State University, San Jose, California.

Stahl, G. 1986. Genetic population structure of Atlantic salmon, in: 'Population Genetics and Fishery Management," N. Ryman and F. Utter, eds., pp. 121-140, University of Washington Press, Seattle.

Swain, D.P. and L.B. Holtby. 1989. Differences in morphology and behavior between juvenile coho salmon (*Oncorhynchus kisutch*) rearing in a lake and its tributary stream. *Can. J. Fish. Aquat. Sci.* 46:1406-1414.

Swales, S., C.D. Levings, and R.M. Lauzier. 1986. Winter habitat preferences of juvenile salmonids in two interior rivers in British Columbia. *Can. J. Zool.* 64:1506-1514.

Taylor, E.B. 1988. Adaptive variation in rheotactic and agonistic behavior in newly emerged fry of chinook salmon, (*Oncorhynchus tshawytscha*), from ocean-and stream-type populations. *Can. J. Fish. Aquat. Sci.* 45:237-243.

Thorpe, J.E. 1989. Developmental variation in salmonid populations. *J. Fish Biol.* 35 (Supplement A):295-303.

Utter, F.M., D. Campton, S. Grant, G. Milner, J. Seeb, and L. Wishard. 1980. Population structures of indigenous salmonid species of the Pacific northwest, in: 'Salmonid Ecosystems of the North Pacific," W.J. McNeil and D.C. Himsworth, eds., pp. 285-304, Oregon State University Press and Oregon State University Sea Grant College Program, Corvallis.

Ware, D.M. and G.A. McFarlane. 1989. Fisheries production domains in the northeast Pacific ocean, in: 'Effects of ocean variability on recruitment and an evaluation of parameters used in stock assessment models," R.J. Beamish and G.A. McFarlane, eds., pp. 359-379, *Can. Spec. Pub. Fish. Aquat. Sci.* 108.

Williams, I.V. and 15 others. 1986. The 1983 early run Fraser and Thompson River pink salmon: morphology, energetics, and fish health. International Pacific Salmon Fisheries Commission *Bulletin* XXII.

Winans, G.A. 1989. Genetic variability in chinook salmon stocks from the Columbia River Basin. *North Amer. J. Fish. Man.* 9:47-52.

Wishard, L.N., J.E. Seeb, F.M. Utter, and D. Steffan. 1984. A genetic investigation of redband trout populations. *Copeia* 1984 (1):121-132.

World Commission on Environment and Development (WCED). 1987. "Our Common Future," Oxford University Press, New York.

Zoltai, S.C. 1988. Ecoclimatic provinces of Canada and man-induced climatic change. *Canadian Committee on Ecological Land Classification Newsletter* 17:12-15, Environment Canada, Ottawa.

Impacts of Fishing on Genetic Structure of Salmonid Populations

J. E. THORPE

1. Introduction

Impacts can only be assessed if there is a baseline against which to compare change. Since it has become possible to characterise the genetic structure of salmonid populations only very recently, their structure prior to exploitation is generally unknown. To date it has been the potential impacts of fisheries on the genetic structure of salmonid populations that have been outlined on the basis of genetic theory (e.g., Lannan et al., 1989). These have shown that it is nearly impossible to exploit a living resource without imparting some genetic change. However, information on the real impacts of fisheries has been elusive and has been correlational rather than causal.

It has been the economic value of the resources rather than their long-term intrinsic genetic value which has determined resource management policies. Management has been perceived to be important only when the numerical strength of a population has begun to decline. Then the practice has been to manage on the concepts of maximum or optimal sustainable yield, to ensure an adequate escapement of spawners to maintain a numerically stable population. This numerical approach has treated all individual fish as of equal genetic potential, and the genetic implications of management have been ignored (Allendorf et al., 1987). But numerical decline has profound genetic implications. As reproductive capacity is influenced genetically (see below), the consequences of overexploitation are genetic drift and potentially increased inbreeding. If the harvest focuses disproportionately on particular genetic segments of the population through gear selectivity, resource allocation, times and places of fishing etc., then directional selection is implicated also. This risk of loss of genetic diversity is exaggerated if the species is also losing reproductive habitat concurrently, and may be even further complicated if numbers are being artificially maintained by hatchery augmentation. Given that genetic variation exists, and that there are phenotypic and genotypic correlations in relevant traits (Gjedrem, 1983), the problem now is to determine the extent to which the phenotypic changes seen in fished populations of salmonids reflect altered genetic status.

To exploit an animal population without altering its genetic constitution requires that all genetic components of that population be harvested in proportion to their relative abundance. Fishing methods vary in the extent to which they are selective, but most commercial fisheries are targeted on a particular range of sizes. This is particularly true of salmonid

SOAFD Freshwater Fisheries Laboratory, Pitlochry, PH16 5LB, Scotland, U.K.

Genetic Conservation of Salmonid Fishes, Edited by J.G. Cloud
and G.H. Thorgaard, Plenum Press, New York, 1993

fisheries, which generally focus on the largest individuals, when they are maturing or mature adults. Since larger fish often command relatively higher prices per unit weight, there is commercial advantage in fishing selectively for those of large size. What are the population consequences of this size-selectivity, and in particular, does it have any genetic relevance?

2. Phenotypic Population Changes following Fishing

Ricker (1981) summarised a series of his detailed studies of changes in the mean age and size of five Pacific salmon (*Oncorhynchus*) species, as evidenced from commercial catches in British Columbia over periods of 25-60 years. In essence, all five species had shown long-term declines in body size, concomitant with decreases in the numerical size of their populations. As these decreases were not generally correlated with oceanic temperature or salinity changes (except for sockeye (*O. nerka*) where the temperature correlation was negative), the changes were interpreted as cumulative genetic effects. The changes were most noticeable after 1950, when the fish became sold by weight rather than by the piece, making

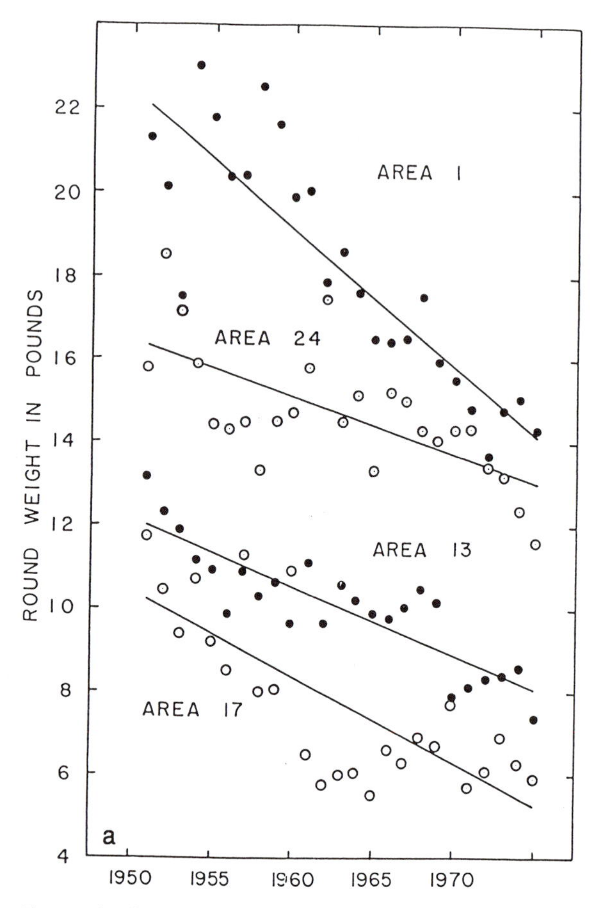

Figure 1. Change in the average size of fish harvested from the ocean since 1950 for: a. chinook (*O. tshawytscha*), b. coho (*O. kisutch*), c. pink (*O. gorbuscha*), d. chum (*O. keta*), and e. sockeye salmon (*O. nerka*) (from Ricker, 1981).

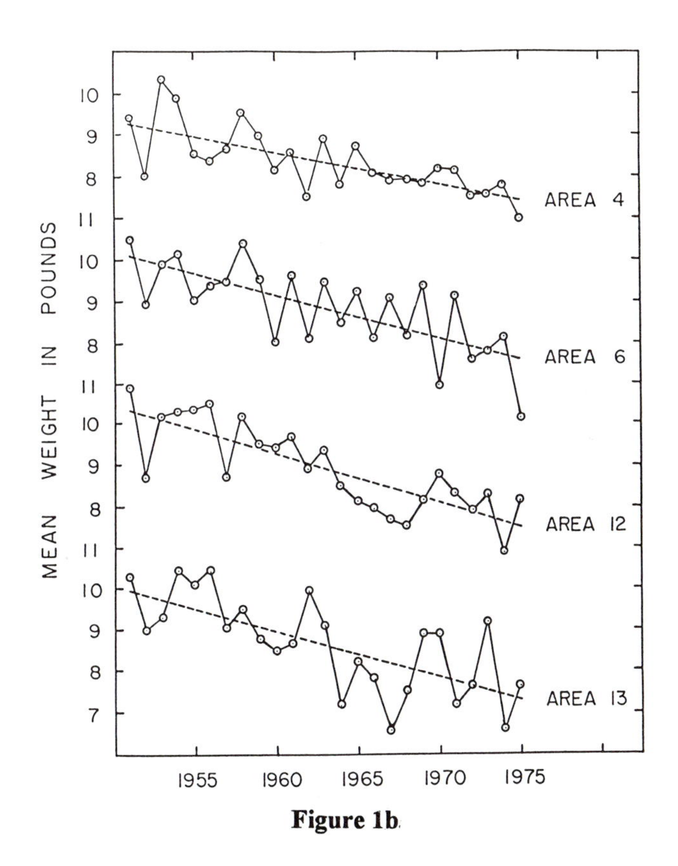

Figure 1b

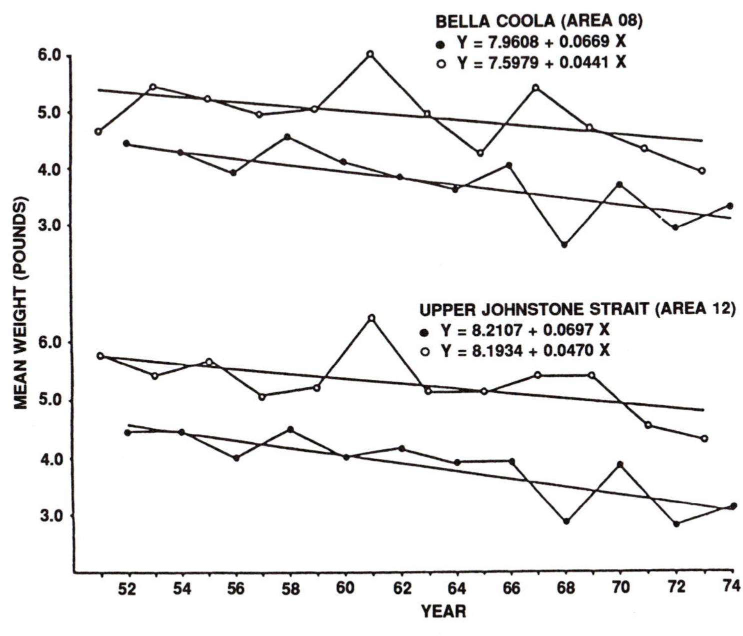

Figure 1c

it more profitable to harvest the larger individuals selectively. Chinook (*O. tshawytscha*) (Fig.1a), coho (*O. kisutch*) (Fig.1b) and pink (*O.gorbuscha*) salmon (Fig.1c) showed the greatest size decreases; chum (*O. keta*) (Fig.1d) and sockeye salmon (Fig.1e) less. The interspecific differences were attributed to differences in fishing practice and to the compensatory effect in sockeye of a positive growth response to a long-term oceanic temperature decline. However, mean age also declined in chinook, coho and sockeye. This did not happen in chum, where the gillnet-fishery took smaller faster maturing individuals, and here mean age actually increased. (All pinks mature at age 2).

Size and age reductions have also occurred in exploited populations of Atlantic salmon (*Salmo salar*) (Schaffer and Elson, 1975; Ritter and Newbould, 1977; Porter et al., 1986).

Before assessing the genetic implications of these data, another question must be posed. What, biologically, does body-size represent? Size is a historical statement. It reflects the accumulated energy and material storage of an individual throughout the period up to the point of measurement. Hence, it reflects the genetically prescribed developmental capacity of the individual under the particular set of environmental opportunities that it has encountered. The

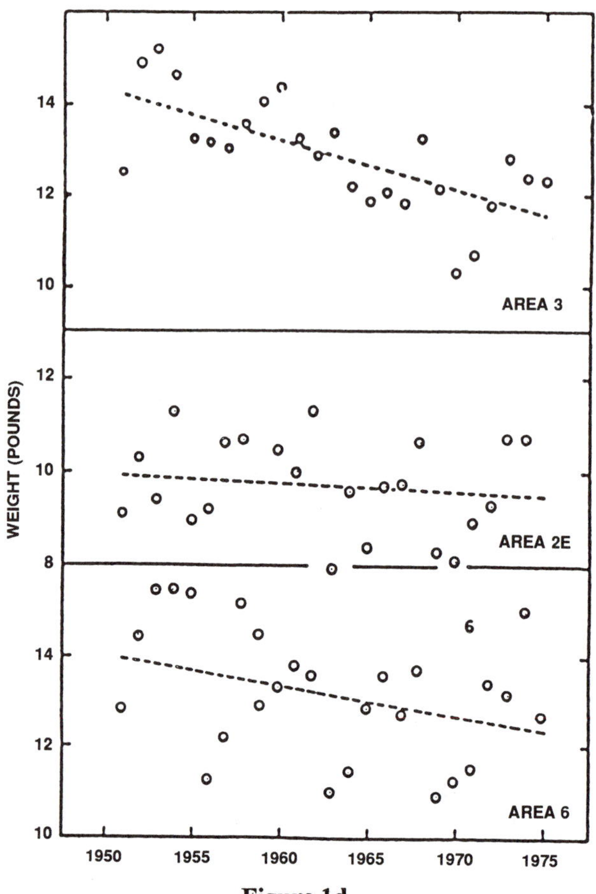

Figure 1d

implications of this statement are twofold. Firstly, size *per se* is not an exclusively genetic trait; and secondly, environmental opportunities change, so do developmental trajectories, and hence, both size and maturation rate.

The fisheries literature is replete with examples of increased growth rates following exploitation, and with simplified age structure due to maturation at earlier ages, and a consequently shortened life-span, including salmonid populations (e.g., Christy and Scott, 1965). The question at issue now is whether these effects are purely phenotypic, or whether they also imply genetic change.

Considering first the phenotypic responses, fishing affects the environmental opportunities. By definition, it reduces the density of the exploited population. Therefore those oceanic fisheries which concentrate on feeding fishes potentially reduce intraspecific competition among the survivors and so increase their environmental opportunity for rapid development. However, many salmonid fisheries concentrate on individuals at the end of their growing life, and their impacts are more on the reduction of density of spawners. Although recruitment of

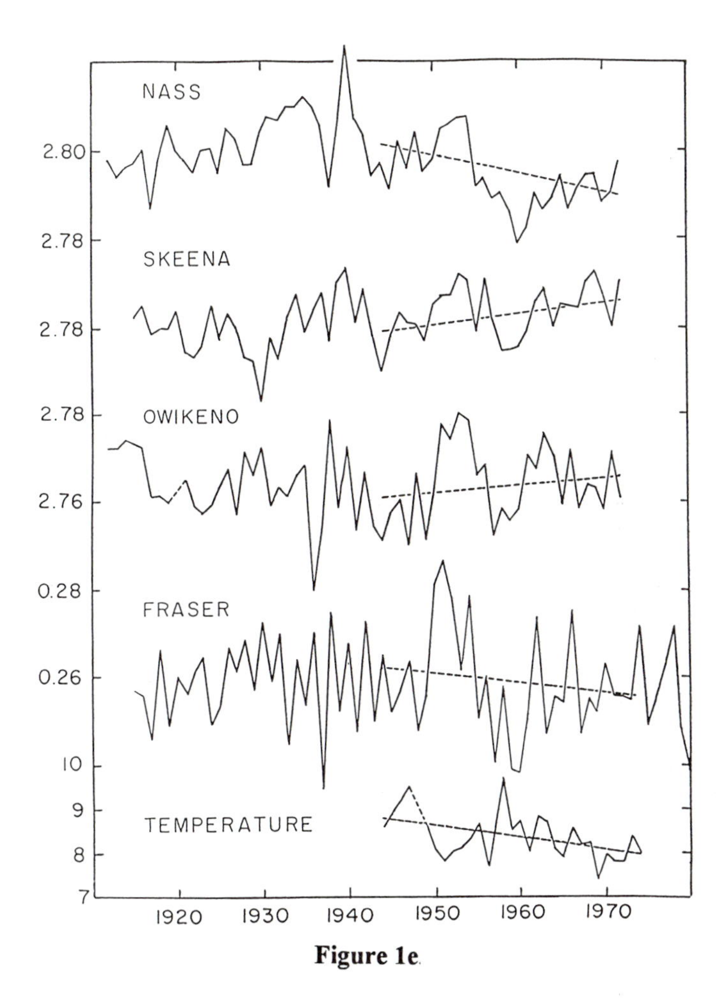

Figure 1e

salmonids is strongly influenced by the limited availability of spawning substrate, at very high exploitation levels competition for this resource becomes reduced among the depleted numbers of survivors. So the fishery may act to reduce the influence of density dependent inhibitions on growth of the progeny (Myers et al., 1986). What is the significance of this?

As Policansky (1983) has pointed out, under conditions of abundant food supply, fishes will mature at their earliest opportunity - that is, maturation has developmental priority over growth. Hence, growth can be viewed as the storage of surplus energy when the individual cannot mature. So the key to understanding the control of growth of individual fishes is an understanding of the control of their maturation.

Traditionally, maturation is thought of as a process which occurs after the fish reaches a certain body size or age - it is customary to characterise species and populations by the size or age at first maturity. However, following the Policansky approach, it can also be thought of as a process which is continuous from fertilisation, but which requires release from physiological inhibition. This model is more realistic, since it allows greater developmental flexibility and appears to represent the real world more closely as it avoids the constraint of assumed threshold size. When under specially favourable circumstances, many fishes show the capacity to reproduce at much younger ages and smaller sizes than customarily seen in the wild. For example, this is true of *Tilapia* spp. (e.g., Fryer and Iles, 1969), and also of salmonids (Thorpe, 1986, 1989).

For salmonids, the gonads are differentiated by the time of first feeding, and maturation represents the further development of gametes. The sequence of biochemical steps in this process is evidently very complex, but the facilitation of these steps (i.e. the release of their inhibition) depends on responsiveness to specific external cues at particular seasons (Scott, 1979; Eriksson and Lundqvist, 1980; Lam, 1983; Lundqvist, 1983; Scott and Sumpter, 1983; Thorpe, 1986; Munro et al., 1990). This implies that for salmonids maturation is fundamentally an annual process, since they are exposed to the specific external cues once every year. It has been shown experimentally that, given favourable feeding opportunities at the critical season for the physiological decision to maintain development on the maturation track, Atlantic salmon (*Salmo salar* L.) can be induced to mature in their first year of life (Adams and Thorpe, 1989). Conversely, given restricted feeding opportunity during such a period in later life, they can be inhibited from maturing that year (Thorpe et al., 1990). This implies that the physiological decision process as to whether or not to release the inhibitions from the next steps of the maturation process depends also on some internal assessment of performance (Thorpe, 1986), which has recently been linked to the animal's fat status at that time (Rowe et al., 1991).

There is evidence that different groups of salmon differ in their maturation responses to the same developmental opportunities (Thorpe, 1975, 1986; Naevdal, 1983; Thorpe et al., 1983; Gjerde and Refstie, 1984; Glebe and Saunders, 1986), thereby implying that there is genetic regulation of the biochemical thresholds which must be overcome to release inhibition to further maturation. I suggest that these are the processes which underlie what is customarily called inheritance of age at maturity. Hence, there is potential for fisheries to act as directionally selective forces. What changes are observed in reality?

The size changes noted by Ricker in British Columbia have also been paralleled in the Western Pacific, but here there have been dramatic changes in age proportions too. Nikulin (1970) and Krogius (1979) noted massive numerical declines in the anadromous component of sockeye salmon returning to Lakes Uyeginsk (from 100,000 to 300) and Dalneye (from 62,000 to 1,600), respectively, over 40 year periods and associated these with increased oceanic fishing pressure on these stocks. Nikulin noted an increased growth rate in freshwater among the depleted Uyeginsk fish, and both authors recorded a dramatic increase in the proportion of fish maturing in freshwater without migrating to sea. This concerned primarily the males (increasing from 13 to 82% of the spawners in Uyeginsk, and from 26 to 92% in Dalneye), but in Uyeginsk the proportion of females maturing as residents rose from 0% in 1930 to 5% in 1968. No mature resident females were recorded in Lake Dalneye in the 1930s, but in 1979, with a continued low population strength there, Varnavskaya and Varnavsky (1988) recorded 43% of the female spawners as residents (Table 1). It is not clear from Krogius' work whether there were no residents among the spawning females up to 1976, or whether they were simply not recorded at the spawning time of the anadromous sockeye. If it were the latter case, the sudden increase in their proportion in 1979 may reflect a change in the method of recording: if the former, then the increase reflects a remarkable change in the reproductive strategy of the females, and could indicate strong action of size-related mate-choice taking place (S. Skulason, personal communication).

Table 1. Changes in life-history patterns of Lake Dalneye sockeye salmon populations under the influence of an oceanic fishery.

Authority		Krogius (1979)		Varnavskaya and Varnavsky (1988)	
Period	1935-46	1947-56	1957-65	1966-76	1979
Spawning migrants	62,000	10,000	5,700	1,600	4,500
% as 1-sea-winter	0.2	0.6	4.3	37.5	?
% females	52	54	59	68	?
Smolt output	197,000	59,000	31,200	27,200	12,500
Residents as % of spawners					
males	26.0	49.4	74.2	88.8	91.8
females	0	0	0	0	43.2

Caswell et al. (1984) noted increases in the proportions of maturing male resident Atlantic salmon in some populations in Quebec, following the increased oceanic exploitation of this species. Schaffer (1974) and Wohlfarth (1986) have suggested that increased mortality of older age classes should select for earlier age at maturity.

Riddell (1986) acknowledged the evidence for genetic variance in maturation rates in salmonids, but in the absence of actual genetic evidence he cautioned against attributing phenotypic change in population ages at maturity to directionally selective effects of fisheries. He noted that life history models developed to predict age at maturity were far more sensitive to juvenile than to adult mortality variation (Jonsson et al., 1984; Stearns and Crandall, 1984), and so fisheries might be expected to exert their greatest influence through reducing density-dependent effects on juveniles (see also Myers et al., 1986). Since that time data have become accessible on the actual genetic status of several sockeye salmon populations in the USSR, subject to different levels of exploitation.

3. Genetic Status of Populations following Fishing

Lewontin (1984) pointed out that no correspondence had been shown between life-history traits and patterns of allelic variation. Hence, differences in breeding structures of stocks may not be related to the capacity of populations to survive and reproduce.

In the anadromous sockeye populations of three neighbouring systems in Kamchatka, Lakes Blizhneye, Nachikinskoye and Dal'neye, the males are larger than the females on average. However, there is also a group of small 3-year-old males which grow and mature rapidly. Altukhov and Varnavskaya (1983) examined the genetic profile of these population components. They found that heterozygosity at the *LDH* and *PGM* loci was maximal in the small males, average in the females, and minimal in the large males. The high heterozygosity of the small males correlated closely with their relatively high rates of growth and maturation (Krogius, 1960, 1972), and this heterozygosity was particularly high in the even smaller and earlier maturing resident males (Altukhov and Varnavskaya, 1983). In Atlantic salmon Jordan et al. (1990) have shown a significant positive association between heterozygosity at the *ME-2* locus and early maturation at sea.

These Asian populations have been fished with gillnets in the ocean since the beginning of the century. Exploitation has been greatest on the largest fish (predominantly males) in the early migrating population components. The severe reduction in numbers, associated with reduced age at maturity, later spawning migration, smaller fish size and an increased proportion of residents among the spawners which characterised the L.Dal'neye population (see above), was reflected also in L.Blizhneye (Altukhov and Salmenkova, in press). The changes in L.

Table 2. Observed heterozygosity at the *PGM* locus in different sex and age groups of sockeye salmon spawners in three Kamchatka lakes (after Altukhov and Salmenkova, 1991). The *PGM* locus accounted for the greatest genetic differences between the stocks, and its total heterozygosity was 45, 49 and 54% in the three lakes, respectively. Although they gave no quantitative data on exploitation levels, the authors stated that the fishery was heaviest on the Dal'neye and least on the Nachikinskoye stocks. They concluded that the genetic effect of the fishery was to select in favour of small, more heterozygous males; to have little selective effect on the intermediate sized females; and to select heavily against the large less heterozygous males. By itself, this might be supposed not to matter from a genetic conservation perspective, since theoretically diversity is not lost.

Groups	Sample size	Small males as % of total males	Heterozygosity
L.Nachinkinskoye			
Large males(57-73cm)	296		0.35
Females (50-69cm)	579		0.45
Small males(43-56cm)	202	40.6	0.55
Total	1077		0.45
L.Blizhneye			
Large males(40-63cm)	104		0.36
Females (43-59cm)	191		0.47
Small males(31-39cm)	248	70.5	0.56
Total	543		0.49
L.Dal'neye			
Total anadromous			
spring race	30		0.30
summer race	26		0.31
Resident males	158	84.0	0.62
Total	214		0.54

Nachikinskoye were less dramatic. The frequency of small individuals among the spawning males was 40.6, 70.5 and 84.0% in Lakes Nachikinskoye, Blizhneye and Dal'neye, respectively (Table 2).

However, in experiments related to the maintenance of polymorphisms in salmonid populations, Altukhov (1989) has recently shown that while high heterozygosity is associated with the highest viability of an individual it may be unfavourable for the fitness of the population as a whole. He suggested that high levels of polymorphism and heterozygosity may result in an increase in genetic recombination frequency, with impaired cooperation of interacting genes. Testing this idea, he studied eight combinations in pink salmon, with genotypes identified by the *MDH-3, 4, PGM, GPD,* and *ME-2* loci. He found that there was heterosis for growth rate, but that the progeny of parents of average heterozygosity survived better than those of high or low (Fig.2, and Table 3).

If this finding is general, then there are long-term implications for the genetic influences of fisheries on the fitness of salmonid populations, since directional selection would appear to drive these populations towards a preponderance of smaller, earlier maturing, but highly heterozygous types. By exaggerating heterozygosity, population stability would be threatened through reduced viability. Hence, such populations might decline faster than can be attributed directly to fishing alone.

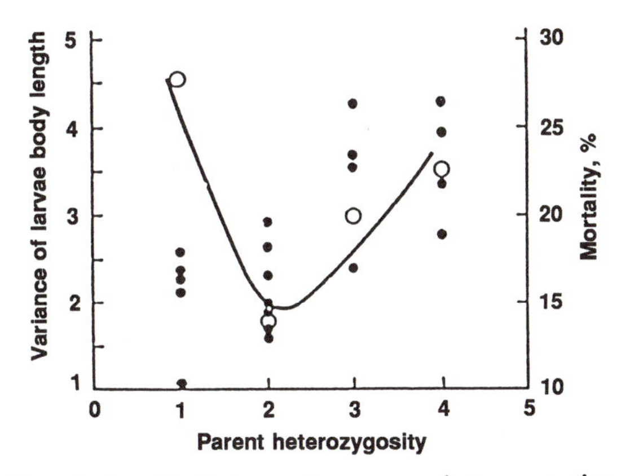

Figure 2. The relationship between the average heterozygosity of pink salmon parents and the survival of their progeny (from Altukhov and Salmenkova, 1991).

Table 3. Distributions of individual heterozygosity for allozyme loci in spawners and larvae of pink salmon (after Altukhov and Salmenkova, 1991).

Group	Frequencies of classes with different numbers of heterozygous loci per fish			Sample size	χ^2 for difference of distributions between spawners and larvae (df=3)
	0	1	>1		
Spawners	0.495±0.016	0.391±0.016	0.114±0.015	606	38.6***
Larvae	0.620±0.020	0.335±0.020	0.044±0.007	901	

Fitness values of the frequency classes in spawners with respect to larvae are 0.080±0.04; 1.117±0.08; and 2.57±0.51, in ascending order of individual heterozygosity.

4. Other Potential Genetic Effects of Fisheries

Salmonid species are made up of thousands of genetic stocks, maintained discrete through precise homing behaviour of adults to spawn in limited stretches of their natal streams (Ricker, 1972; STOCS, 1981). While the pattern is consistent with the view that present stocks evolved through small founding populations colonising different river basins following glaciations, general electrophoretic analyses by themselves do not allow the inference that such breeding organisation is adaptive. However, in a recent review Taylor (1991) has presented positive evidence for local adaptation in salmon, including six different associations of biochemical gene frequency variation and temperature-dependence. Verspoor and Jordan

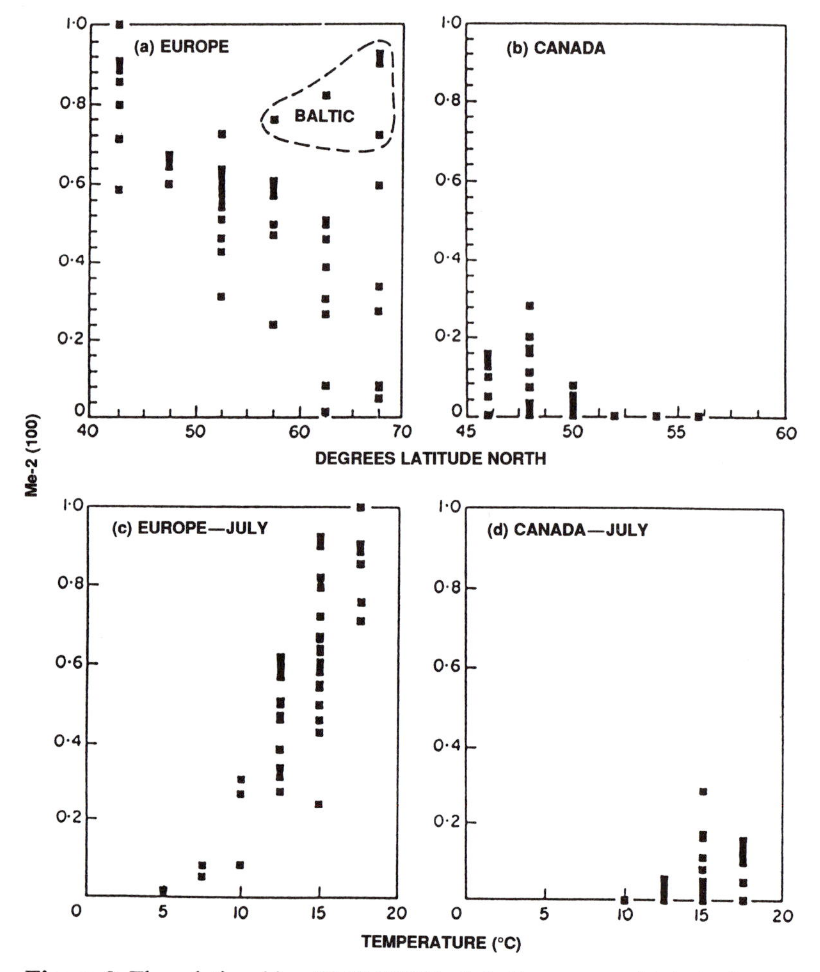

Figure 3. The relationship of *ME-2(100)* allele frequency with latitude and with mean July air temperature in Europe and Canada (from Vespoor and Jordan, 1989).

(1989) found evidence of latitudinal clines in malic enzyme-2 gene frequences in Atlantic salmon. These authors showed that the *ME-2 (100)* allele decreased in frequency with increasing latitude in both America and Europe and increased in frequency with summer stream temperature (Fig.3). The reality of this environmental association was strengthened on finding the same temperature relationship within single river systems. Since fishing tends to favour the survival of more rapid developers, and since Jordan et al. (1990) found that the earlier maturing Atlantic salmon at sea had an excess of heterozygotes, fisheries would tend to reduce the frequency of the *ME-2 (125)* variant in the north and the *ME-2 (100)* variant in the south. The impact of this would be to reduce the degree of physiological adaptation of these populations, most especially at the extremes of the species range.

Natural selection has probably operated to maintain the timing of critical physiological and reproductive events, to correspond with local climatic seasons. For example, Rich and Holmes (1928) showed that fall and spring adult migration timing in chinook salmon was genetically controlled, and Hansen and Jonsson (1991) have provided experimental evidence for genetic control of seasonal return timing in different stocks of Atlantic salmon.

Oceanic and many coastal fisheries do not discriminate between stocks, but harvest them in mixtures on feeding grounds or along migration paths before they have segregated into separate genetic components. Heavy harvesting may result in over-exploitation of numerically small stocks, driving them below viable effective numbers in the breeding units (Hilborn, 1985). However, this conclusion needs close examination, since some apparently stable spawning populations of Atlantic salmon may number fewer than 40 individuals (Saunders and Schom, 1985). Since as many as 9 generations may overlap in these spawnings, the real spawning unit would include elements spread over not one but nine years.

In conclusion, while resource managers must achieve some level of conservation in balancing social and economic pressures, they must nevertheless act on the prudent assumption that stock structure is adaptive and that genetic conservation of salmonid species would require that fishing be done on an individual stock basis.

References

Adams, C.E. and J.E. Thorpe. 1989. Photoperiod and temperature effects on early development and reproductive investment in Atlantic salmon (*Salmo salar* L.). *Aquaculture* 79:403-409.

Allendorf, F.W., N. Ryman, and F.M. Utter. 1987. Genetics and fishery management: past, present and future, in: "Population genetics and fishery management," N. Ryman and F. Utter, eds., pp. 1-19, University of Washington Press, Seattle.

Altukhov, Yu. P. 1989. "Genetic processes in populations." Nauka, Moscow.

Altukhov, Yu. P., and E.A. Salmenkova. 1991. The genetic structure of salmon populations. *Aquaculture* 98:11-40.

Altukhov, Yu. P., and N.V. Varnavskaya. 1983. Adaptive genetic structure and its connection with intrapopulation differentiation for sex, age and growth rate in sockeye salmon, *Oncorhynchus nerka* (Walbaum). *Genetika* 19:796-807.

Caswell, H., R.J. Naiman, and R. Morin. 1984. Evaluating the consequences of reproduction in complex salmonid life-cycles. *Aquaculture* 43:123-134.

Christy, F.T. Jr. and A. Scott. 1965. "The common wealth in ocean fisheries." Johns Hopkins University Press, Baltimore, Maryland.

Eriksson, L.O. and H. Lundqvist. 1980. Photoperiod entrains ripening by its differential effect in salmon, *Naturwissenschaften* 67:202-203.

Fryer, G. and T.D. Iles. 1969. Alternative routes to evolutionary success as exhibited by African cichlid fishes of the genus *Tilapia* and the species flocks of the Great Lakes. *Evolution* 23:359-369.

Gjedrem, T. 1983. Genetic variation in quantitative traits and selective breeding in fish and shellfish. *Aquaculture* 33:51-72.

Gjerde, B. and T. Refstie. 1984. Complete diallel cross between five strains of Atlantic salmon. *Livestock Production Science* 11:207-226.

Glebe, B.D. and R.L. Saunders. 1986. Genetic factors in sexual maturity of cultured Atlantic salmon (*Salmo salar*) parr and adults reared in sea cages. *Canadian Special Publication in Fisheries and Aquatic Sciences* 89:24-29.

Hansen, L.P. and B. Jonsson. 1991. Evidence of a genetic component in the seasonal retum pattern of Atlantic salmon, *Salmo salar* L. *Journal of Fish Biology* 38:251-258.

Hilbom, R. 1985. Apparent stock-recruitment relationships in mixed stock fisheries. *Canadian Journal of Fisheries and Aquatic Sciences* 42:718-723.

Jonsson, B., K. Hindar, and T.G. Northcote. 1984. Optimal age at sexual maturity of sympatric and experimentally allopatric cutthroat trout and Dolly Varden charr. *Oecologia (Berlin)* 61:319-325.

Jordan, W.C., A.F. Youngson, and J.H. Webb. 1990. Genetic variation at the malic enzyme-2 locus and age at maturity in sea-run Atlantic salmon. *Canadian Journal of Fisheries and Aquatic Sciences* 47:1672-1677.

Krogius, F.V. 1960. Growth rate and the age grouping of sockeye salmon, *Oncorhynchus nerka* (Walbaum) in the sea. *Voprosy Ikhtiologii* 17:67-88.

Krogius, F.V. 1972. The linear growth of *Oncorhynchus nerka* fry in Lake Dal'neye. *Izvestia TINRO* 82:19-31.

Krogius, F.V. 1979. On the relationship between freshwater and marine life-spans of sockeye salmon from Lake Dal'neye. *Biologiya Morya* 3:24-29.

Lam, T.J. 1983. Environmental influences on gonadal activity in fish, in: "Fish physiology: reproduction, vol. IXB," W. S. Hoar, D. J. Randall, and E. M. Donaldson, eds., pp. 65-116, Academic Press, New York.

Lannan, J.E., G.A.E. Gall, J.E. Thorpe, C.E. Nash, and B.E. Ballachey. 1989. Genetic resource management of fish. *Genome* 31:798-804.

Lewontin, R.C. 1984. Detecting population differences in quantitative characters as opposed to gene frequencies. *American Naturalist* 123:115-124.

Lundqvist, H. 1983. Precocious sexual maturation and smolting in Baltic salmon (*Salmo salar* L.): photoperiodic synchronisation and adaptive significance of annual biological cycles, Ph.D. thesis, University of Umeå, Sweden.

Munro, A.D., A.P. Scott, and T.J. Lam. 1990. "Reproductive seasonality in Teleosts: environmental influences." CRC Press, Boca Raton.

Myers, R.A., J.A. Hutchings, and R.J. Gibson. 1986. Variation in male parr maturation within and among populations of Atlantic salmon, *Salmo salar* L.. *Canadian Journal of Fisheries and Aquatic Sciences* 43:1242-1248.

Naevdal, G. 1983. Genetic factors in connection with age at maturation. *Aquaculture* 33:97-106.

Nikulin, O. 1970. On the relationship between the decreased absolute numbers of sockeye *Oncorhynchus nerka* (Walb.) and the increased relative numbers of dwarfs among young fish feeding in Lake Uyeginsk (Okhotsk Region). *Izvestia TINRO* 71:205-217.

Policansky, D. 1983. Size, age and demography of metamorphosis and sexual maturation in fishes. *American Zoologist* 23:57-63.

Porter, T.R., M.C. Healey, M.F. O'Connell, E.T. Baum, A.T. Bielak, and Y. Cote. 1986. Implications of varying the sea age at maturity of Atlantic salmon (*Salmo salar*) on yield to the fisheries. *Canadian Special Publication in Fisheries and Aquatic Sciences* 89:110-117.

Rich, W.H. and H.B. Holmes. 1928. Experiments with marking young chinook salmon on the Columbia River, 1916-1927. *Bulletin of the United States Bureau of Fisheries* 44:215-264.

Ricker, W.E. 1972. Hereditary and environmental factors influencing certain salmonid populations, in: "The stock concept in Pacific salmon," R. C. Simon and P. A. Larkin, eds., pp. 19-160, University of British Columbia Press, Vancouver.

Ricker, W.E. 1981. Changes in the average size and average age of Pacific salmon. *Canadian Journal of Fisheries and Aquatic Sciences* 38:1636-1656.

Riddell, B.E. 1986. Assessment of selective fishing on the age at maturity in Atlantic salmon (*Salmo salar*): a genetic perspective. *Canadian Special Publication in Fisheries and Aquatic Sciences* 89:102-109.

Ritter, J.A. and K. Newbould. 1977. Relationship of parentage and smolt age to first maturity of Atlantic salmon (*Salmo salar*). *International Council for the Exploration of the Sea* CM 1977/M.32.

Rowe, D.K., J.E. Thorpe, and A.M. Shanks. 1991. Role of fat stores in the maturation of male Atlantic salmon (*Salmo salar*) parr. *Canadian Journal of Fisheries and Aquatic Sciences* 48, 405-413.

Saunders, R.L. and C.B. Schom. 1985. Importance of the variations in life-history parameters of Atlantic salmon (*Salmo salar*). *Canadian Journal of Fisheries and Aquatic Sciences* 42:615-618.

Schaffer, W.M. 1974. Selection for optimal life histories: the effect of age structure. *Ecology* 55:291-303.

Schaffer, W.M. and P.F. Elson. 1975. The adaptive significance of variations in life history among local populations of Atlantic salmon in North America. *Ecology* 56:577-590.

Scott, A.P. and J. Sumpter. 1983. The control of trout reproduction: basic and applied research on hormones, in: "Control processes in fish physiology" J.C. Rankin, T.J. Pitcher, and R.T. Duggan, eds., pp. 200-220, Croom Helm, London.

Scott, D.B.C. 1979. Environmental timing and the control of reproduction in teleost fish. *Symposia of the Zoological Society of London* 44:105-132.

Skulason, S. 1991. Personal communication.

Stearns, S.C. and R.E. Crandall. 1984. Plasticity for age and size at sexual maturity: a life-history response to unavoidable stress, in: "Fish reproduction: strategies and tactics" G. W. Potts and R. J. Wootton, eds., pp. 13-33, Academic Press, London.

STOCS. 1981. Proceedings of the Stock Concept International Symposium, Alliston, Ontario, 1980. *Canadian Journal of Fisheries and Aquatic Sciences* 38:1457-1921.

Taylor, E.B. 1991. A review of local adaptation in Salmonidae, with particular reference to Pacific and Atlantic salmon. *Aquaculture* 98:185-207.

Thorpe, J.E. 1975. Early maturity in male Atlantic salmon. *Scottish Fisheries Bulletin* 42:15-17.

Thorpe, J.E. 1986. Age at first maturity in Atlantic salmon. *Salmo salar*: freshwater period influences and conflicts with smolting. *Canadian Special Publication in Fisheries and Aquatic Sciences* 89:7-14.

Thorpe, J.E. 1989. Developmental variation in salmonid populations. *Journal of Fish Biology* 35 (Supplement A):295-303.

Thorpe, J.E, R.I.G. Morgan, C. Talbot, and M.S. Miles. 1983. Inheritance of developmental rates in Atlantic salmon, *Salmo salar* L. *Aquaculture* 33:119-128.

Thorpe, J.E., C. Talbot, M.S. Miles, and D.S. Keay. 1990. Control of maturation in cultured Atlantic salmon, *Salmo salar*, in pumped seawater tanks, by restricting food intake. *Aquaculture* 86:315-326.

Varnavskaya, N.V. and V.S. Varnavsky. 1988. On the biology of dwarf forms of sockeye in Lake Dal'neye (Kamchatka). *Biologiya Morya* 1988:16-23.

Verspoor, E. and W.C. Jordan. 1989. Genetic variation at the *Me-2* locus in Atlantic salmon within and between rivers: evidence for its selective maintenance. *Journal of Fish Biology* 35 (Supplement A):205-213.

Wohlfarth, G.W. 1986. Decline in natural fisheries - a genetic analysis and suggestion for recovery. *Canadian Journal of Fisheries and Aquatic Sciences* 43:1298-1306.

Genetic Change in Hatchery Populations

GRAHAM A. E. GALL

1. Introduction

Artificial propagation of salmonid fish has a long and varied history (Gall and Crandell, 1992) with both its critics and its proponents. However, most criticism has not been particularly constructive due to a lack of appreciation or knowledge by the critics of the hatchery programs or their objectives.

Frequently, critics charge that "hatcheries" per se have specific detrimental effects, primarily loss of genetic variation. Nothing could be further from the truth. What is true is that "hatchery management" often has been faulty relative to proper genetic management of the stocks. There are few physical hatchery facilities that have been responsible directly for poor genetic management of the fish reared in the hatchery. While critics may be correct in the nature of their complaints, the focus of comment is most often toward an assumed purpose of a hatchery program. It is essential that the mission and objectives of a hatchery program be clearly assessed, including whether or not the program has carefully defined genetic and breeding goals. Only then can the real potential impact of hatchery operations be properly assessed and criticized.

However, it does seem fair to ask those involved in all aspects of fisheries biology to accept a fair share of responsibility for the existing hatchery program design and operation. In particular, it should be realized that seldom is the discipline of genetics considered as an integral component of fisheries biology, either in training or practice. This is the most likely reason that genetic and breeding goals tend not to exist as part of the stated mission of most hatchery programs.

There are four major considerations of a genetic nature that must be integrated into the design and operation of a hatchery program to insure proper genetic management of the resource cultured in the hatchery. These are: (1) the initial sampling of fish to form the broodstock, (2) domestication of the broodstock, (3) management and operation of the hatchery facility, and (4) defining genetic and breeding goals. It must be noted that the fourth item must be determined first; otherwise, it is impossible to competently address the first three. With defined goals, it is expected that facilities and their mode of operation can be developed that will facilitate the program goals.

Department of Animal Science, University of California, Davis, CA 95616, U.S.A.

Genetic Conservation of Salmonid Fishes, Edited by J.G. Cloud
and G.H. Thorgaard, Plenum Press, New York, 1993

2. Background

The history of hatchery operations designed for a variety of specific purposes indicates that carefully defined genetic considerations generally have not played a part in program design and development. Unfortunately, most programs tend to an after-the-fact approach. This often takes on a pattern of identifying some fish that are easily obtained, specifying a general use facility with fish production as the prime motivator to design, and hurrying to produce fish so the program will be effective as early as possible.

Setting goals requires knowledge of both genetic management tools and the inheritance of characters. Unfortunately, a look at history shows that geneticists with such knowledge are seldom consulted until after the fact. It also is unfortunate that most fishery biologists and many fish geneticists refuse to accept or acknowledge the existence of over fifty years of experience accumulated by animal and plant breeders. There is a myth among fisheries folk that fish are unique and must never be confused with animals used for agricultural production. This attitude is so pervasive that most fish genetic programs are training students in isolation from this rich heritage of science. A sound theory and practice of animal breeding exists and has been shown experimentally to be founded on solid principles. A careful search of the scientific literature will show that essentially everything of interest to a fish geneticist has been examined in some animal breeding experiment. There may be biological differences among organisms that require knowledge obtained from farm animals be "fine-tuned," something a good biologist should find easy and rewarding. It is unfortunate when the "wheel must be re-invented."

To help overcome the difficulty many find in addressing basic ideas, it may be useful to briefly review some straightforward animal breeding theory and practice. The observed "value" of an individual, its phenotypic value, is the result of an interplay between the genotype of the individual (genotypic value) and the environment under which the individual developed (environmental effects). Many appear to forget that often the largest fish in a population, or the population with the largest average-sized fish, was the fish, or the population of fish, that was able to obtain the most food. Our interest is in the genetic value of the individuals. But the observed value, the phenotypic value, will reflect the genotypic value of an individual accurately only to the extent that environmental effects are absolutely equal for all individuals.

The notion of heritability is used as a guide in determining the degree to which the phenotypic value can be expected to reflect an individual's genotypic value. The heritability coefficient (h^2) is a simple regression coefficient relating genotypic value to phenotypic value. We can refer to genotypic value simply as genetic value; many books on animal breeding abbreviate even further to use the descriptive term, breeding value, particularly when discussing selection. Since the heritability coefficient is in reality a regression coefficient, it is convenient to describe the relationship between genetic value (G) and phenotypic (P) value with a simple regression equation,

$$G = h^2P + (1-h^2)\overline{P}$$

where $\overline{P}$ represents the mean of the population. From this formulation, we see that any estimate of the genetic value of an individual will be a function of both the individual's phenotypic value and the average phenotypic value of the population in which it resides. Clearly, if individuals from different populations are to be compared, it is imperative that the heritability coefficient used in the second term be one which accounts for environmental differences between populations, a fact seldom acknowledged (see Crandell and Gall, 1992a). For individuals within a single population, the genotypic value of individuals can only be predicted accurately from phenotypic values when the heritability coefficient is unity, that is, when environmental effects are exactly the same for all individuals.

Characters can differ widely in the degree to which phenotypic values accurately reflect genetic values of individuals. A summary of heritability coefficients for a number of traits of rainbow trout is presented in Table 1.

These numbers can be considered examples of the level of disparity expected between observed phenotypic values for traits and the breeding value of individuals with respect to the specific trait. Note that juvenile body weight and survival traits tend to have low heritabilities. Heritabilities for egg production and age at maturity are higher indicating that differences between individuals within a population more closely reflect differences in genetic values than is expected for other traits.

Table 1. Heritability Estimates for Traits of Rainbow Trout Qualitatively Summarized from the Published Literature.

Trait	Estimate
Juvenile body weight	.05 - .10
Market body weight	.15 - .20
Embryo survival	.05 - .20
Egg size	.20 - .30
Egg number	.20 - .30
Age at maturity	.20 - .40
Carcass lean	.10 - .15
Carcass fat	.25 - .45

Information about the nature of phenotypic values also can be gained from comparing heritability coefficients estimated by various methods, for example, by examining estimates from sire and dam families. Estimates for body weight at various ages of rainbow trout are given in Table 2.

Table 2. Estimates of Heritability for Various Measures of Body Weight Based on Half-sib (Sire) and Full-sib (Dam) Components of Variance for a Rainbow Trout Stock Maturing for the First Time at 30 Months of Age (Gall and Huang, 1988).

Body Weight	Estimate[1]	
Trait	Sire	Dam
End nursery period	.54	.37
One year of age	.20	.97
25 months of age	.19	.63
Mature males	.17	.76
Mature females	.21	.43

1Standard error of Sire estimates: 0.11 to 0.14.

Heritability estimates obtained from comparisons among sire (half-sib) families provide information about additive genetic differences (true breeding values) since common environmental effects are removed by the estimation procedure and there is little evidence of non-additive genetic effects for body weight of animals. In contrast, the estimates derived from comparisons among dam (full-sib) families, using the identical data set, include the effects of environmental factors common to all members of a full-sib family. The examples provided in Table 2 indicate the existence for fairly large common environmental effects on body weight of rainbow trout at all ages except the nursery period. These common environmental effects probably are of maternal origin. The results outlined in Table 2 suggest care in comparing individuals because brothers and sisters are likely to be more similar phenotypically for body weight than two individuals chosen at random from the population.

Finally, it is generally necessary to obtain multiple estimates of heritability coefficients to establish reliability. One trait of interest in fisheries for which there have been numerous estimates is age at maturity within season (these values may also apply equally well to maturity measured as year-of-age). Estimates from six separate studies are high and very similar (Table 3); a similar value of 0.39 for pink salmon was presented at these proceedings by Gharrett (these proceedings). One estimate for rainbow trout was lower than the other six which could

Table 3. Summary of Estimates of Heritability for Age at Maturity for Various Species.

Species	Estimate	Reference
Atlantic Salmon	.48	Gjerde (1986)
Atlantic Salmon	.39	Gjerde and Gjedrem (1984)
Rainbow Trout	.21	Gjerde and Gjedrem (1984)
Rainbow Trout	.37	Gall et al. (1988)
Rainbow Trout	.35	Gall and Huang (1988)
Mosquitofish	.41	Busack and Gall (1983)
Mosquitofish	.46	Campton and Gall (1986)

be due to sampling or the genetic nature of the particular stock used in the study. This high consistency among estimates from a variety of sources indicates that confidence can be placed in the estimates and the conclusion that age at maturity is a very heritable phenotype.

Many are probably wondering what all this discussion of estimation and theory has to do with considerations of genetic changes in hatchery populations of fish since all the data were obtained from domestic fish used in fish production programs. The answer is very simple; this old animal breeding methodology allows for intelligent thinking and planning. As an example, consider a prediction of response to selection to increase number of eggs per female for rainbow trout that I prepared about 20 years ago. In absence of data, other than phenotypic mean and variance, it was postulated that a change of about 200 eggs per female per generation could be achieved, an improvement of 1000 eggs per female in ten years. Sixteen years later, the selection program resulted in a total genetic change of about 600 eggs per female and a total phenotypic change of 1500 eggs (Table 4). The large phenotypic change could be attributed to management improvements resulting from the availability of data collected by hatchery personnel and improvements in the environment such as feed quality.

Table 4. Summary of Average Egg Number Per Female for One of Two Replicates (Even Year Class) of an Age-Two Spawning Rainbow Trout Stock (Hot Creek) Held under Multiple Trait Index Selection for Egg Production and Yearling Weight.

	Egg Number	
Year Spawned	Phenotypic	Genotypic[1]
1974	2790	2210
1976	2460	2510
1978	2470	2700
1980	2980	2800
1982	3275	2980
1984	3820	3490
1986	4110	3910
Estimated Total Change	1500	600

1Estimated using linear models techniques to estimate genetic trend.

3. Initial Sampling

The main concern in assessing a sampling strategy is the phenomenon known as founder effect; the genetic material possessed by the fish used to "found" a broodstock. It is generally accepted that potential source populations will differ in level and kind of genetic variation. Thus, the first consideration in developing a sampling strategy is the goal of the culture program.

Three general goals can be used to demonstrate this point. The goals outlined also represent the likely range of the genetic objectives for most hatchery programs.

1. Interest could be in a high degree of genetic variation, rapid growth rate, and high fecundity essential for a fish production program. This goal will require a specific set of sampling strategies.
2. Interest could be in supplementation of a native stock to enhance a fishery, a goal indicating concern for both production and native stock conservation. This goal will require a second set of sampling strategies.
3. Interest could be purely in conservation of a wild stock threatened with extinction due to habitat degradation or some other environmental problem. This goal will require a third set of sampling strategies.

Note that, in general, the more the goal of the program tends toward production effectiveness, the less it will tend toward conservation of the genetic integrity of a specific stock. It is likely that the two extreme goals (1 and 3) will be incompatible genetically. For the remainder of this discussion, it will be assumed that conservation is at least a major component of program missions.

Sampling for conservation purposes is the most difficult. All sampling decisions must be based on phenotypic estimates for all characters since genetic testing, specifically quantitative genetic stock evaluation, is not possible. Consequently, the only alternative is to sample the full range of observed phenotypic variability. This approach will result in excessive sampling effort since a great proportion of the observed variability will be primarily environmental in origin.

The approach of sampling all observed variation presents two serious problems. First, a very large number of fish must be sampled, meaning a large broodstock facility is essential.

If the value of sampling a range of phenotypes is to be realized, individual groups of fish must be reared to retain identity of the various types. The second problem is that a method to carefully weight sample sizes must be devised that will reflect the density distribution of each phenotypic character while balancing this effort for a large number of characters. For example, in sampling an anadromous salmon stock, consideration should be given to the density of individuals returning over the duration of the anadromous run while balancing this weighting for the density distribution at spawning time. There are few, if any, hatchery programs designed with sufficient physical facilities or management rigor to accomplish this level and type of sampling strategy.

4. Domestication

There is only a brief literature on how domestication occurs and how it might be managed. It has not been of interest even to zoological gardens and aquaria since these institutions generally are constrained to sampling a few individuals for display or preservation. Doyle (1983) has provided a discussion of the theory of 'domestication selection' along with a few experimental results. Given this paucity of experience, it is fairly simple to substitute good animal breeding and population genetic logic to assess a number of relevant considerations.

A species can undergo significant phenotypic change when placed under artificial culture conditions (Robinson and Doyle, 1990). Most often the changes will be in response to the new environment, including program management. It must be remembered that (a) program managers determine the environment, (b) management of the animals is part of the management system, (c) not all observed phenotypic changes will reflect genetic change; and (d) no evaluation method can be used to remove environmental effects (Busack and Gall, 1983; Campton and Gall, 1988).

The most obvious component of the system is the physical environment. The type of physical facilities utilized, as well as the management of the fish within the limitations of the physical facilities, establishes the environmental regime for the fish. These two factors, the physical environment and fish management, will determine the compatibility of the sample of genotypes contained in the broodstock to the environmental conditions. The actual level of incompatibility will impart limitations on the rate of survival of various genotypes. An environment that results in high mortality increases the risk of genetic selection or drift or both. An environment that requires artificial induction of spawning is likely to select for response to spawning induction. A standardized spawning procedure, such as identifying and spawning ripe females every 10 days, can impart a selection pressure on reproductive behavior or simply change genotypic frequencies due to sampling (genetic drift). How many hatchery managers spawn rainbow trout every day of the week?

5. Program Management

The management of the broodstock can have subtle effects that require greatest attention. What must be recognized is that some highly heritable traits can be genetically correlated (Table 5).

For example, rainbow trout body weight is correlated with age at spawning (maturation of females) and with most egg production traits. In addition, spawning time is correlated with age of female and spawn date (day within the season). Since spawn date is highly heritable, spawning only those females that mature early in the season will not only result in a change in spawn time but should also cause a correlated response toward earlier maturation (Table 6).

Thus, we would expect a management program that derives a major portion of the egg take from the first portion of the spawning season to result in a trend toward higher proportions of younger mature fish in addition to a trend toward an earlier spawning season.

Discarding (culling) small sized juvenile fish because they don't fit program objectives, such as a minimum size at a specified release date, will result in selection. The larger fish will reproduce in the native habitat while the smaller fish will be artificially spawned in the hatchery to produce the next generation of broodstock. Unfortunately, there is no information upon which to assess the nature of the selection response. For anadromous species, such selection is likely to effect age of maturity since size of fish is correlated with age of smoltification.

Table 5. Some Examples of Genetic Correlations Among Traits of Rainbow Trout (Taken from Huang and Gall, 1989).

Body Weight at Spawning with		Egg Size with	
Age	.48	Egg Number	-.40
Egg Size	.38	Spawn Age	.25
Egg Number	.42	Egg Volume	-.04

Table 6. Relationship between Week of the Spawning Season and Age at Spawning for Rainbow Trout Females Examined over Three Consecutive Generation (Davis Stock).

Week of	Generation (year spawned)		
Spawning	78	80	82
8/1 - 8/26	723	714	719
8/27 - 9/3		719	728
9/3 - 9/9	723	705	733
9/10 - 9/16		709	735
9/17 - 9/23	751	715	739
9/24 - 9/30		721	743
10/1 - 10/7	762		747
10/8 - 10/14	762		
10/15 - 10/28	772		767
Average Age at Spawning	761	713	743

Thus, naturally returning adults will tend toward an earlier age of maturation. For freshwater or marine fish, the effect may be random since the skewness generally observed for body size tends to disappear with age.

A particularly critical management consideration is the control of inbreeding, even at mild levels. Unfortunately, this cannot, *I repeat, cannot* be accomplished accurately without some level of pedigree information on the broodstock. Since random mating is the desired mating scheme, simply isolating groups by spawn time will not work effectively. Neither is it feasible to achieve randomness by mixing gametes prior to the fertilization event — there is adequate evidence that fertilization of eggs with mixed sperm simply results in most of the eggs being fertilized by sperm from a dominant male. McKay et al. (1992) provided an assessment of the effect of various mating systems on combined rates of inbreeding and selection response. However, they acknowledge the inherent difficulty in determining the effectiveness of fertilizing eggs with mixed sperm from several males. Consequently, it is imperative that effective physical marking techniques be developed and implemented.

Finally, not all observed phenotypic changes will reflect major genetic change. An example is the reduction of variation in body size usually observed after three or four generations in the hatchery. Unfortunately, studies have not been performed to provide evidence on this point. Unpublished data on the flour beetle (*Tribolium castaneum*; Gall, 1975) provides an indication of the nature of the effects. The flour beetle, a natural pest in wheat storage facilities, has been used extensively as a model organism for the study of animal breeding theory (see Gall, 1971). The organism is cultured at room temperature in a simple artificial medium of dry wheat flour supplemented with 10% dried brewers yeast. If a sample of flour beetles is brought into the laboratory from the wild, it takes about three generations for the population to stabilize in terms of mean pupa weight and variance in pupae weight (pupa weight is an excellent measure of final larval growth). No genetic changes have been identified. The response appears to be a simple phenotypic adjustment to the uniform and improved artificial environment. A larger number of generations are required before other environmental effects stabilize, such as those that affect body size of parents and fecundity and age at reproduction. These traits have a strong maternal component that effects subsequent progeny generations. Therefore, it is important to consider that not all phenotypic changes observed as a stock is brought under artificial culture will reflect genetic changes.

6. Genetic and Breeding Goals

The method used to choose brood fish to reproduce the population is a critical step. All management decisions, beginning with fertilized eggs, that influence which fish are allowed to reach maturity will influence the gene pool available to form gametes. Which fish from among those that survive to maturity are allowed to spawn (contribute to the gamete pool) determines the final genetic make up of the progeny generation. The goals governing these choices must be carefully defined in terms of the purpose of the breeding program. Too often, the choices are determined on a *ad hoc* basis, are based on maintaining efficient operation of the hatchery, or are determined as simple preferences of the manager.

There have been cases of managers replacing broodfish simply by retaining what is assumed to be a random sample of fish from production ponds. One problem with this approach is that the sample is not likely to be random with regard to the genotypic make-up of the population. The fish may have been segregated by size at some earlier age or some fish may have been removed for some other purpose. In addition, the approach ignores the contribution of different progeny families to the sample of parents so some families can contribute disproportionately large numbers of future parents. Often in the selection of parents, spawning time is ignored in favor of ensuring reproduction at a most opportune time for management or to meet the needs for specific production goals. Spawning season is one of the most heritable of the important traits so is easily modified. Thus, this simple procedure of sampling from fish available in production ponds is not likely to achieve complete genetic sampling.

Even random sampling within only a portion of the spawn season can cause severe genetic drift due to a non-random distribution of individuals within families. The effective population size is reduced because the high heritability of spawn time results in a high correlation of spawn time among family members. Without adequate pedigree information, it is not possible to detect this problem. Another source of non-random selection occurs when some brood fish are discarded because they do not fit the manager's ideal of a typical fish. For example, early maturing males often are discarded, small fish are not spawned, or fish with phenotypic abnormalities are discarded.

It is well established that directional selection is as effective with salmonids (rainbow trout, brook trout, coho, Atlantic salmon) as it is with other animal species. Kincaid et al. (1977) achieved significant gains in nursery weights after three generation of selection (Table 7). It is possible to increase weight at spawning by over 10% per generation through selection for increased fecundity and high yearling weight (Table 8; Gall, unpublished data). Age at sexual maturity also is easily manipulated at least within 1 and 2 year age range and within season (Siitonen and Gall 1989; Gall and Crandell, 1993b).

Whether early maturing individuals should be discarded is an open question, if the goal is to retain maximum genetic material in the population. Clearly, any environmental factor that increases growth rate (nutrition, water temperature, good husbandry) will reduce age at maturity. Consequently, it is arguable that conservation programs should spawn all maturing fish (or an accurate random sample) to minimize genetic loss. This conclusion is based on the high likelihood that the observed reduction in age or time of maturity is environmentally induced, rather than a response to selection.

Table 7. Response to Selection for 147-d Weight of Rainbow Trout (Adapted from Kincaid et al., 1977).

	Mean 147-day weight (g)		
Generation	Selected	Control	Response
0	2.02	2.02	—
1	3.40	3.08	0.32
2	4.44	3.77	0.99
3	5.27	4.30	1.94

Table 8. Summary of Correlated Response in Body Weight at Spawning for Two Year-Classes of a Domestic California Rainbow Trout Stock (Mt. Shasta) under Multitrait Selection to Improve Egg Production and Yearling Weight (Data from 1976 to 1988).

	Mean Body Weight at Spawning (g)	
Generation	Even Year	Odd Year
1	550	610
2	920	625
3	1125	800
4	975	1110
5	1260	1140
Average Response/Gen.	149	153
	(11.2 %/yr)	(13.8 %/yr)

Consideration of the question of how to deal with changes in age at maturity provides an excellent example of the naivety of many programs—they assume that any phenotypic change must be bad because it reflects a response to some selection pressure imparted by hatchery rearing. The real question that should be addressed is how to minimize the conflict among the effects of rearing in an artificial environment, the desire for cost effective production, and a goal of culturing a fish that fits into a particular natural ecosystem. If one breeding goal is to stabilize age at maturity, then changes in the culture regime for the hatchery program may be the most important consideration, not genetic management. For example, reducing the water temperature and restricting feed intake (Table 9) are likely to retard sexual development; the challenge is to find the regime that will produce mature fish at the age desired, given sound genetic management of the stock. Such management changes almost always will increase the cost of producing fish. If conservation is the purpose of the program, then one mission of the program must be to produce quality fish for conservation rather than fish at a minimum unit cost.

Table 9. Proportion of Males Maturing before One Year (First Year Males), Proportion of Males Not Maturing until the Second Year (Second Year Males), and Proportion of Females for a Rainbow Trout Stock Reared at Two Water Temperatures (Adapted from Crandell and Gall, 1993b)

Water Temperature (C)	First Year Males (%)	Second Year Males (%)	Females (%)
16 constant	16.2	31.0	52.8
9-17 ambient	6.9	45.8	47.3

7. Conclusion

Experience and logic demonstrate that it is possible to design hatchery management regimes and breeding programs to achieve most genetic goals. There will always be some genetic change resulting from artificial production that is different than changes expected in the natural environment, but the changes can be managed. And it must be remembered that natural populations do not exist in a static genetic condition. If local adaptation is important, then natural selection is capable of nullifying modest genetic changes caused by hatchery production. If hatchery production is essential for supplementation or recovery of stocks suffering from environmental damage to the habitat, a quality hatchery program it can be designed and managed, at least for the short-term, if the will is strong and common genetic sense and experience prevail.

References

Busack, C.A. and G.A.E. Gall. 1983. An initial description of the quantitative genetics of growth and reproduction in the mosquitofish, *Gambusia affinis*. *Aquaculture* 32:123-140.

Campton, D.E. and G.A.E. Gall. 1988. Response to selection for body size and age at maturity in the mosquitofish, *Gambusia affinis*. *Aquaculture* 68:221-241.

Crandell, P. and G.A.E. Gall. 1993a. The effect of sex on heritability estimates of body weight for individually tagged rainbow trout (*Oncorhynchus mykiss*). *Aquaculture* (in press).

Crandell, P. and G.A.E. Gall. 1993b. The genetics of age and weight at sexual maturity for individually tagged rainbow trout (*Oncorhynchus mykiss*). *Aquaculture* (in press).

Doyle, R.W. 1983. An approach to the quantitative analysis of domestication selection in aquaculture. *Aquaculture* 33:167-185.

Gall, G.A.E. 1971. Replicated selection for 21-d pupa weight of *Tribolium castaneum*. *Theor. App. Gen.* 41:164-173.

Gall, G.A.E. and P. Crandell. 1992. The rainbow trout. Aquaculture 100: (1-10).

Gall, G.A.E. and N. Huang. 1988. Heritability and selection schemes for rainbow trout: reproductive performance. *Aquaculture* 73:57-66.

Gall, G.A.E., J. Baltodano and N. Huang. 1988. Heritability of age at spawning for rainbow trout. *Aquaculture* 68:93-102.

Gjerde, B. 1986. Growth and reproduction in fish and shellfish. *Aquaculture* 57:37-55

Gjerde, B. and T. Gjedrem. 1984. Estimates of phenotypic and genetic parameters for carcass traits in Atlantic salmon and rainbow trout. *Aquaculture* 36:97-110.

Huang, N. and G.A.E. Gall. 1989. Correlation of body weight and reproductive characteristics in rainbow trout. *Aquaculture* 86:101-200.

Kincaid, A.L., W.R. Bridges, and B. von Limbach. 1977. Three generations of selection for length and weight in fall-spawning rainbow trout. *Trans. Am. Fish. Soc.* 106:621-628.

Kinghorn, B.P. 1983. A review of quantitative genetics in fish breeding. *Aquaculture* 31:283-304.

McKay, L.R., I McMillan, E. Sadler, and R.D. Moccia. 1992. Effects of mating system on inbreeding levels and selection response in salmonid aquaculture (Abst.). *Aquaculture* 100: (in press).

Robinson, B.W. and R.W. Doyle. 1990. Phenotypic correlations among behavior and growth variables in tilapia: implications for domestication selection. *Aquaculture* 85:177-186

Siitonen,L. and G.A.E. Gall. 1989. Response to selection for early spawn date in rainbow trout *Salmo gairdneri*. *Aquaculture* 78:153-161.

Potential Impacts of Transgenic and Genetically Manipulated Fish on Natural Populations: Addressing the Uncertainties through Field Testing

E. HALLERMAN[1] and A. KAPUSCINSKI[2]

"In this decade, the success of biotechnology will rely on obtaining the public's confidence that these new products of biotechnology can be used in the environment safely and beneficially."

Robert A. Roe (1991), U.S. House of Representatives, sponsor of a bill regulating development and use of genetically modified organisms

1. Introduction

Fishes can be subjected to many types of genetic modification (Chourrout, 1987), including gene transfer, chromosome set manipulation, and chimera production. Fish so manipulated are the subject of substantial experimentation for scientific, aquaculture, and fisheries management applications.

1.1. Gene Transfer

Following the demonstration that quantitative phenotypes could be favorably altered through expression of introduced genes (Palmiter et al., 1982), gene transfer experiments have been undertaken in a wide range of vertebrate species. Successful gene transfers have been reported for at least 14 species of fishes, including Atlantic salmon (e.g., Fletcher et al., 1988; Rokkones et al., 1989) and rainbow trout (e.g., Chourrout et al., 1986; Penman et al., 1990). Gene transfer experiments involving fish are most frequently aimed at improving aquacultural performance, with increased growth rate or improved freezing resistance being the traits targeted to date (Kapuscinski, 1990). A lesser number of gene transfer efforts have been directed at species of prime value for sport fisheries.

Breeding programs for transgenic fishes are in rather early stages of development, with introduction of genetic constructs, rapid growth of original generation transgenic individuals, and germ line transmission having been demonstrated in various studies (Moav et al., 1990). The performance of transgenic lines of fish under field conditions has not yet been reported.

[1]Department of Fisheries and Wildlife Sciences, Virginia Polytechnic Institute and State University, Blacksburg, VA 24061, U.S.A. [2]Department of Fisheries and Wildlife, University of Minnesota, St. Paul, MN 55108, U.S.A.

Genetic Conservation of Salmonid Fishes, Edited by J.G. Cloud and G.H. Thorgaard, Plenum Press, New York, 1993

Transgenic lines will not be ready for use in aquaculture until the late 1990s. Despite the early stage of development of transgenic lines, some experimenters advocate gene transfer as a means of genetic improvement of fish.

1.2. Chromosome Set Manipulation

Chromosomal manipulation of fish genomes involves suppression of normal meiosis or mitosis to retain additional haploid sets of chromosomes, prevention of genetic contribution of spermatozoa or egg nuclei, or both. A large number of experiments have evaluated chromosomally manipulated fishes for aquaculture, sport fishing, or aquatic weed control purposes (Thorgaard and Allen, 1987; Chourrout, 1987; Ihssen et al., 1990). The utility of diploid gynogens includes rapid production of inbred lines and production of monosex cultures, and such fish have found utility in production aquaculture (Bye and Lincoln, 1986). Diploid androgens can be utilized to rapidly produce inbred lines or, of interest to this Study Institute, to produce diploid individuals of extinct stocks or species through fertilization of inactivated eggs by cryopreserved sperm derived from the extirpated group (Thorgaard and Cloud, 1992). Interest in triploids stems from their reproductive sterility, the possibility of rapid growth, eased production of interspecific hybrids, and possible production of trophy fish (Thorgaard and Allen, 1987).

1.3. Chimera Production

Recent studies in fish development have demonstrated that cells isolated from a trout embryo will become established and participate in embryogenesis when microinjected into a recipient embryo at the same stage (Nilsson and Cloud, 1989). Since donor cells in mammalian chimeras have become established in the germ line, it is anticipated that donor fish cells could similarly become established. Should this prove true, chimera production would open the possibility of producing individuals bearing genetic material from salmonid stocks extinct in the wild using embryonic cells stored in liquid nitrogen (Thorgaard and Cloud, 1993).

1.4. Ecological Risk

Whether intentionally stocked or whether escaping from aquaculture facilities, genetically modified fish of any sort could enter natural ecosystems, where they might survive, disperse, and reproduce. Phenotypic alteration induced by a particular genetic modification could be the basis for undesirable ecological or genetic impacts of genetically modified fishes in natural ecosystems. Different types of genetic modification could give rise to ecological impacts through different mechanisms. For instance, reproduction of transgenic fishes could lead to introgression of transgenes into native populations of fish, whereas broods of triploid fish could include percentages of nontriploid, fertile individuals. The likelihood and ecological impacts of these respective events is uncertain with present knowledge.

Judgments regarding the threat posed by genetically modified fish to natural ecosystems are currently based on little empirical evidence. Rational assessment of both utility and ecological risk can be approached only through generation of reliable data through well conceived and executed field testing programs. Given the ecological and genetic uncertainties posed by the different classes of genetically modified fish, the objectives of this review and synthesis study were:

- to present the scientific questions to be addressed in field tests of genetically modified fish, and
- to outline proactive roles for scientists involved in genetic modification of fishes within the context of genetic conservation of salmonids.

2. Ecological and Genetic Uncertainties Posed by Transgenic Fishes

We have previously reviewed the anticipated ecological and genetic impacts of transgenic fishes in natural aquatic systems in detail (Kapuscinski and Hallerman, 1990, 1991; Hallerman and Kapuscinski, 1992a, 1992b). The discussion below will draw upon our earlier work, presenting the ecological and genetic uncertainties that should be approached through field testing of transgenic fishes. For a more complete discussion of the anticipated ecological impacts of transgenic fishes, the interested reader is referred to our earlier papers.

2.1. Phenotypic Alteration

As the basis for ecological impacts of a genetically modified organism, neither the source of the transferred gene nor the presence or absence of a homologous gene in the host genome will prove important, but rather the type and magnitude of phenotypic alteration of the modified organism (Tiedje et al., 1989; Kapuscinski and Hallerman, 1990). Although we cannot predict the full range of phenotypic changes which might be expressed among transgenic fishes, we have anticipated five classes of phenotypic changes which could lead to such ecological impacts (Kapuscinski and Hallerman, 1991):

- **Changes in metabolic rates**. For example, expression of a transferred growth hormone gene could affect many processes in the host, including partitioning of energy between growth, respiration, and reproduction.
- **Changes in tolerance to physical factors**. For example, expression of an introduced antifreeze polypeptide gene could extend the range of tolerance to low environmental temperatures.
- **Changes in behavior**. Expression of an introduced gene for a reproductive hormone could affect a range of traits, including reproductive behavior.
- **Changes in resource use**. Expression of a transgene closing a genetic lesion in a key biochemical pathway could eliminate a dietary requirement.
- **Changes in resistance to parasites or pathogens**. Expression of viral epitope genes or antigen receptor genes could effect novel resistance to disease.

Phenotypic alterations consequent to transfer of a given gene are not limited to the direct effects anticipated. Prediction of the types and magnitudes of phenotypic alterations is complicated by the possibility of several types of indirect effects:

- Expression of introduced genes is typically under novel regulation; frequently, promoter or enhancer elements for introduced structural genes are specifically chosen so that transgene expression would not be subject to the regulatory signals controlling expression of the host's homologous gene.
- Unanticipated pleiotropic effects of transgenes on expression of host genes may occur.
- Expression of the transgene beyond the control of the host genome, novel pleiotropy, or the presence of the transgene product itself may disrupt homeostatic mechanisms of the host.

- Integration of the transgene into a functional gene would constitute insertional mutagenesis of the gene.

These possibilities refute the assertions of molecular biologists that the phenotypic effects of transfers of particular, well-characterized genes can be predicted with confidence.

In some cases, transgene expression giving rise only to the anticipated direct effects might yield a phenotype no different in kind than in non-transgenic individuals, but out of the normal phenotypic range. However, the objective of most gene transfer experiments is production of phenotypes outside the normal range for the species.

2.2. Phenotypic Alteration as the Basis for Ecological Impacts

Transgenic lines useful for aquaculture or fisheries management would be the product of several generations of breeding, and would have been subject to domestication effects. Because of this, impacts of transgenic fish in natural ecosystems would include those associated with non-transgenic hatchery fish (Hallerman and Kapuscinski, 1992a; Goodman, 1991; Steward and Bjornn, 1990). However, phenotypic alterations stemming from transgene expression would form the basis for *additional* ecological impacts *unique* to transgenic fishes.

Although our ability to make predictions is limited, the different classes of phenotypic alterations would be expected to form the basis for different mechanisms giving rise to ecological impacts. Impact mechanisms for different phenotypic modifications would include the following examples:

- Taking introduced growth hormone genes as an example of genes effecting changes in *metabolic rates*, the large size at age attained by transgenic individuals might favor them in competition for food, habitat resources, or spawning sites.
- For transgenes extending *environmental tolerances*, expression of antifreeze polypeptides in salmon would extend the range of areas where salmon could be cultured, leading to the range of ecological and genetic impacts attendant to intraspecific contact between cultured and wild stocks and to unnatural interspecific interactions.
- Expression of introduced hormone genes could affect *behavior* regarding migration, territoriality, or mating.
- Regarding transgene-mediated change in *substrate use*, increased mouth gape of fish at a given age bearing growth factor transgenes and achieving extraordinary size might enable them to utilize prey items until then too large to handle.
- Transgenic individuals exhibiting improved *disease resistance* would be favored in natural communities relative to native, non-transgenic individuals of the same and of closely related species.

This list is not exhaustive, but instead is intended to be illustrative.

Given (1) the possibility of having transgene-based phenotypes which are truly novel or outside the normal range for a species, and (2) the complex nature of aquatic ecosystems, it is difficult to predict the specific mechanisms by which altered phenotypes among transgenic salmonids might perturb aquatic communities. Further, it is difficult to anticipate long-term responses of conspecific populations or aquatic communities to a perturbation, and whether such responses would jeopardize the self-perpetuation of community components.

In previous publications, we have considered at greater length possible impact-causing mechanisms within lake (Kapuscinski and Hallerman, 1990) and stream (Hallerman and Kapuscinski, 1992a) communities. For purposes of this review, we note that salmonids, as predators, play an important role in ordering their aquatic community, both in terms of the abundance of potential prey species, and in terms of community energy flow. It is possible that large numbers of transgenic individuals in a system could, through changes in their resource utilization, affect the structure or function of the community and so affect not only conspecifics, but also other species. Such changes could be subtle or exhibit time lags, complicating their detection. It is important that community-level effects of transgenic fish be investigated through field tests in systems representative of the wild before transgenic fish gain access to the wild thorough intentional stocking or accidental escape. Because we cannot predict the full range of ecological impacts of transgenic fish, our purpose here is limited to stimulating critical thinking regarding both the possible mechanisms giving rise to ecological impacts of transgenic fishes and the means of quantifying such impacts.

2.3. Possible Genetic Impacts of Transgenic Fishes

Should transgenic fish reproduce within natural systems, any ecological or genetic impacts that they pose would be perpetuated. In determining the extent of such impacts, the key issue is the fitness of transgenic fish in natural ecosystems. Anticipating the fitness of transgenic fish is problematic, as key unknowns would affect both the reproductive success and viability components of fitness.

Transgenic lines will be developed from a very limited number of founders. It seems likely that genetic drift will limit the amount of variability in a particular line, perhaps reducing the likelihood that individual members would be adaptable to natural conditions. However, large size conferred through transgene expression could favorably affect the access of adapting transgenic males to mates. The reproductive success of transgenic females could be favored by positive body size effects on egg production, acquisition of prime spawning sites, or nest defense.

Should transgenic individuals reproduce, their descendants would include individuals with a diversity of genotypes—non-transgenic, hemizygous, and homozygous—with a corresponding diversity of fitness values. Effects of transgene expression on the viability of progeny from matings involving different combinations of such genotypes will be the most critical unknown factor in determining the fitness of transgenic genotypes in the wild. Because we lack data on both viability and reproductive success of transgenic fish in systems reflective of natural ecosystems, we cannot rule out either the possibility of wild populations becoming dominated by transgenic genotypes or the possibility of transgenic genotypes rapidly becoming extinct.

2.4. The Receiving Ecosystem as the Context for Ecological or Genetic Impacts

Potential ecological or genetic impacts posed by a particular line of transgenic fish depend not only upon phenotypic alteration exhibited by the fish, but also upon the receiving ecosystem. Important aspects of ecosystem structure and function include, but would not be limited to the following:

- How does the number of transgenic individuals compare to that of non-transgenic individuals in the receiving conspecific stock?
- What factors limit the flow of material and energy in the ecosystem, and how vulnerable are they to perturbation?
- Does the ecosystem contain depleted conspecific stocks vulnerable to demographic perturbation or depleted stocks of any species vulnerable to heightened predation or competitive effects?

These considerations underline the importance of considering the implications of release of transgenic lines of fish on a site-specific basis.

3. Field Testing of Transgenic Fishes

"Before deciding whether to allow a field test, we are first obliged to define what scientific information and issues must be considered, and then we must ask whether we know enough scientifically to be able to determine the relative safety or risk of the introduction."
National Research Council (1989)

Faced with a large number of key unknowns, it is clear that only field testing will provide data needed to characterize and quantify any ecological risks posed by transgenic fishes. Key unknowns for a given transgenic line include (1) quantification of transgene-induced phenotypic alterations, (2) quantification of fitness, i.e., viability and reproductive success, and (3) possible ecological mechanisms giving rise to community level impacts of transgenic fish in natural environments. However, before considering specific hypotheses and experimental designs, we must address issues pertinent to the conduct of experiments with organisms bearing recombinant DNA constructs.

3.1. Effective Confinement

Field testing of transgenic fishes must be carried out in secure confinements. The philosophy embodied within public policies regulating field testing is to permit development of genetically modified organisms while minimizing environmental risk (Hallerman and Kapuscinski, 1990). The issue of containment of genetically altered fish is considered at length by Donaldson, 1993. For purposes of this review, it will suffice to observe (1) that confinement protocols required for environmental release permits of recombinant DNA-bearing animals in the U.S. have become more strict, and (2) that development of requests for field testing permits now involves environmental risk assessment.

The first outdoor confinement of transgenic fish involved common carp (*Cyprinus carpio*) broodstock at Auburn University in 1989 and utilized ponds outfitted with, among other features, double screening of inlets and outlets (Cooperative State Research Service, 1990). However, the ponds were alongside a drainage ditch and were located in the floodplain of a large stream. Controversy associated with the broodstock release led to an improved confinement protocol for a planned experiment with transgenic carp fry. A new, more secure facility well above the floodplain was constructed with an elaborate discharge filter system containing several backups for catchment of any escaped fish and on-site security. The Office of the Secretary of the U.S. Department of Agriculture (1990) found that conduct of the planned experiments in this facility posed no significant environmental impacts.

U.S.D.A.'s experience in regulating outdoor releases of over 100 genetically altered plant lines, over a dozen microbial lines, and this one fish line led to promulgation of proposed field testing guidelines (Office of the Secretary, 1991). The guidelines are intended as points to be considered by principal investigators and institutions in designing research that can be conducted safely outside contained facilities. A four-step process set out for determining appropriate confinement measures includes: (1) determination of the level of safety concern for the parental organism, (2) determination of whether the genetic modification increases, decreases, or has no effect on the level of safety concern, (3) determination of the level of safety concern for the modified organism, and (4) determination of the confinement level appropriate for the modified organism and development of a safety protocol to meet that level of confinement. Determination of the level of concern associated with a modified organism entails consideration of whether it is or is not sexually mature, fertile, or able to cross with related species, and whether the modification might increase its competitive ability. Specific measures which might be taken within the four confinement levels are set out. In following the guidelines to develop a confinement protocol for a permit request, the principal investigator is essentially carrying out a preliminary environmental assessment for the field test of the modified organism.

We note that the proposed guidelines were presented without an implementation plan. It is not clear at present when or by whom the plan will be prepared (Marshall, 1991). As with any new regulatory policy, the success of the proposed guidelines will depend largely on procedural and substantive details of its implementation.

3.2. Questions to Be Considered in Field Tests of Transgenic Salmonids

Questions of practical utility or ecological impact posed by a line of transgenic fish will be a function of expression of the genetic construct with which they were transformed. The range of questions raised by the universe of potential gene transfers is beyond the scope of this review. Observing that, to date, the majority of gene transfer experiments involving salmonids entail the introduction of growth promoting genes, we focus here on issues of practical utility and ecological impacts raised by development of such lines. We aim to identify areas of needed research and to encourage discussion of experimental designs useful in addressing key questions.

A first set of critical unknowns regards characterization of the transgenic line at issue. Key questions to be addressed for a transgenic line bearing an introduced growth promoting gene include:

- To what degree has the targeted modification, increased growth rate, been achieved?
- Recognizing that growth is but one component of the energy budget of a fish, how has alteration of the metabolic allocation to growth affected other components of the energy budget, such as respiration or reproduction?
- Will energy requirements of transgenic individuals exceed those of non-transgenic individuals?
- Expression of the introduced growth promoting gene will be under the transcriptional control of a different regulatory element than the host's native copy. Recognizing (1) that expression of the introduced gene will not be regulated by the host's normal homeostatic feedbacks and (2) that growth promoting hormones affect metabolic processes other than growth (e.g., lipid metabolism, possibly osmoregulation), to what degree have other processes been affected, and is the host's metabolic profile still in homeostasis?
- Following gene transfer, have the reproductive performance and viability of the genetic line been substantially maintained?
- Is the transgene stably transmitted to subsequent generations?
- If measures to achieve sterility have been attempted, to what degree have they been successful?

It is clear that the broad range of issues raised will require a series of laboratory and field experiments entailing a large commitment of time and resources. Experiments addressing the questions posed will require the participation of geneticists, physiologists, and aquaculturists.

We propose that these experiments be conducted in contained laboratories or in confined microcosms (e.g., ponds or channels). We emphasize that these questions simultaneously address issues of practical utility, basic science, and environmental impact.

Characterization of growth promoting transgene-bearing lines will guide development of hypotheses for environmental impact studies. Expression of an introduced growth promoting gene suggests several experiments addressing possible mechanisms of ecological or genetic impact of transgenic lines:

- Would rapid growth of transgenic individuals favor them competitively in terms of (1) rapid achievement of minimum size needed to capture key prey items, or (2) competition with cohort members for habitat resources?
- Would the large size of transgenic individuals allow them access to otherwise unexploited prey items?
- Noting that transgenic pigs bearing an introduced growth hormone gene proved lethargic (Marx, 1988), has the behavior of transgenic fish been altered?
- Would rapid growth affect the propensity of transgenic salmon to mature as jacks or affect the overall age structure of a population?
- Would the large size of transgenic individuals at spawning favorably affect their access to mates or to prime spawning sites?
- Would offspring of transgenic spawners come to dominate the population? How would genetic diversity levels and long-term viability of the population and other intra- or interspecific populations with which it interacts be affected?

We anticipate that hypotheses about possible ecological or genetic impacts of transgenics would be addressed in iterative fashion, involving first a simplified ecosystem in microcosm ponds or channels, and later in more complex mesocosms. These experiments would share the feature of monitoring interactions of co-stocked transgenic and non-transgenic individuals over time periods ranging from weeks to several years. Experiments could involve field observation of individuals marked with color tags. Monitoring reproductive success of transgenic fish would be facilitated if the introduced growth promoting construct contained an easily detected marker gene. Although this program of field tests would be targeted to addressing environmental impacts of a growth-enhanced transgenic line, its results would have broader interest than the consideration of whether or not to allow expanded utilization of that transgenic line. Because environmental impacts are the consequence of phenotypic alteration regardless of cause, experimental results would be of interest to ecologists because they reflect size-mediated ecological interactions applicable to any salmonid population.

3.3. Review and Synthesis of Field Testing Program Results

After a large number of field tests of transgenic fishes have been carried out, it would be appropriate to convene key investigators with representatives of regulatory agencies and concerned sectors of the public to review and synthesize performance and ecological impact data. The purposes of such a meeting would be two-fold, (1) to identify any gaps in knowledge needed for quantitative risk assessment, and (2) to determine whether confinement measures required to that date should be maintained as is, increased, or possibly relaxed.

Over 200 field tests have been conducted around the world with genetically modified plants and microorganisms (Office of Agricultural Biotechnology, 1990). A conference on "Biosafety Results of Field Tests of Genetically Modified Plants and Organisms" was held in November 1990, when roughly 200 people from about a dozen nations gathered to share first-hand experience on biosafety issues in field testing. This conference should serve as a model for a future transgenic fish biosafety meeting.

4. Ecological and Genetic Uncertainties Posed by Chromosomally Manipulated Fishes

Chromosomally manipulated fishes include triploids, tetraploids, gynogens, and androgens. The utility of these respective types and risks they potentially pose to natural ecosystems differ as a function of their sterility or fertility, the degree of phenotypic alteration they exhibit, and the degree of genetic variability they retain.

4.1. Triploid Fishes

Much of the utility of triploid fishes to aquaculturists or to fisheries managers is a consequence of their sterility, which precludes the possibility of unwanted reproduction. Many of the ecological risks posed by genetically modified fishes would be obviated if concerns about their reproduction were overcome. To that degree, culture or stocking of triploid fish would limit potential impacts to those with ecological bases, precluding those with genetic ones. This would have the effect of limiting any realized impacts in terms of time. Three relevant practical applications can be cited. Reproductive sterility would limit potential genetic impacts of the frequently large numbers of domesticated, but otherwise unaltered, escapees from commercial aquaculture operations on wild genetic stocks in receiving ecosystems (Anonymous, 1989a, 1989b; Sattaur, 1989; Hallerman and Kapuscinski, 1992a). Reproductive sterility would be particularly important in cases where interspecific hybrid triploids were cultured or stocked, because the possibility of introgressive hybridization through backcrosses to one of the parent species could be eliminated. Similarly, culture of sterile triploid transgenic fish would be preferable to culture of fertile diploids. The practical utility of the sterile nature of triploid fish is complicated, however, by three factors.

The percentage of triploidy induced among treated individuals has varied considerably for different protocols and different species of fish (Ihssen et al., 1990). Although recent refinements of technique have led to consistent achievement of near 100% triploidy induction (Benfey and Sutterlin, 1984; Chourrout, 1984), the need for ploidy verification remains. Because there are a number of methods for ploidy verification, this is not a technical problem, but merely an economic one, adding an additional expense to the cost of seed stock for aquaculture or fisheries management.

Triploid individuals have abnormal gonadal development (Thorgaard, 1983) and are functionally sterile because they produce aneuploid gametes (Allen et al., 1986; Benfey et al., 1989). However, differences in morphology and physiology between diploids and triploids differ between species, and generalities cannot be made regarding the reproductive development of triploids. For example, despite abnormal gonadal development, triploid rainbow trout do exhibit normal external sexual differentiation, and at least some triploid males do produce sperm (Thorgaard and Gall, 1979; Lincoln and Scott, 1984). Also, triploid males of some species exhibit testosterone levels comparable to those of diploid males (Lincoln and Scott, 1984; Benfey et al., 1989). Should courtship and spawning behavior of triploid males in natural ecosystems sufficiently duplicate that of diploids, they could successfully mate with diploid females. No viable progeny would result, as the embryos would be aneuploids. However, were many triploids to secure matings, the loss of entire broods could reduce the reproductive success of the population. In turn, this could increase two types of genetic risk for natural populations (Busack, 1990): (1) loss of within-population genetic variation; and (2) extinction arising from a final blow of a demographic catastrophe.

A third potential problem stemming from sterility of triploids is that they might survive and continue to grow for an indeterminate number of years beyond the normal lifespan of their species (Robertson, 1961; Hunter and Donaldson, 1983). Results of energetics modelling (Kitchell and Hewett, 1987) indicated that old, large, sterile salmon consumed far more prey than fertile individuals which matured, and that a stocking of sterile chinook salmon would consume about 40% more forage than an equivalent stocking of fertile salmon over their life span. Sterile fish would compete with fertile fish for food items during the normal life span of the species, which could limit the growth rate of the entire population in energy-limited systems. More troublesome possibilities are (1) that large, sterile individuals might change prey preferences to include otherwise invulnerable size classes of prey, and (2) that this alteration of the community food web could prove destabilizing for community function.

Concern over the occurrence of diploid individuals among broods subject to triploidy induction has led several U.S. states to require triploidy verification for every individual intended for stocking. Although we appreciate this rigor in the case of grass carp (*Ctenopharyngodon idella*), a highly invasive exotic species, this approach does not seem justifiable on economic or practical grounds for the culture of salmonids. It may be more rational to verify ploidy among individuals sampled from larger groups and to accept entire groups if the frequency of triploids is at or near 100%, reasoning that leakage of a low percentage of fertile diploids would still be preferable to large scale culture of fertile diploids exclusively.

Potential impacts of reproductive behaviors on the part of triploids could be assessed in a field study with several treatments:

1. Isolation of triploid males with diploid females in spawning channels would determine whether the spawning of triploid males could be 'forced' by the absence of diploid males.
2. Observation of reproductive behavior of triploid males co-stocked in spawning channels with diploid males and females would determine whether triploids gain access to females or interfere with normal spawning behaviors of diploids. Comparisons of fry emergence rates among such treatments with those of controls containing only diploid fish would address the issue of whether group reproductive success was impacted by spawns involving triploids.
3. In mesocosms, observations of movements of marked triploids would determine whether they joined in pre-reproductive and spawning behaviors.

Basic questions regarding possible impacts of non-maturing triploids on trophic function in aquatic communities could be investigated in a series of field studies. The potential for triploid salmonids to survive and grow beyond the normal lifespan could be assessed by monitoring the survival and growth of a cohort in a culture facility, or better, in an isolated mesocosm. Knowing lifespan, age-specific survival rates, weight at age, and food conversion rates, the forage demand of a triploid cohort can be estimated and compared with that of diploids. To assess predation of extraordinarily large triploids on otherwise invulnerable size classes of prey, a first laboratory experiment might isolate triploids with particular size classes of potential prey to determine whether the triploids *could* prey upon them. To estimate potential impacts of such predation in natural systems, a second, mesocosm experiment might involve observation of predation-mediated depletion upon known numbers of potential prey items of different size classes.

4.2. Tetraploid Fishes

The major practical interest in tetraploid fish is for production of triploids through matings with diploids (Chourrout et al., 1989). Advantages of this approach over triploidy induction are production of 100% triploid progeny and higher viability among the progeny, which eliminate two factors constraining large-scale culture of triploids. From the standpoint of potential ecological impacts, however, entry of tetraploid individuals into natural ecosystems is cause for concern.

Tetraploid individuals in natural systems pose a potential risk because tetraploid females could breed with diploid males. This would yield all triploid broods, resulting in large numbers of sterile individuals in the systems, posing competition with and reducing the recruitment of normal diploids. The reciprocal cross, that of tetraploid males with diploid females, seems to pose less concern. Diploid sperm produced by tetraploid male rainbow trout exhibited low fertilizing ability, possibly because the larger diploid sperm could not penetrate the micopyle of a normal, haploid egg (Chourrout and Nakayama, 1987). Should such crosses become frequent in a natural ecosystem, they could reduce overall reproductive success of and incur related genetic risks in natural salmonid populations.

Because the utility of tetraploids is for production of triploids, and valuable tetraploid females are likely to be carefully confined, the level of concern posed by tetraploids would be limited. Still, the level of concern which should be attached to the possible problem of escaped tetraploid individuals backcrossing with diploids can be measured with an appropriately designed field test. In isolated, experimental spawning channels, some combination of the following treatments might be set up:

1. Tetraploid males with diploid females.
2. Tetraploid females with diploid males.
3. Tetraploid males with tetraploid females. Treatments 1, 2, and 3 would determine whether matings in natural settings could be forced by a limited choice of prospective mates.
4. Both tetraploid and diploid females and diploid males. This treatment is most representative of conditions of practical concern.
5. A mixed group of tetraploid males and females and diploid males and females, and
6. Diploid males with diploid females, a control.

This combination of treatments would demonstrate the likelihood of crosses of different types under a range of ecological conditions.

4.3. Gynogenetic and Androgenetic Fishes

Gynogenetic and androgenetic fishes are produced by manipulations which preclude the genetic contribution to the embryo of sperm or egg, respectively. The haploid gamete or embryo is restored to diploidy through a well-timed physical or chemical shock. Because the utility and possible environmental impacts of these classes of genetically manipulated fishes are similar, they will be considered together below.

Depending upon whether meiosis or mitosis was disturbed, two types of gynogenetic diploids can be produced:

- Meiotic gynogens may be useful for the production of inbred lines, which can in turn be used to produce crossbred lines which might exhibit superior performance due to heterotic effects. Through successive generations, inbreeding achieved through development of gynogenetic lines at first exceeds, but later trails that achieved through sibmating (Ihssen et al., 1990). However, meiotic gynogens are useful for the production of isogenic lines, that is, lines in which every individual has the same genotype, but is not homozygous at every locus. Isogenic lines will prove more useful than inbred lines for some applications because they retain significant levels of heterozygosity, and hence, higher viability. Gynogenetic diploids can also be used to produce monosex cultures. All female broods of pink salmon (Maximovich and Petrova, 1980), rainbow trout and coho salmon (Refstie et al., 1982) have been produced by gynogenesis in one generation. Additionally, monosex sperm obtained via hormonal sex reversal of first generation gynogens can be used to perpetuate all-female stocks (e.g., Bye and Lincoln, 1986).

- Although more difficult to produce, mitotic gynogens are of practical interest because completely homozygous lines can be produced in two generations, compared to ten or more for meiotic gynogens or sibmatings (Nagy and Csanyi, 1982).

Androgenetic diploids could be used to produce fully homozygous lines within two generations, which are in turn useful for production of crossbred lines. They might also have some utility for rehabilitation of extirpated stocks because diploid individuals could be produced by fertilization of inactivated eggs by cryopreserved sperm, making possible the generation of individuals bearing some of the extinct stock's original genetic material. Such genetic techniques will be useful in rehabilitation if they are coupled with primary actions to remove or circumvent the original causes of stock extirpation.

Culture or stocking of inbred lines, or to a lesser degree, of resultant crossbred lines, presents the likelihood of large numbers of fish genetically identical to one another, but different from natural stocks, coming into contact with natural stocks. Reproduction of fish from these inbred lines with natural stocks would give rise to a cohort with reduced genetic variability and a strongly altered array of allele frequencies relative to the unaffected natural stock. These conditions suggest decreased ability of natural populations to persist in a changing environment and possible disruption of any local adaptation they may have.

These possibilities are of somewhat greater concern for the case of aquaculture of monosex stocks. Large numbers of fish escape from aquaculture operations (Hallerman and Kapuscinski, 1992a), and the entry of large numbers of cultured females into spawning runs could (1) increase competition for prime nesting sites, or (2) contribute to losses of earlier spawns through superimposition of their nests on established redds, in addition to the possible genetic impacts cited above.

Populations of fish produced by androgenesis using gene-banked sperm from extinct stocks would be subject to strong founder effects and probably to subsequent genetic drift and inbreeding. However, because the choice in such a situation is between reduced genetic diversity or total loss of the genetic stock, acceptance of the former is clearly preferable.

Although reductions in genetic variability suggested above can be modelled, the experimental designs needed for quantifying the ecological and genetic uncertainties posed by gynogens and androgens would require unseemingly large commitments of resources. We suggest that culture of such fishes be conducted in secure facilities, using sterile fish wherever possible.

4.4. Perspectives on Ecological Impacts Posed by Chromosomally-Manipulated Fishes

As a class, chromosomally manipulated fishes do not seem to pose as varied a range of impacts to the environment as transgenic fish. This is because any phenotypic alterations at issue are not novel, and so mechanisms of possible ecological impact can, to some degree, be anticipated. However, with regard to the magnitude of anticipated impacts, chromosomally manipulated fishes are no different than transgenic fishes inasmuch as it is the degree of phenotypic alteration which will affect the magnitude of any impact, not the source of the alteration, whether gene transfer or chromosomal manipulation. Therefore, significant uncertainties which are posed, especially concerning possible effects of ploidy manipulated fish on reproductive success of natural conspecific populations, should be investigated before large numbers of triploids come into contact with wild populations, for example, following mass escape from net pen aquaculture facilities. Execution of field tests to quantify unknown aspects of behavior and fitness of ploidy manipulated fish under natural conditions seem justifiable for both scientific and fisheries management reasons.

Chromosomal manipulation techniques could play an important role in minimizing the potential impacts of transgenic fish. Crossing of diploid transgenics with tetraploids would produce 100% triploids, but the environmental safety of such fish would be subject to the reservations noted above. Alternatively, triploid induction could be combined with gynogenesis and hormonal sex reversal to achieve a higher level of safety (Kapuscinski, 1990). All female gynogens in the parental generation would be sex reversed by methyltestosterone treatment. Outcrossing of the resultant functional males to normal females of another line and triploidy induction would produce a second generation of all-female, sterile triploids. Integration of gene transfer into the protocol at the time of gynogenesis in the parental generation or at the time of outcrossing in the second generation would effect the production of a functionally sterile transgenic line.

The issue of whether chromosomally manipulated fishes are "organisms with deliberately modified hereditary traits" (Office of the Secretary, 1991) in a legal sense may affect the course of their development and utilization. Under the proposed guidelines regulating outdoor release of genetically modified organisms (Office of the Secretary, 1991), it is unclear whether chromosomally manipulated fish are within the regulatory scope of purview. Chromosomal manipulation of animals is not among the methodologies of conventional breeding which are specifically excluded from oversight. Should chromosomally manipulated fish be subject to such oversight, permits for outdoor releases and assessments of environmental impact may be required.

5. Scientific and Ecological Uncertainties Posed by Chimeric Fishes

Production of piscine chimeras is still very much a technical coup. Many scientific questions with bearing upon their utility for rehabilitation of extirpated stocks, not to mention ecological or genetic implications of their release, have yet to be approached. Key among these are:

- What is the viability of such fish?
- To what degree will chimeric individuals be mosaics of the respective parent embryos?
- Assuming that donor-derived cells become incorporated into the germ line, what percentage of germ cells, and more importantly, gametes, will come from the donor? How can we manipulate production conditions to maximize the proportion of gametes derived from the donor?
- What is their fertility or fecundity?

These questions can all be approached experimentally in laboratory studies.

It has been suggested that, through use of female triploid embryos as recipients, resultant fertile chimeras would produce only gametes with the genetic composition of the injected, diploid donor cells (J. Cloud, personal communication). Should this technique succeed, it would eliminate uncertainty over the genetic lineage of any offspring of the chimeras. Fertilizing these eggs with cryopreserved sperm would facilitate rehabilitation of an extinct stock, subject only to the loss of genetic diversity consequent to sperm and embryonic cells having necessarily been taken from a limited number of individuals. Chimera production would clearly conserve genetic material. Nonetheless, a number of practical questions will have to be approached:

- How would genetically-based behaviors of a chimeric individual be expressed, given that its somatic cells were derived from both donor and recipient? Would the fish act like a member of the donor stock as we might wish?
- How would it be best to handle the husbandry of chimeric broodstock? It may prove advisable to use the chimeras as captive broodstock and plant only their progeny in the wild. Should we maintain chimeric individuals in a hatchery to sidestep this potential problem?
- Will enough genetic variability be maintained for the re-founded stock to adapt to a changing environment?

These questions can be addressed in well designed field tests.

6. Decision Making and Risk Management

We have presented the case that genetically modified fishes may pose both benefits and risks to the environment. Our arguments are intended to promote quantitative research into the ecological risks as well as the benefits posed by genetically modified fishes. Beyond their immediate scientific interest, quantitative data are useful as input for risk management analysis aimed at reaching defensible decisions regarding releases of genetically modified fish. Risk management is aimed at reducing the risk associated with achievement of anticipated benefits, and for choosing from among alternative options that yield different types or amounts of risks and benefits. Execution of a risk assessment-based decision making process, such as that outlined below, would result in a decision of whether or not to go ahead with a proposed release of a genetically modified organism.

Gregory (1988, in press) outlined a general framework for reaching decisions about risks of a deliberate release of a genetically altered aquatic organism which can be applied to a particular proposed introduction. The established structure of risk assessments for expected damages of established and novel technologies is applicable to risk assessments for genetically modified organisms (Fiksel and Covello, 1986). The approach requires: (1) a clear presentation of the basis for environmental effects of a proposed release, combined with treatment of associated uncertainties, and (2) analysis of the values of relevant societal groups potentially affected by the release. The decision process for managing the risks of deliberate releases into aquatic systems (Gregory, in press) includes: (1) identification of key groups of people who would be affected by the proposed release; (2) identification of the technical alternatives that might be adopted; (3) identification of the consequences of each alternative, notably in our case, ecological consequences; (4) consideration of the likelihood of these effects in the form of probability estimates or frequency distributions; (5) consideration of the values of the various stakeholder groups regarding the various consequences, and finally; (6) reaching a decision between alternatives. With all information in hand, the decision maker can link values to the consequences of different technical alternatives.

In the context of risk management regarding genetically modified organisms, knowledge about some of the more common fallacies in thinking about risks may help to anticipate the basis for controversy and to diminish it (Gregory, in press):

- No risk-free alternatives are available. All options, including the option of doing nothing, i.e., releasing no genetically modified fish, will involve risks.
- Risk decisions will involve conflicting objectives. For example, culture of genetically modified fishes and environmental protection will often conflict.
- Foreknowledge of ecological effects of genetically modified fishes will not be absolute. Risk decisions will involve statistical consideration of the likelihoods of particular effects.
- Analysis of risks is never objective, starting with the very framing of the problem. Acknowledgement of subjectivity and explicit statement of the bases of a decision will be important to the stakeholder groups.

Case studies of field tests of genetically modified organisms suggest that these potential problems and the six steps in decision making have often been ignored (Gregory, in press), to a large degree affecting whether the proposed test was eventually carried out (Office of Technology Assessment, 1988). For example, approval of a planned experiment involving environmental release of transgenic carp fry was subject to a two-year delay due to failure to recognize possible consequences of proposed actions, to threatened lawsuits from environmental organizations, and to changes in federal regulatory policy. We feel that improved adherence to all six steps in the decision making process outlined above and recognition of the noted factors which complicate decision making will result in expedited approval of field testing experiments.

The model outlined above was developed in the context of reaching a decision for going forward with a particular field test of a particular genetically modified organism. However, the general approach will also be applicable in future cases where data from many such releases are already in hand and decisions are to be made regarding (1) whether a particular type of genetically modified fish is indeed attractive for a proposed more widespread application, (2) whether significant ecological risk is posed by utilization of such fish, and (3) whether and what measures might be required to confine such fishes in order to both enjoy benefits and minimize risk.

7. Perspectives on Genetic Modification of Fishes

7.1. Genetically Modified Fishes and Genetic Conservation of Salmonids

The production and use of genetically modified salmonids has some bearing upon genetic conservation of salmonid stocks. A recent survey of the status of native Pacific salmonid stocks in California, Oregon, Idaho, and Washington (Nehlson et al., 1991) revealed 214 which were depleted to varying degrees. Recognizing that the list was incomplete and that British Columbia, Alaska, and Asian stocks were not surveyed, it is clear that the number of declining Pacific salmonid stocks is considerably higher. Further recognizing that a large proportion of Atlantic salmon (*Salmo salar*), trout, and char stocks are also in decline underlines the urgency of genetic conservation of salmonid stocks. This decline of native salmonid stocks resulted from habitat loss or damage, inadequate passage and flows caused by hydropower, agriculture, logging and other developments, overfishing, and negative interactions with other fishes, including non-native hatchery salmon (Nehlson et al., 1991). An overall strategy for successful conservation of salmonids must include as its major elements measures which address these factors.

Within the context of this larger conservation strategy, genetic modification of salmonids has the potential both to increase the threat to native stocks and to provide techniques to help conserve them (see also Miller, 1991). The potential threats to native stocks posed by genetically modified fishes can be minimized (1) by thorough evaluation of genetically altered stocks through well designed programs of field testing, and (2) through careful judgment of which, where, and how genetically modified lines of fish are utilized. Genetic manipulation can play an important supporting role in the overall conservation strategy, for example, in use of chimeras to partly reconstitute extinct stocks, or in culture of sterile triploids to minimize impacts of aquaculture escapees on native populations. In any case, the role for genetically modified fish within the larger salmonid genetic conservation strategy will be relatively narrow.

7.2. Roles of Scientists Practicing Genetic Manipulation of Fishes

Within the context of a comprehensive strategy for conservation of salmonid stocks, scientists practicing genetic manipulation of salmonids have two broad roles to play. The first role is largely technical. Scientists must strive to develop and appropriately utilize techniques of genetic manipulation to serve the goal of genetic conservation. We have argued that development of genetically manipulated lines must include thorough characterization of both performance traits and environmental impacts posed. Moreover, development and effective implementation of rational public policies regulating ecologically sound development and use of genetically modified organisms will require the active participation of concerned scientists.

Recognizing that genetic conservation of salmonids will have to be practiced within a societal milieu, scientists will have to devote some of their efforts to working in the public arena. Genetic conservation of salmonids will introduce a host of issues into public debate. With few exceptions, however, public debate on key issues is shaped and controlled by opinion makers and civic leaders who are not scientists and who rarely have a comprehensive perspective on the relevant body of scientific knowledge (Langenberg, 1991). Scientists must ensure that there is effective communication between those who understand the issues and those who must act upon them. The second important role of scientists, therefore, is to engage fully in public debate and act on conservation issues. This necessitates professional integrity, including carefully identifying and articulating one's subjective biases, and honestly distinguishing between verified knowledge, professional judgment, and critical unknowns.

ACKNOWLEDGMENTS: Support for E.M.H. while preparing the manuscript was provided in part through U.S.D.A.-C.S.R.S. Hatch Program Project No. 6129270 and Aquaculture Research Grant No. 89-34123-4956. Support for A.R.K. was provided in part by the Minnesota Agricultural Experiment Station, the Minnesota Sea Grant College Program supported by the NOAA Office of Sea Grant, Department of Commerce under Grant No. NOAA-90AA-D-SG149, Projects No. RA/5, J.R. 280, the Legislative Commission on Minnesota Resources, and the Greater Minnesota Corporation. This is article 19196 of the Minnesota Agricultural Experiment Station Scientific Journal Article Series. The authors are grateful to the organizers of the NATO Advanced Study Institute for the opportunity to present our views and for coverage of E.M.H.'s expenses in attending the meeting.

References

Allen, S.K., Jr., R.G. Thiery, and N.T. Hagstrom. 1986. Cytological evaluation of the likelihood that triploid grass carp will reproduce. *Transactions of the American Fisheries Society* 115:841-848.

Anonymous. 1989a. "Possible development of codes of practice to minimize threats to wild stocks." North Atlantic Salmon Conservation Organization, Commission Paper CNL(89)23, Edinburgh, U.K.

Anonymous. 1989b. "Draft protocols dealing with ecological concerns respecting Atlantic salmon due to introductions and transfers of fishes." North Atlantic Salmon Conservation Organization, Commission Paper WAC(89)16, Edinburgh, U.K.

Benfey, T.J., H.M. Dye, I.I. Solar, and E.M. Donaldson. 1989. The growth and reproductive endocrinology of triploid Pacific salmonids. *Fish Physiology and Biochemistry* 6:113-120.

Benfey, T.J. and A.M. Sutterlin. 1984. Triploidy induced by heat shock and hydrostatic pressure in Atlantic salmon (*Salmo salar* L.). *Aquaculture* 36:359-367.

Busack, C. 1990. Yakima/Klickitat production project genetic risk assessment. Yakima/Klickitat production project preliminary design report, appendix A, Washington Department of Fisheries and Bonneville Power Administration (Division of Fish and Wildlife - PJ, P.O. Box 3521, Portland, OR 97208).

Bye, V.J. and R.F. Lincoln. 1986. Commercial methods for the control of sexual maturation in rainbow trout (*Salmo gairdneri*). *Aquaculture* 57:299-309.

Chourrout, D. 1984. Pressure-induced retention of second polar body and suppression of first cleavage in rainbow trout: Production of all-triploids, all-tetraploids, and heterozygous and homozygous diploid gynogenetics. *Aquaculture* 36:111-126.

Chourrout, D. 1987. Genetic manipulations in fish: Review of methods, in: "Proceedings of the Word Symposium on Selection, Hybridization and Genetic Engineering in Aquaculture, Vol. II," R. Tiews, ed., pp. 111-126, G. Heenemann, Berlin.

Chourrout, D., B. Chevassus, F. Krieg, A. Happe, G. Burger, and P. Renard. 1986. Production of second generation triploid and tetraploid rainbow trout by mating tetraploid males and diploid females—potential of tetraploid fishes. *Theoretical and Applied Genetics* 72:193-206.

Chourrout, D., R. Guyomard, and L.M. Houdebine. 1986. High efficiency gene transfer in rainbow trout (*Salmo gairdneri* Rich.) by microinjection into egg cytoplasm. *Aquaculture* 51:143-150.

Chourrout, D. and I. Nakayama. 1987. Chromosome studies of progenies issued from tetraploid females of rainbow trout. *Theoretical and Applied Genetics* 74:687-692.

Cooperative State Research Service, U.S. Department of Agriculture. 1990. Research proposed on transgenic fish; Publication of environmental assessment; Notice of opportunity for public comment. *Federal Register* 55:5751-5772.

Donaldson, E. 1993. Containment of genetically altered fishes, in: "Genetic conservation of salmonid fishes," J. G. Cloud, ed., pp. 113-129, Plenum Press, New York.

Fiskel, J. and V. Covello. 1986. "Biotechnology risk assessment: Issues and methods for environmental introduction," Pergamon Press, New York.

Fletcher, G. L., M.A. Shears, M.J. King, P.L. Davies, and C.L. Hew. 1988. Evidence for antifreeze protein gene transfer in Atlantic salmon (*Salmo salar*). *Canadian Journal of Fisheries and Aquatic Sciences* 45:352-357.

Goodman, B. 1991. Keeping anglers happy has a price: Ecological and genetic effects of stocking fish. *BioScience* 41:294-299.

Gregory, R.S. 1988. A framework for managing the risks of deliberate releases of genetic material into aquatic ecosystems. *Journal of Shellfish Research* 7:557.

Gregory, R.S. In press. A decision framework for managing the risks of deliberate releases of genetic materials, in: 'Dispersal of living organisms and genetic materials into aquatic ecosystems," A. Rosenfield and R. Mann, eds., University of Maryland Press.

Hallerman, E.M. and A.R. Kapuscinski. 1990. Transgenic fish and public policy: Regulatory concerns. *Fisheries* 15(1):12-20.

Hallerman, E.M. and A.R. Kapuscinski. 1992a. Ecological implications of using transgenic fishes in aquaculture, in: 'Effects of Introduced Organisms on Aquatic Communities," ICES Marine Science Symposium 194:56-66.

Hallerman, E.M. and A.R. Kapuscinski. 1992b. Ecological and regulatory uncertainties associated with transgenic fish, in: 'Transgenic Fishes," C. Hew and G. L. Fletcher, eds., pp 209-228, World Science Publishing, Singapore.

Hunter, G.A. and E.M. Donaldson. 1983. Hormonal sex control and its application to fish culture, in: 'Fish physiology," W.S. Hoar, D.J. Randall, and E.M. Donaldson, eds., Vol 9B, pp.223-303, Academic Press, New York.

Ihssen, P.E., L.R. McKay, I. McMillan, and R.B. Phillips. 1990. Ploidy manipulation and gynogenesis in fishes: Cytogenetic and fisheries applications. *Transactions of the American Fisheries Society* 119:698-717.

Kapuscinski, A.R. 1990. Integration of transgenic fish into aquaculture. *Food Reviews International* 6:373-388.

Kapuscinski, A.R. and E.M. Hallerman. 1990. Transgenic fish and public policy: Anticipating environmental impacts of transgenic fish. *Fisheries* 15(1):2-11.

Kapuscinski, A.R. and E.M. Hallerman. 1991. Implications of introduction of transgenic fish into natural ecosystems. *Canadian Journal of Fisheries and Aquatic Sciences*, 48 (suppl. 1): 99-107.

Kitchell, J.F. and S. W. Hewett. 1987. Forecasting forage demand and yield of sterile chinook salmon (*Onchorynchus tshawytscha*) in Lake Michigan. *Canadian Journal of Fisheries and Aquatic Sciences* 44(Suppl. 2):384-389.

Langenberg, D.N. 1991. Science, slogans, and civic duty. *Science* 252:361-363.

Lincoln, R F. and A.P. Scott. 1984. Sexual maturation in triploid rainbow trout. *Salmo gairdneri* Richardson. *Journal of Fish Biology* 25:385-392.

Marshall, E. 1991. How to regulate environmental releases. *Science* 251:1023-1024.

Marx, J.L. 1988. Gene-watchers feast served up in Toronto. *Science* 242:32-33.

Maximovich, A.A. and G.A. Petrova. 1980. Production of radiation-induced diploid gynogenetic pink salmon, in: 'Proceedings of the First International Conference on Biology of Pacific Salmon," pp. 151-155, TINRO, Vladivostok.

Miller, J.A. 1991. Biosciences and ecological integrity: Do the advances of modern biology threaten the ecosystem, or do they provide the tools to conserve it? *BioScience* 41:206-210.

Moav, B., Z. Liu, N.L. Moav, M.L. Gross, A.R. Kapuscinski, A.J. Faras, K.S. Guise, and P.B. Hackett. 1990. Expression of heterologous genes in transgenic fish, in: 'Transgenic Fishes", C. Hew and G.L. Fletcher, eds., pp. 120-141, World Science Publishing, Singapore.

Nagy, A. and V. Casanyi. 1982. Changes in genetic parameters in successive gynogenetic generations and some calculations for carp gynogenesis. *Theoretical and Applied Genetics* 63:105-110.

National Research Council. 1989. Field testing genetically modified organisms: Framework for decisions. National Academy Press, Washington, D.C.

Nehlson, W., J.E. Williams, and J.A. Lichatowich. 1991. Pacific salmon at the crossroads: Stocks at risk for California Oregon, Idaho, and Washington. *Fisheries* 16(2):4-21.

Nilsson, E. and J.G. Cloud. 1989. Production of chimeric embryos of trout (*Salmo gairdneri*) by introducing isolated blastomeres into recipient blastulae. *Biology of Reproduction* 40(Suppl. 1):186.

Office of Agricultural Biotechnology (United States Department of Agriculture). 1990. International conference on biosafety planned. *Biotechnology Notes* 3(8):4.

Office of Technology Assessment, U.S. Congress. 1988. New developments in biotechnology, 3, Field testing engineered organisms: Genetic and ecological issues. OTA-BA-350, U.S. Government Printing Office, Washington, D.C.

Office of the Secretary, U.S. Department of Agriculture. 1990. Finding of no significant impact; Research on transgenic carp in confined outdoor ponds to be conducted at the Alabama Agricultural Experiment Station (AAES), Auburn University, Auburn, Alabama, With cover letter dated November 15, 1990.

Office of the Secretary, U.S. Department of Agriculture. 1991. Proposed guidelines for research involving the planned introduction into the environment of organisms with deliberately modified hereditary traits; Notice. *Federal Register* 56:4134-4152.

Palmiter, R.D., R.L. Brinster, R.E. Hammer, M.E. Trumbauer, M.G. Rosenfeld, N.C. Birnberg, and R. M. Evans. 1982. Dramatic growth of mice that develop from eggs microinjected with metallothionein-growth hormone fusion genes. *Nature* 300:611-615.

Penman, D.J., N.J. Beeching, S. Penn, and N. MacLean. 1990. Factors affecting the integration of microinjected DNA into the rainbow trout genome. *Aquaculture* 85:35-50.

Refstie, T.J., J. Stoss, and E. Donaldson. 1982. Production of all female coho salmon (*Onchorynchus kisutch*) by diploid gynogenesis using irradiated sperm and cold shock. *Aquaculture* 29:67-82.

Robertson, O.H. 1961. Prolongation of the lifespan of kokanee salmon (*Onchorynchus nerka kennerlyi*) by castration before the beginning of gonad development. *Proceedings of the National Academy of Sciences U.S.A.* 47:609-621.

Roe, R.A. 1991. Testimony before the Subcommittee on Department Operations, Research, and Foreign Agriculture of the Committee on Agriculture, U.S. House of Representatives, October 2, 1990, pages 101-107, in: Serial No. 101-75, U.S. Government Printing Office, Washington, D.C.

Rokkones, E., P. Alestrom, H. Skjervold, and K.M. Gautvik. 1989. Microinjection and expression of a mouse metallotheionein human growth hormone fusion gene in fertilized salmonid eggs. *Journal of Comparative Physiology B* 158:751-758.

Sattaur, O. 1989. The threat of the well bred salmon. *New Scientist* 29 April:54-58.

Steward, C.R. and T.C. Bjornn. 1990. Supplementation of salmon and steelhead stocks with hatchery fish: A synthesis of published literature, part 2, in: "Analysis of salmon and steelhead supplementation," W. H., Miller, ed., parts 1-3, Technical report 90-1, Bonneville Power Authority, U.S. Department of Energy, Portland.

Thorgaard, G.H. 1983. Chromosome set manipulation and sex control in fish, in: 'Fish physiology," W.S. Hoar, D.J. Randall, and E.M. Donaldson, eds., Vol. 9B, pp. 405-434, Academic Press, New York.

Thorgaard, G.H. 1986. Ploidy manipulation and performance. *Aquaculture* 57:57-64.

Thorgaard, G.H. and S.K. Allen, Jr. 1987. Chromosome manipulation and markers in fishery management, in: 'Population Genetics and Fishery Management," N. Ryman and F. Utter, eds., pp. 319-331, University of Washington Press, Seattle.

Thorgaard, G.H. and J.G. Cloud. 1993. Reconstruction of genetic strains of salmonids using biotechnical approaches, in: "Genetic Conservation of Salmonid Fishes," J.G. Cloud ed., pp. 184-191, Plenum Press, New York.

Thorgaard, G.H., and G.A.E. Gall. 1979. Adult triploids in a rainbow trout family. *Genetics* 93: 961-973.

Tiedje, J.M., R.K. Colwell, Y.C. Grossman, R.E. Hodson, R.N. Mack, and P.J. Regal. 1989. The planned introduction of genetically engineered organisms: Ecological considerations and recommendations. *Ecology* 70:298-315.

The Reproductive Containment of Genetically Altered Salmonids

EDWARD M. DONALDSON, ROBERT H. DEVLIN, IGOR I. SOLAR, AND FRANCESC PIFERRER*

1. Introduction

Genetically altered fish may be categorized according to the process by which they became genetically altered *vis-à-vis* the wild stock, e.g., through: (1) interuption of the process of natural selection by maintenance in captivity through part or all of the life cycle, (2) selective removal of a specific segment of the population by the fishery, (3) selective breeding, (4) chromosome set manipulation, (5) transgenesis, and (6) intra- and inter-specific hybridization. While there is a tendency to categorize genetically altered fish according to the process by which they became altered, alternatively and perhaps more appropriately, they may also be categorized according to the nature and degree of the resultant genetic alteration. This latter form of categorization is hampered by our ability to properly characterize and quantify the degree of alteration to the genome and to interpret these changes at the genomic level in terms of the resultant phenotype and its potential impact on wild stocks.

Current concerns regarding genetically altered salmonids include the potential impact of hatchery fish and escaped farm fish on wild stocks, the introduction of exotic salmonids and the development and implementation of transgenic fish. Escapes of genetically altered salmonids can occur from hatcheries, aquaria, tanks, ponds, raceways and lake, river and sea cage sites. They can also occur during transfer at any developmental stage by truck, boat or helicopter. Primary causes of escapes include human error, storms, floods, ice, etc., structural failure, wear and tear, predator damage, navigational accidents and vandalism.

Containment methodologies for genetically altered salmonids and other fishes fall into two general categories: first, physical containment, i.e., maintenance of the fish in a physical quarantine structure which prevents their escape and interaction with other fish, and second, biological containment, i.e., prevention or inhibition of reproduction by sterilization or monosex culture depending on the circumstance.

While the focus of this paper is on the second category of containment methodology

Biotechnology, Genetics and Nutrition Section, West Vancouver Laboratory, Biological Sciences Branch, Department of Fisheries and Oceans, 4160 Marine Drive, West Vancouver, B.C. V7V lN6 Canada.
*Current address: Dept. Reproductive Medicine, Univ. of California-San Diego, 9500 Gilman Drive, LaJolla, CA 92093-0947, U.S.A.

Genetic Conservation of Salmonid Fishes, Edited by J.G. Cloud and G.H. Thorgaard, Plenum Press, New York, 1993

there will also be an increasing need for improved means of physical containment as reproductively viable broodstock for genetically altered fish will have to be maintained even though the offspring grown by aquaculture enterprises may be prevented from reproducing by appropriate biological containment methods.

Physical containment can be implemented with varying degrees of stringency. It is generally accepted that total containment or quarantine is only feasible in a land based facility. Total containment has recently been implemented for the field testing of transgenic carp in the USA and at several facilities for the initial quarantine (for disease control purposes) of Atlantic salmon imported into British Columbia. Land based physical containment systems may include some or all of the following measures: a 24 hr security system, triple screening of out flows with appropriate mesh sizes, treatment of the effluent (e.g., by chlorination), appropriate covers, surrounding of field trial ponds with dry ponds, and provision for removal or poisoning of fish in the event that a flood is predicted. Total physical containment is not feasible in lake or ocean based net pens; however, recent concern regarding the potential genetic impact of escaped aquaculture stocks on wild stocks of salmonids has prompted consideration of an improved code of practice concerning the physical integrity of net pen systems and the husbandry practices associated with the transport and handling of fish. Physical containment measures which can be implemented at net pen sites include: 24 hr security, the use of pens with strong and durable mesh, complete sewn on covers, predator nets and measures to discourage predators, improved anchoring systems, navigational markers, and provision for recapture of escaped fish. There is also the possibility of locating net pen sites in aquaculture zones which are geographically separate from areas set aside for the conservation of natural stocks.

A number of techniques and potential techniques have been described for the sterilization of salmonids. These techniques are in various stages of development, some are available for pilot trials with little further research and development required, while others are in earlier stages of development or can be eliminated as being impractical for economic, technical, or environmental reasons. Reviews on sex control and/or chromosome set manipulation which include reference to sterilization in salmonids include, Donaldson and Hunter, 1982a; Hunter and Donaldson, 1983; Thorgaard, 1983; Utter et al. 1983; Yamazaki, 1983; Donaldson, 1986; Bye and Lincoln, 1986; Purdom, 1986; Donaldson and Benfey, 1987; Secombes et al. 1987; Chourrout, 1987; Piferrer and Donaldson, 1988, 1989a; Benfey and Donaldson, 1988; Johnstone, 1989; Dunham, 1990. The proceedings of a workshop on non-maturing salmonids were recently published (Pepper, 1991), which contain several reviews including Benfey, 1991, Donaldson et al., 1991, Johnstone et al. 1991 and Jungalwalla, 1991.

Until recently the induction of sterility in salmonids was largely considered as a means of preventing sexual maturation in aquacultured fish thus permitting the harvesting of high quality fish on a year round basis and also permitting growth to a large size, as large salmon currently have a consistently higher market value per unit weight than small salmon. These facts have been known for some considerable time as Watson (1755) noted that castrated fish including trout "grew much larger than their usual size, were more fat, and, which is no trifling consideration, were always in season."

Now, however, with concern being raised about first, the potential for reproductive interaction between escaped aquacultured salmonids and wild salmonids (Skaala et al., 1990; NASCO, 1990; Egidius et al., 1991) and the consequent potential impact on the genome of the wild salmonids and second, the potential or hypothetical risks to the genome of wild salmonid populations associated with the development and application of transgenic salmonids for aquaculture purposes (Kapuscinski and Hallerman, 1990), there is heightened interest in the development and implementation of viable sterilization technologies.

2. Sterilization Techniques

In this section we describe each technique and discuss the merits and demerits of each procedure, the degree of sterility obtained, where they are being developed and whether they are at the laboratory, pilot demonstration, or commercial stages of development. The known techniques are summarized in Table 1 and are subsequently described and assessed. Where possible, references quoted in the following sections are to research on salmonids.

Table 1. Methods and potential methods for sterilization of salmonids.

1. Surgical removal of gonads.	5. Treatment with androgen.
2. Induction of autoimmunity.	6. Induction of female triploidy.
3. Chemosterilization.	7. Production of sterile hybrids.
4. Exposure to X or gamma irradiation.	8. Production of sterile transgenics.

2.1. Surgical Removal of Gonads

This is the oldest technique of all having been used by the Chinese in cultured carps many centuries ago. It was also the first technique to be used in salmonids and was described in some detail over two centuries ago (Watson, 1755). Its use in Pacific salmon was described by Robertson et al. (1961) and McBride et al. (1963) and in rainbow trout *(Oncorhynchus mykiss)* by Wunder (1977). Its use in Atlantic salmon *(Salmo salar)* in Scotland has been investigated by Brown and Richards (1979) and Brown (1982). The procedure involves (1) anaesthesia in a suitable anaesthetic such as tricainemethanesulphonate, 2-phenoxyethanol, or metomidate (it should be noted that these and most other anaesthetics have not yet been cleared for veterinary use in food fish), (2) making a longitudinal mid-ventral incision posterior of the pectoral fins, (3) making cuts at the anterior and posterior insertion of the gonads, taking particular care to leave no gonadal fragments underneath the liver at the anterior point of connection, (4) removal of the gonads, and (5) administration of antibiotic and closure of the wound using surgical suture or surgical staples.

The procedure is effective, the recovery rate can be high and the fish remain in silver bright immature status until their ultimate death. If a surgical production line is set-up, many fish can be operated upon; however, there are two major limitations, (1) the procedure is not regarded as economic, and (2) the procedure is only feasible on fish of reasonable size (100 g plus) owing to the need to visualize the gonads during surgery. The procedure could, however, be used to generate true sterile fish for research purposes or pilot scale trials.

2.2. Induction of Autoimmunity

The concept that it may be possible to sterilize Atlantic salmon by induction of gonadal autoimmunity through injection of an antigen, consisting of macerated ovarian and testicular tissue from mature salmon, combined with Freunds complete adjuvant (FCA) has been investigated by Laird et al. (1978 and 1980). Initially the results looked promising; however, a field trial (Ellis, 1981) did not confirm the initial results. In this field trial S1 (one-year) and potential S2 (two-year) smolts were injected, just prior to sea water transfer of the Sl smolts, with the above antigen combined with either saline, FCA, alum or saponin. The fish were examined the following September. Gonads were absent in 7 out of 21 male parr which had received saponin and 2 out of 18 male parr which received FCA. Ellis (1981) concluded that the gonads in these fish may have been missed during histological examination and noted that the treatments failed to prevent maturation in precocious male parr. The induction of gonadal autoimmunity in Atlantic salmon thus appears to be unattainable at the present time. Furthermore, if the technique did work, questions could be raised concerning the presence of residues from the adjuvant. The autoimmune technique has recently been further investigated and reviewed (Secombes et al., 1987). In this study, it was concluded that autoimmune responses could be induced in the rainbow trout testis, but that the autoantigens only appeared on the post-meiotic cells which appear when the fish is already exhibiting secondary sexual characteristics. Occlusion of the seminiferous tubules with granulomatous tissue occurred in some fish and prevented sperm release, but not sexual maturation. This technique after further development may therefore provide a means of preventing reproductive interaction with other fish, but will not prevent the loss of quality associated with sexual maturation in salmonids.

In mammals the feasibility of inducing antibodies to modified gonadotropin releasing hormone (GnRH) peptides has been investigated as a means of immunocastration in livestock. In one such study, sheep were vaccinated with cysteine substituted GnRHa conjugated to a carrier molecule with a cross linking agent. Immunized animals generally had lower gonadotropin and testosterone concentrations and demonstrated testicular atrophy (Goubau et al., 1989). As yet there are no reports of similar studies having been conducted in fish; however, it appears to be a promising area for further research.

3. Chemosterilization

The use of chemosterilization agents to interfere with gonadal development has been investigated to a limited degree. These include mutagens (Tsoi, 1969), gonadotropin antagonists (Donaldson, 1973; Flynn, 1973), antiestrogens and antiandrogens (Schreck and Fowler, 1982). The preferred objective of such treatments would be to interfere with the early stages of gonadal differentiation. The use of such chemicals, in particular mutagens, in fish destined for human consumption would most probably be unacceptable and for this reason they have received little attention from investigators. One area, however, that deserves further investigation is the possibility of developing gonadotropin releasing hormone (GnRH) antagonists that are effective in inhibiting gonadotropin synthesis and/or release in fish. There is as yet insufficient information to determine whether mammalian GnRH antagonists are effective in fish or whether specific piscine GnRH antagonists can be developed.

4. Exposure to X or Gamma Irradiation

Extensive studies in the medaka *(Oryzias latipes)*, reviewed by Egami and Ijiri (1979) have shown that acute or chronic irradiation during early development can interfere with the development of both testis and ovary. Studies on the low level gamma irradiation of chinook salmon *(Oncorhynchus tshawytscha)* during early development resulted in delayed sexual maturation in some individuals (Bonham and Donaldson, 1972; Hershberger et al., 1978). In the rainbow trout gamma irradiation has also been shown to interfere with gonadal development (Konno, 1980) and produce sterile trout (Konno and Tashiro, 1982).

Recently, the feasibility of sterilizing Atlantic salmon by gamma irradiation from a Cobalt 90 source has been investigated by Thorpe et al. (1987a,b). In these studies gamma irradiation at a dosage of 10 Sv at the eyed embryo stage i.e., after Gorodilov's embryo development stage 25, was shown to destroy germ cells; in fact, in fish irradiated at stage 28 there was no ovarian development in any fish examined 26 months after fertilization. Structural deformities of the vertebral column and mandible in irradiated fish were relatively low, 5%

compared to 1.1% in control fish (Thorpe et al., 1987a). However, in many fish the endocrine tissues of the gonads, which synthesize reproductive steroids, were not eliminated and these fish later underwent the changes in secondary sexual characteristics which are typical of normal maturing salmon (Thorpe et al., 1987b). These irradiated fish which exhibit secondary sexual characteristics, but produce no gametes would thus be unable to reproduce, although it is conceivable that they would demonstrate reproductive behavior in the presence of an intact sexually mature salmon. A second drawback concerning this technique is the accessibility of the gamma source. It would normally be necessary to transport the developing eggs to the gamma source unless it were possible to develop a transportable gamma source. An alternative would be to develop a suitable X irradiation source, however, further research would be required to define the dosage parameters.

5. Treatment with Androgen

The utilization of androgens to regulate sex differentiation in salmonids has been investigated by a number of groups and the early research has been reviewed by Donaldson and Hunter, 1982a, and Hunter and Donaldson, 1983. Most research has focused on the optimization of techniques to masculinize genotypic female salmonids and thus obtain monosex female sperm for the generation of monosex female salmonid stocks (Donaldson and Benfey, 1987; Piferrer and Donaldson, 1988); however, a number of studies have also focused on the utilization of androgen treatment to induce permanent sterility in salmonids. Masculinization of genotypic female salmonids can be achieved by minimal treatment with androgen if presented at the appropriate developmental stage. Thus, in the coho salmon *(Oncorhynchus kisutch)*, it was possible to increase the proportion of males to 73% from control levels of 42-54% by a single 2 hr immersion in 400 μg/L 17α–methyltestosterone 6 days after hatching (Piferrer and Donaldson, 1989) and from control levels of 49.5% males to 89% males after a single immersion in 1600 μg/L 17α–methyl-dihydrotestosterone (Piferrer and Donaldson, 1991). In monosex female chinook salmon over 90% masculinization was achieved with one or two immersions in 17α–methyltestosterone around the time of hatching (Baker et al., 1988). Sterilization, on the other hand, occurs when androgen is administered not just during the labile period when male/female differentiation occurs, but over a longer period of time during the histological differentiation of the gonad. There is now information available on the androgen sterilization of Atlantic salmon and several Oncorhynchids including the rainbow trout, coho salmon and chinook salmon. In the Atlantic salmon, Simpson (1975-76) immersed eyed eggs and alevins at 250 μg/L followed by dietary treatment at 30 mg/kg 17α–methyltestosterone for 120 days. At 9 months no ovarian development was apparent. In a subsequent study, a similar treatment resulted in sterile gonads at 6 months while dietary treatment alone at 3 mg/kg for 90 days produced both male and sterile Atlantic salmon (Johnstone et al., 1978). Recently it has been reported that dietary 17α–methyltestosterone administered at >20 mg/kg for 600 degree days from first feeding is extremely effective in sterilizing Atlantic salmon (Johnstone, 1989). Further research and development would be required before androgen induced sterilization could be utilized on a production basis for Atlantic salmon. At the West Vancouver Laboratory, we have investigated the androgen induced sterilization of Pacific salmon and trout. Immersion of coho salmon eyed eggs and alevins in 25-400 μg 17α–methyltestosterone/L followed by dietary administration of 20 mg 17α–methyltestosterone/kg diet for 3 months from first feeding resulted in 94-100% sterile salmon. The 25 μg/L immersion group had the lowest percentage of steriles. In coho that received 17α–methyltestosterone in the diet alone only 52% of the fish were sterile (Goetz et al., 1979). When coho were immersed twice at the eyed egg and alevin stages in 400 µg 17α–methyltestosterone/L and then fed a lower, 10 mg/kg, dose of 17α–methyltestosterone, the sterility rate was 94% at three yrs of age. At this time all except 3% of the control salmon and all but 6% of a group directly feminized with estradiol-17ß had matured (Hunter et al., 1982). Sterile coho salmon from this study lived 3-4 years in captivity beyond the normal time of death, i.e., to an age of 6 or 7 years at which time they probably underwent a natural aging process quite different from the rapid aging process which normally occurs in this species at 3 or occasionally 4 years of age, directly following sexual maturation and spawning. Coho salmon sterilized by androgen treatment under experimental study certificates issued by Health and Welfare Canada, have been released from both Capilano Salmon Hatchery and Big Qualicum Salmon Hatchery into British Columbia coastal waters. Sterile fish were harvested by the commercial and recreational fisheries within the Canadian and United States economic zones at age classes 2, 3, 4 and 5 years. Owing to the intensity of the fishery 73.6% of the total salmon harvested were captured in year class 3; approximately 24.7% in year class 4 and 1.3% in year class 5 (Solar et al., 1986). The only androgen treated fish which underwent the normal anadromous migration to the hatchery of origin were a relatively small percentage of incompletely sterilized fish. No sterilized fish returned to the hatchery of origin. This provided, for the first time, experimental confirmation for the hypothesis that sexual maturation is essential in salmonids for the initiation of the anadromous migration. We concluded that the sterile fish remained in the marine environment until they were either harvested in the fishery or died of natural causes.

Recent studies in this laboratory have demonstrated that chinook salmon can also be sterilized using synthetic androgen treatment. Chinook salmon also require both immersion at the alevin stage and dietary administration from first feeding to achieve a high percentage of sterility; however, in this species the dose of dietary androgen required, up to 80 mg/kg, is higher than in coho (Solar et al., unpublished). We are currently interested in determining whether an immersion regime can be developed which results in a high level of sterility without the necessity for subsequent dietary androgen administration. In an intial study immersion of coho for 2 hrs on 10 occasions during the alevin stage at a dosage of 10 mg 17α–methyltestosterone/L resulted in the production of 51% sterile fish and 15% partially sterile fish. Immersion of alevins on a continuous basis for 30 days in 100 µg 17α–methyltestosterone/L resulted in a sterility rate of 43% (Piferrer and Donaldson, 1988; unpublished in Piferrer and Donaldson, 1988; Piferrer, 1990). These early results indicate the potential, after further manipulation of critical variables including dose, duration, timing and form of androgen, to induce a degree of sterility in salmonids by immersion during the incubation process which is high enough to be of practical use.

When conducted properly by trained personnel, sterilization by androgen administration can be a very effective means of inducing a high percentage of sterility in salmonids. However, there are several aspects to the procedure which require caution.

1. The procedure currently requires several months and must be carried out carefully to avoid under or overtreatment.
2. The sterilization process must preferably be completed two months or more prior to smoltification and sea water transfer, as androgen is known to interfere with sea water adaptability. This would not be a problem with 1 + Atlantic or coho smolts, but poses a potential problem for the sterilization of zero age coho or chinook smolts depending on the incubation and rearing temperature.
3. Androgens are potent steroids which are biologically active in humans. Operators must therefore utilize appropriate protective equipment and avoid self exposure.

4. In general, authority for the use of androgens must be obtained from the appropriate regulatory agency.
5. Androgen solutions must be disposed of after use in an environmentally sound manner.
6. While androgen residues are completely eliminated in a matter of two to three weeks after the termination of treatment and the sterilized fish at the time of marketing contain undetectable levels of natural androgen, the potential exists for consumer resistance to fish which have received androgen treatment during early development.

6. Induction of Female Triploidy

The induction of female triploidy is probably the most promising sterilization technique developed to date. Reviews on chromosome set manipulation which include the induction of triploidy in salmonids include those of Thorgaard, 1983, 1986; Utter et al., 1983; Bye and Lincoln, 1986; Purdom, 1986; Chourrout, 1987; Donaldson and Benfey, 1987; Benfey and Donaldson, 1989; Johnstone, 1989; Johnstone et al., 1991; and Jungalwalla, 1991. In addition, a bibliography on triploid teleosts has recently been published (Benfey, 1989). When triploidy induction is contemplated as a means of sterility induction there are two separate technological issues, the methodology for the induction of triploidy and the technique for ensuring that the triploids are all female. Triploidy in salmonids was originally induced by applying heat shock shortly after fertilization to prevent the separation of the second polar body. In Atlantic salmon, eggs can be exposed to a high temperature shock, 30C for 6-10 minutes at 20-30 minutes after fertilization, to a moderate temperature shock, 28C for 10-15 minutes, at 15-25 minutes after fertilization or to a lower temperature shock at 26C for 15-20 minutes, at 10-20 minutes after fertilization (Johnstone, 1985, 1989). Johnstone (1989) proposed that the heat shocks have the effect of accelerating the fertilized eggs through the limited period when retention of the second polar body can be effected, thus making it difficult to obtain maximal levels of triploidy particularly when dealing with ova from different individuals or stocks which may exhibit genetic differences or may differ in time from ovulation.

In the last decade there has been increasing interest in the application of pressure shock to induce triploidy in salmonids. This technique which involves application of a pressure shock shortly after fertilization was first tested on Atlantic salmon using a landlocked strain in Newfoundland (Benfey and Sutterlin, 1984). Since that time large scale studies have been initiated on the induction of triploidy in Atlantic salmon in Scotland (Johnstone, 1989; Johnstone et al., 1991). In these trials a two litre pressure vessel is being used which is capable of pressure shocking 50,000 ova per hour with 100% triploidy rate and 90% survival rate relative to controls. Triploid Atlantic salmon are also being produced in Tasmania by means of a two litre pressure chamber (Jungalwalla, 1991). In British Columbia a one litre pressure vessel has been developed for triploidy induction in salmonids (Benfey et al., 1988).

Although temperature and pressure are the main triploidy induction techniques which have been utilized in salmonids, trials have also been conducted with nitrous oxide and other anaesthetics as a means of triploidy induction in Atlantic salmon (Johnstone et al., 1989). This technique must be applied immediately after fertilization and the eggs must be in a monolayer thus making the procedure less feasible on a production basis (Johnstone, 1989). Nitrous oxide was most successful when eggs were exposed to it from 0-30 minutes after fertilization at a pressure of 11 atmospheres (Johnstone et al., 1989). A recent study has shown that application of electrical shocks in conjunction with temperature shocks may enhance the induction of triploidy (Teskeredzic et al., 1991). It has also been shown possible to induce triploidy by dispermy (Ueda et al., 1986) and by high pH-high calcium treatment (Ueda et al., 1988).

Another procedure for triploidy production which obviates the need for post-fertilization shock treatments involves the generation of tetraploid salmonids which in turn generate diploid spermatozoa. These diploid spermatozoa when used to fertilize normal haploid ova produce triploid zygotes. This procedure has been conducted on an experimental basis in rainbow trout *(Oncorhynchus mykiss)* (Chourrout et al., 1986) but has yet to be tested in other salmonids. One potential problem with the procedure is associated with the ability of the larger diploid spermatozoa to penetrate the micropyle.

Several alternative procedures are available for the production of monosex female triploids (Donaldson, 1986). The two main alternatives are either the use of monosex female sperm in the initial fertilization or the direct feminization of the triploid embryo during early development. Procedures have been reviewed recently (Donaldson, 1986; Donaldson and Benfey, 1987; Piferrer and Donaldson, 1988) for both the generation of monosex female sperm and for the direct feminization of salmonids. As indicated above there has been some success in rainbow trout in the generation of diploid spermatozoa from tetraploid fish. If this becomes a feasible procedure, then it would also be desirable to manipulate these to produce monosex female diploid sperm, i.e., the tetraploids would be genotypic females with a male phenotype. Regardless of the initial means of production, once genotypic monosex female embryos have been generated a portion can be treated with androgen during early development to generate phenotypic males which when mature produce additional monosex female spermatozoa for the production of monosex female or monosex triploid populations (Donaldson and Benfey, 1987). In Atlantic salmon there have been reports of the presence of occasional oocytes of normal appearance within the otherwise sterile ovary (Benfey and Sutterlin, 1984; Johnstone, 1989; Johnstone et al., 1991). In the latter study viable ova were obtained from two triploid females. After fertilization only 10% survived to the eyed stage. These were shown to be hypo or hyperdiploid and were not expected to survive hatching (Johnstone et al., 1991). Similar observations have not been reported in Pacific salmonids although we have seen degenerate ova in 4-5 yr old triploid female *Oncorhynchus mykiss* during autopsy (Donaldson and Solar, 1990, unpublished). In general, as indicated above, the production of female triploids appears to be a promising procedure for sterilization in salmonids; however, as with other sterilization procedures further information is needed on the performance of these fish under production grow out conditions. Current evidence in triploid coho salmon indicates that triploid salmon grow at a similar rate to immature salmon, but slower than maturing salmon (W.C. Clarke et al. unpublished, 1989, 1991). This is also true for Atlantic salmon where during spring and early summer grilse grow faster than either immature salmon or triploid salmon of the same age (Johnstone et al., 1991).

7. Production of Sterile Hybrids

While a number of investigators have investigated the generation of interspecific hybrids between salmonids (Dangel et al., 1973; Chevassus, 1979, 1983) most studies have encompassed only the early development phase and few investigators have grown hybrids to maturity and determined whether they can undergo gonadal maturation and generate viable gametes. Three forms of sterility have been defined in hybrid teleosts: (1) zygotic sterility, where the gametes are viable but embryonic death occurs after fertilization, (2) gametic sterility, where the gonads are of normal size but produce gametes which are inviable, (3) gonadal sterility, where the gonad is smaller and fails to produce viable gametes (Chevassus, 1983). The degree of sterility in female and male hybrids may not be the same. Thus the *Salmo trutta* x *Salvelinus fontinalis* hybrid has been referred to as an example of gonadal sterility, however, some males are capable of sperm production (Suzuki and Fukuda, 1973). The *Salmo salar* x *Salmo trutta* hybrid is an example of zygotic sterility, as gametes are produced but the resultant embryos are inviable. The *Salvelinus fontinalis* x *namaycush*, *Oncorhynchus keta* x *nerka* and the *Oncorhynchus keta* x *gorbuscha* hybrids, on the other hand, each produce viable offspring. Viable offspring have also been obtained from *Oncorhynchus kisutch* x *tshawytscha* hybrids in this laboratory (I. Baker et al., 1986, unpublished). If it were possible to produce a sterile interspecific hybrid salmonid, one would then be faced with the question as to whether this hybrid had suitable production characteristics and whether it would be acceptable for culture purposes. In the *Oncorhynchus kisutch* x *tshawytscha* hybrid for example, growth rates were satisfactory in many of the fish; however, a proportion of the hybrids exhibited morphological deformities, especially in the area of the caudal peduncle (I. Baker et al., 1986, unpublished).

Another hybrid option is the production of triploid hybrids. Some triploid hybrids have been shown to have a higher survival rate than the equivalent diploid hybrid (Chevassus, 1983). In a triploid hybrid the maternal genome should predominate and in female triploid hybrids sterility should be a certainty. Studies involving triploid hybrids have been listed by Benfey (1989).

8. Production of Sterile Transgenics

The last decade has seen the generation of transgenic fish carrying additional genes coding for factors which are involved in growth (Zhang et al., 1990) and freezing resistance (Fletcher et al., 1988). The manipulation of other production characteristics such as disease resistance, flesh colour, etc., is also possible as is the regulation of sex. In the initial stages of transgenic fish development, there appears to be a consensus building in the fisheries research community that transgenic fish, especially those with altered production characteristics, should be sterilized by one means or another prior to transfer outside the laboratory (Kapuscinski and Hallerman, 1990; Maclean and Penman, 1990; Devlin and Donaldson, 1992). In the context of this review the potential for the production of inherently sterile transgenic salmonids is of interest. This could be achieved by linking a gene which generates a toxic gene product to the regulatory portion of a gonad specific gene in such a way that gonad development is blocked at the stage when the regulatory portion of the gonad specific gene is activated (Maclean and Penman, 1990). Inhibition of gonadal development could also be achieved by producing transgenic salmon in which a specific enzyme-regulated step in the endocrine control of reproductive development has been compromised. As the resultant transgenic fish derived from these procedures would be sterile, it would be necessary to maintain a special monosex or bisexual broodstock in which the gene construct inducing sterility is not expressed.

9. Monosex Techniques

While sterilization is the primary means of biological containment, in certain circumstances the utilization of monosex stocks may provide a sufficient degree of containment. For example, if a monosex stock of cultured salmonids escaped into the natural environment and survived to reproduce, the only possible offspring would be wild x wild or wild x cultured and there would be no cultured x cultured, an important factor if the escape was large relative to the size of the natural stock. The utilization of monosex stocks would be of even greater use for the containment of salmonids which are transplanted out of their natural range, e.g., *Oncorhynchus* species to countries bordering the Atlantic ocean or *Salmo* species to the Pacific rim countries. Thus, if hybridization between the exotic and indigenous salmonids is shown to be unlikely or impossible on the basis of biological incompatibility then monosex culture would ensure that the released exotic salmonids would form neither selfsustaining populations nor interspecific hybrids.

The implementation where feasible of monosex culture of genetically altered fish offers certain advantages over sterilization. These mainly relate to performance. Monosex female or male salmonids would be expected to have production characteristics which are similar to those of the same sex in a mixed sex population. To date only monosex female salmonid populations have been produced on a commercial scale. The production and growout of monosex male salmonid populations has not yet been seriously investigated. Such a population would offer the possibility of rapid growth but earlier maturity. Methods for the production of monosex female populations fall into two main categories: direct feminization and indirect feminization. Direct feminization involves the treatment of salmonids at about the time of hatching with a natural or synthetic estrogen which directs sex differentiation toward the production of females. In recent studies we have produced 100% female *Oncorhynchus tshawytscha* after a single 2 hr treatment immediately after hatch (Donaldson and Piferrer, 1991; Piferrer and Donaldson, 1991). Directly feminized fish have the normal female phenotype; however, half of them are genetically male and at maturity these would produce ova half of which would carry the Y chromosome and half the X chromosome.

The utilization of the indirect method of monosex female production, on the other hand, results in the production of females which are phenotypically and genotypically normal. The indirect method involves the fertilization of normal ova with monosex female sperm obtained from salmon which have a female genotype but a male phenotype. These latter fish can be produced by a variety of methods which involve the direct masculinization of embryos having a female genotype. Masculinization levels of 90% have been achieved in *Oncorhynchus kisutch* after a single two hour treatment at the time of hatching (Piferrer and Donaldson, 1991) and 100% masculinization has been achieved in *Oncorhynchus tshawytscha* using a similar procedure (Piferrer and Donaldson, unpublished). Until recently the production of monosex female sperm has involved the masculinization of embryos of mixed sex followed by progeny testing combined with repeat masculinization in the second generation (Donaldson, 1986) or alternatively the utilization of dietary androgen treatment and separation of putative genotypic males and females on the basis of testicular morphology, with the testes of genotypic females generally having no sperm ducts and having an abnormal appearance.

Currently two new procedures are being investigated for the production of monosex female sperm in a single generation. One involves the production of gynogenes, which are inherently female in salmonids, followed by masculinization at hatching. The second, which is expected to have wide application, is the utilization of a Y specific DNA probe to separate genotypic males and females after the direct masculinization of embryos of mixed sex. This procedure is now available for *Oncorhynchus tshawytscha* (Devlin et al., 1991).

10. Discussion

In this review we have considered various methods for the containment of genetically altered salmonids. The type of physical and/or biological containment that is appropriate in a particular circumstance depends on the degree of containment required. If no genetic interaction is acceptable, then the fish either have to be under strict quarentine or one hundred percent sterilized. If a small to moderate degree of genetic interaction is acceptable then a lesser degree of containment can be implemented. However, if full genetic interaction is acceptable then no containment is required. As indicated in the introduction, the current practice is to classify genetically altered fish by the means or process by which they became altered rather than by an objective measure of the degree of genetic alteration and its potential impact. Thus, transgenic fish are currently maintained in quarantine and it is likely that they could only be used in aquaculture if fully sterilized (Devlin and Donaldson, 1992). On the other hand, farmed fish with varying degrees of domestication and/or genetic selection are normally maintained under moderate physical containment with either full reproductive capacity, as monosex female stocks or as sterile stocks. The use of monosex female and sterile stocks in salmonid aquaculture has to date been for economic reasons rather than as means for containment; however, this could change depending on the outcome of current studies on the impact of interbreeding between escaped farm fish and wild fish on the fitness of the wild population (NASCO, 1990). A recent study in which mature cultured brown trout *(Salmo trutta)* were introduced into a stream containing mature wild brown trout has demonstrated that the hybrid offspring had a decreased fitness level in the natural environment (Skaala, Jorstad and Borgstron, unpublished, 1991). Salmonids which are produced under controlled conditions in hatchery facilities for release into the natural environment normally fall into the category of requiring no containment although the impact of these potentially genetically altered fish on wild stocks has been the subject of recent discussion (Department of Fisheries and Oceans, 1991).

The various technologies for the induction of sterility, which we have discussed above, result in differing degrees of sterility. For the enhancement of product quality by prevention of sexual maturation in aquacultured fish, it is essential that the sterile fish contain neither gametes nor gonadal steroidogenic tissue capable of elaborating androgens, estrogens or progestins, as production of these steroids will cause the fish to undergo changes in secondary sexual characteristics even in the absence of gamete development. On the other hand, prevention of mature gamete formation and/or release is sufficient to prevent the occurrence of direct reproductive interaction between genetically altered and other salmonids. However, even in this latter situation, it would be preferable to produce completely sterile fish as these have been shown not to undergo the characteristic anadromous migration to the river and hatchery of origin (Donaldson and Hunter, 1982b; Donaldson and Hunter, 1985; Solar et al., 1986; Baker et al., 1989). Fish in which the gonadal steroidogenic tissue is still functional would undergo the typical migration to freshwater and may well exhibit reproductive behavior which could interfere with the spawning ritual in wild fish even though the former fish may be unable to contribute gametes either through lack of gametes or inability to externalize the gametes.

Of the techniques described above two sterilization techniques and two techniques for the production of monosex females are currently available for use for the reproductive containment of salmonids. These are sterilization by androgen treatment during early development, sterilization by production of female triploids, production of monosex females by utilization of monosex female sperm and production of females by direct feminization with estrogen during early development. Of these techniques the production of monosex female salmonid populations by the indirect technique has proven to be the most successful from the standpoint of acceptability, ease of application, 100% success rate and lack of impairment of production characteristics such as growth. The direct feminization technique is also effective; however, it does require the use of a natural or synthetic estrogen on production fish albeit at a very early life stage. There is as yet no perfect means of inducing sterility in salmonids. All sterile fish in which the steroidogenic cells are compromised fail to show the growth spurt which characterizes the performance of both male and female intact salmon.

Sterilization by androgen treatment currently involves treatment during the alevin stage and for the first two months after first feeding. Sterilization by induction of female triploidy, through temperature or pressure shock, involves some mortality if the treatment is of sufficient intensity to ensure production of 100% triploids. In the future we can expect the further improvement of these methods and the development of new technologies such as the production of inherently sterile transgenic salmonids. There has also been considerable recent progress in the development of technologies to accelerate growth in salmonids either through the administration of recombinant (Down et al., 1989; McLean et al., 1990) or through transgenesis (Du et al., 1992; Devlin et al., 1992 unpublished) and it is our expectation that these technologies will be implemented in conjunction with sterilization technologies to produce salmonids which grow rapidly, but never reach sexual maturity, thus providing an optimal fish for culture purposes which is also reproductively contained and therefore incapable of reproductive or genetic interaction with wild salmonids.

References

Baker, I.J., I.I. Solar, and E.M. Donaldson. 1988. Masculinization of chinook salmon *(Oncorhynchus tshawytscha)* by immersion treatments using 17α–methyltestosterone around the time of hatching *.Aquaculture* 72: 359-367.

Baker, I.J., I.I. Solar, K. Mulji, E.M. Donaldson, G.A. Hunter, and E.T. Stone. 1989. Coded wire tag recoveries from the second release of sterile coho salmon *(O. kisutch)* into the marine environment. *Can. Data Rep. Fish. Aquat. Sci.* No. 775. 21 pp.

Benfey, T.J. 1989. A bibliography of triploid fish, 1943 to 1988. *Can. Data Rep. Fish. Aquat. Sci.* No. 1682. 33 pp.

Benfey, T.J. 1991. The physiology of triploid salmonids in relation to aquaculture. *Can. Tech. Rep. Fish. Aquat. Sci.* No. 1789: 73-80.

Benfey, T.J. and E.M. Donaldson. 1988. Triploidy in the culture of Pacific salmon, in: Proc. Aquaculture International Congress (Vancouver, Canada: Sept. 6-9, 1988), pp. 549-554. British Columbia Pavilion Corp., Vancouver.

Benfey, T.J. and A.M. Sutterlin. 1984. Growth and gonadal development in triploid landlocked Atlantic salmon *(Salmo salar). Can. J. Fish. Aquat. Sci.* 41: 1387-1392.

Benfey, T.J., P.G. Bosa, N.L. Richardson, and E.M. Donaldson. 1988a. Effectiveness of a commercial-scale pressure shocking device for producing triploid salmonids. *Aquac. Engineer.* 7: 147-154.

Benfey, T.J., H.M. Dye, I.I. Solar, and E.M. Donaldson. 1989b. The growth and reproductive endocrinology of adult triploid Pacific salmonids. *Fish Physiol. Biochem.* 6: 113-120.

Bonham, K. and L.R. Donaldson. 1972. Sex ratios and retardation of gonadal development in chronically gamma-irradiated chinook salmon smolts. *Trans. Am. Fish. Soc.* 101: 428-434.

Brown, L.A. 1982. Surgical gonadectomy of salmonid fish and subsequent prevention of diseases associated with sexual maturation. *Proc. R. Soc. Endinburgh, Sect.* B. 81: 211-219.

Brown, L.A. and R.H. Richards. 1979. Surgical gonadectomy of fish: a technique for veterinary surgeons. *Vet. Rec.* 104: 215.

Bye, V.J. and R.F. Lincoln. 1986. Commercial methods for the control of sexual maturation in rainbow trout *(Salmo gairdneri R.). Aquaculture* 57: 299-309.

Chevassus, B. 1979. Hybridization in salmonids: results and perspectives. *Aquaculture* 17: 113-128.

Chevassus, B. 1983. Hybridization in fish. *Aquacultue* 33: 245-262.

Chourrout, D. 1987. Genetic manipulations in fish: review of methods, in: "Selection, Hybridization, and Genetic Engineering in Aquaculture, Vol. II," Tiews, K., ed., pp. 111-126. Heenemann Verlags. mbH, Berlin.

Chourrout, D. and I. Nakayama. 1987. Chromosome studies of progenies of tetraploid female rainbow trout. *Theor. Appl. Genet.* 74: 687-692.

Chourrout, D., B. Chevassus, F. Krieg, A. Happe, G. Burger, and P. Renard. 1986. Production of second generation triploid and tetraploid rainbow trout by mating tetraploid males and diploid females - potential of tetraploid fish. *Theor. Appl. Genet.* 72: 193-206.

Dangel, J.R., P.T. Macy, and F.C. Withler. 1973. Annotated bibliography of interspecific hybridization of fishes of the subfamily Salmoninae. *NOAA Tech. Mem.*, NMFS NWFC. 1:48.

Department of Fisheries and Oceans. 1991. International Symposium on Biological Interactions of enhanced and wild salmonids, June 17-20, 1991. Nanaimo, B.C. Book of Abstracts. 80 pp.

Devlin, R.H. and E.M. Donaldson. 1992. Containment of genetically altered fish with emphasis on salmonids, in: "Transgenic Fish," C.L. Hew and G.L. Fletcher, eds. pp. 229-265. World Scientific, Singapore.

Devlin, R.H., B.K. McNeil, T.D.D. Groves, and E.M. Donaldson. 1991. Isolation of a Y-chromosome DNA probe capable of determining genetic sex in chinook salmon *(Oncorhynchus tshawytscha). Can. J. Fish. Aquat. Sci.* 48:1606-1612.

Donaldson, E.M. 1973. Reproductive endocrinology of fishes. *Am. Zool.* 13: 909-927.

Donaldson, E.M. 1986. The integrated development and application of controlled reproduction techniques in Pacific salmonid aquaculture. *Fish Physiol. Biochem.* 2: 9-24.

Donaldson, E.M. and G.A. Hunter. 1982a. Sex control in fish with particular reference to salmonids. *Can. J. Fish. Aquat. Sci.* 39: 99-110.

Donaldson, E.M. and G.A. Hunter. 1982b. The ocean release and contribution to the fishery of all-female and sterile groups of coho salmon *(Oncorhynchus kisutch)*, in: "Proceedings of the International Symposium on Reproductive Physiology of Fish, Wageningen, The Netherlands, 2-6 August, 1982," H.J.Th. Goos and C.J.J. Richter, eds., p. 78. Pudoc, Wageningen.

Donaldson, E.M. and G.A. Hunter. 1985. Sex control in Pacific salmon: Implications for aquaculture and resource enhancement, in: "Proc. of Salmonid Reproduction an Int. Symp., Review Papers," R.N. Iwamato and S. Sower, eds., pp. 26-32. Wash. Seagrant Program, Bellevue, Wash.

Donaldson, E.M. and T.J. Benfey. 1987. Current status of induced sex manipulation, in: "Proc. Third Int. Symp. on the Reproductive Physiology of Fish" Idler, D.R., Crim, L.W. and Walsh, J.M., eds., pp. 108-119. Memorial Univ. of Nfld., St. John's.

Donaldson, E.M. and Piferrer. 1991. Direct feminization of chinook salmon utilizing 17-ethynylestradiol. Proc. 4th Int. Symp. on the Reproductive Physiology of Fish, Norwich, U.K., July 7-12, 1991, p. 274.

Donaldson, E.M., G.A. Hunter, I.J. Baker, and E.T. Stone. 1984. The first release of hormonally sterilized coho salmon *(Oncorhynchus kisutch)* into the marine environment. Int. Conf. on the Biology of Pacific salmon, Victoria/Agassiz, B.C. Sept. 5-12, 1984. Abstr.

Donaldson, E.M., F. Piferrer, I.I. Solar, and R.H. Devlin. 1991. Studies on hormonal sterilization and monosex technologies for salmonids at the West Vancouver Laboratory. *Can. Tech. Rep. Fish. Aquat. Sci.* No. 1789: 37-45.

Down, N.E., P.M. Schulte, E.M. Donaldson, and H.M. Dye. 1989. Growth acceleration of seawater-adapted female chinook salmon *Oncorhynchus tshawytscha* by constant infusion of recombinant bovine growth-hormone under ambient summer conditions. *J. World Aqua. Soc.* 20: 181-187.

Du, S.J., Z. Gong, G.L. Fletcher, M.A. Shears, M.J. King, D.R. Indler, and C.L. Hew. 1992. Growth enhancement in transgenic Atlantic salmon by the use of an "all fish" chimeric growth hormone gene construct. *Bio/Technology*. 10:176-181.

Dunham, R.A. 1990. Production and use of monosex or sterile fishes in aquaculture. *Reviews in Aquatic Sciences.* 2: 1-17.

Egami, N. and K.-I. Ijiri. 1979. Effects of irradiation on germ cells and embryonic development in teleosts. *Int. Rev. Cytol.* 59: 195-248.

Ellis, A.E. 1981. A field trial of a method to induce auto-immune castration in *Salmo salar*. *Int. Coun. Explor. Sea.* CM 1981/F:37. 10 pp.

Egidius, E., L.P. Hansen, B. Jonsson, and G. Naevdal. 1991. Mutual impact of wild and cultured Atlantic salmon in Norway. *J. Cons. Int. Explor.* Mer, 47: 404-410.

Fletcher, G.L., M.A. Shears, M.J. King, P.L Davies, and C.L. Hew. 1988. Evidence for antifreeze protein gene transfer in fish: The first step toward a more freeze-resistant salmon. *Can. J. Fish. Aquat. Sci.* 45: 352-357.

Flynn, M.B. 1973. The effect of methallibure and a constant 12-hours-light : 12-hours-dark photoperiod on the gonadal maturation of pink salmon (*Oncorhynchus gorbuscha*), in: Masters Thesis UBC. Vancouver B.C., *70 pp.*

Glebe, B.D., J. Delabbio, P. Lyon, R.L. Saunders, and S. McCormick. 1986. Chromosome engineering and hybridization of Arctic char *Salvelinus alpinus* and Atlantic salmon *Salmo salar* for Aquaculture, in: EIFAC Symposium E35: 14pp.

Goetz, F.W., E.M. Donaldson, G.A. Hunter, and H.M. Dye. 1979. Effects of estradiol-17ß and 17α–methyltestosterone on gonadal differentiation in the coho salmon, *Oncorhynchus kisutch. Aquaculture 17: 267-278.*

Goubau, S., D.W. Silversides, A. Gonzalez, B. Laarveld, R.J. Mapletoft, and B.D. Murphy. 1989. Immunization of sheep against modified peptides of gonadotropin releasing hormone conjugated to carriers. *Domest. Anim. Endocrinol.* 6: 339-347.

Hanson, L.H. and P.S. Manion. 1980. Sterility method of fertility control and its potential role in an integrated sea lamprey *(Petromyzon marinus)* control program. *Can. J. Fish. Aquat. Sci.* 37: 2108-2117.

Hershberger, W.K., K. Bonham, and L.R. Donaldson. 1978. Chronic exposure of chinook salmon eggs and alevins to gamma irradiation: effects on their return to freshwater as adults. *Trans. Am. Fish. Soc.* 107: 622-631.

Hunter, G.A. and E.M. Donaldson. 1983. Hormonal sex control and its application to fish culture, in: "Fish Physiology, Vol. IX, Reproduction, Part B, Behavior and Fertility Control," W.S. Hoar, D.J. Randall and E.M. Donaldson, eds., Chap. 5. p. 223-303. Academic Press, New York.

Hunter, G.A., E.M. Donaldson, F.W. Goetz, and P.R. Edgell. 1982. Production of all female and sterile groups of coho salmon *(Oncorhynchus kisutch)* and experimental evidence for male heterogamety. *Trans. Am. Fish. Soc.* 111: 367-372.

Hunter, G.A., E.M. Donaldson, J. Stoss, and I. Baker. 1983. Production of monosex female groups of chinook salmon *(Oncorhynchus kisutch)* by the fertilization of normal ova with sperm from sex reversed females. *Aquaculture* 33: 355-364.

Johnstone, R. 1985. Induction of triploidy in Atlantic salmon by heat shock. *Aquaculture* 49: 133-139.

Johnstone, R. 1989. Maturity control in Atlantic salmon. A review of the current status of research in Scotland, in: M. Carrillo, S. Zanuy and S. Planas, Compilers, XIth Int.

Symp. Comp. Endocrinol., Proc. of the Satellite Symp. on Applications of Comp. Endocrinol. to Fish Culture, Almunecar, Spain. May 22-23, 1989. p. 89-94.

Johnstone, R., T.H. Simpson, and A.F. Youngson. 1978. Sex reversal in salmonid culture. *Aquaculture* 13: 115-134.

Johnstone, R., R.M. Knott, A.G. Macdonald, and M.V. Walsingham. 1989. Triploidy induction in recently fertilized Atlantic salmon ova using anaesthetics. *Aquaculture* 78: 229-236.

Johnstone, R., H.A. McLay, and M.V. Walsingham. 1991. Production and performance of triploid Atlantic salmon in Scotland. *Can. Tech. Rep. Fish. Aquat. Sci.* No. 1789, 15-36.

Jungalwalla, P.J. 1991. Production of non-maturing Atlantic salmon in Tasmania. *Can. Tech. Rep. Fish. Aquat. Sci.* No. 1789, 47-71.

Kapuscinski, A.R. and E.M. Hallerman. 1990. Transgenic fish and public policy: anticipating environmental impacts of transgenic fish. *Fisheries* 15:2-11.

Konno, K. 1980. Effects of γ-irradiation on the gonads of the rainbow trout, *Salmo gairdneri irideus,* during embryonic stages, in: "Radiation Effects on Aquatic Organisms," N. Egami, ed., p. 129-133. Japan Soc. Sci. konPress, Tokyo/Univ. Park Press, Baltimore.

Konno, K. and F. Tashiro. 1982. The sterility of rainbow trout *(Salmo gairdneri)* irradiated with cobalt-60 gamma rays. *J. Tokyo Univ. Fish.* 68: 75-80.

Laird, L.M., A.E. Ellis, A.R. Wilson, and F.G.T. Holliday. 1978. The development of the gonadal and immune systems in the Atlantic salmon *(Salmo salar L.)* and a consideration of the possibility of inducing autoimmune destruction of the testis. *Ann. Biol. Anim. Biochim. Biophys.* 18: 1101-1106.

Laird, L.M., A.R. Wilson, and F.G.T. Holliday. 1980. Field trials of a method of induction of autoimmune gonad rejection in Atlantic salmon *(Salmo salar L.). Reprod. Nutr. Dev.* 20: 1781-1788.

McLean, E., E.M. Donaldson, H.M. Dye, and L.M. Souza. 1990. Growth acceleration of coho salmon *(Oncorhynchus kisutch)* following oral administration of recombinant bovine somatotropin. *Aquaculture* 91: 197-203.

Maclean, N. and D. Penman. 1990. The application of gene manipulation to aquaculture. *Aquaculture* 85: 1-20.

McBride, J.R., U.H.M. Fagerlund, M. Smith, and N. Tomlinson. 1963. Resumption of feeding by and survival of adult sockeye salmon *(Oncorhynchus nerka)* following advanced gonad development. *J. Fish. Res. Board Can.* 20: 95-100.

NASCO. 1990. Report on the Norwegian Meeting on impacts of aquaculture on wild stocks. North Atlantic Salmon Conservation Organization. Paper CLN(90)28: 9 pp.

Pepper, V.A. 1991. Proceedings of the Atlantic Canada Workshop on methods for the production of non-maturing salmonids. *Can. Tech. Rep. Fish. Aquat. Sci.* No. 1789, 152 pp.

Piferrer, F. 1990. Hormonal manipulation of the process of sex differentiation in Pacific salmon. Ph.D. thesis, University of Barcelona. 399 pp.

Piferrer, F. and E.M. Donaldson. 1988. Progress in the development of sex control techniques for the culture of Pacific salmon. Aquaculture International Congress & Exposition, Sept. 6-9, 1988, Vancouver, B.C. Publ. Aqua. Int. Congr. B.C. Pavilion Corp. p. 519-530.

Piferrer, F. and E.M. Donaldson. 1989a. Hormonal sex control in Pacific salmon: The importance of treatment timing, in: M. Carrillo, S. Zanuy and S. Planas, Compilers, XIth Int. Symp. Comp. Endocrinol. Proc. Satellite Symp. Applications of Comparative Endocrinology to Fish Culture. Almunecar, Spain, May 22-23, 1989. p. 81-88.

Piferrer, F. and E.M. Donaldson. 1989b. Gonadal differentiation in Coho salmon *Oncorhyn-*

chus kisutch, after a single treatment with androgen or estrogen at different stages during ontogenesis. *Aquaculture* 77: 251-262.

Piferrer, F. and E.M. Donaldson. 1991. Dosage-dependent differences in the effect of aromatizable and non-aromatizable androgens on the resulting phenotype of coho salmon *(Oncorhynchus kisutch)*. *Fish Physiol. Biochem.* 9: 145-150.

Piferrer, F. and E.M. Donaldson. 1992. The comparative effectiveness of the natural and a synthetic estrogen for the direct feminization of chinook salmon (*Oncorhynchus tshawytscha*). *Aquaculture* 106: 183-193.

Purdom, C.E. 1986. Genetic techniques for control of sexuality in fish farming. *Fish Physiol. Biochem.* 2: 3-8.

Robertson, O.H. 1961. Prolongation of the life span of kokanee salmon *(Oncorhynchus nerka kennerlyxi)* by castration before beginning of gonad development. *Proc. Natl. Acad. Sci.* 47: 609-621.

Schreck, C.B. and L.G. Fowler. 1982. Growth and reproductive development in fall chinook salmon: effects of sex hormones and their antagonists. *Aquaculture* 36: 253-263.

Secombes, C.J., L.M. Laird, and I.G. Priede. 1987. Immunological approaches to control maturation in fish. II. A review of the autoimmune approach. *Aquaculture* 60: 287-302.

Simpson, T.H. 1975-76. Endocrine aspects of salmonid culture. *Proc. R. Soc. Edinburgh* (B). 17: 241-252.

Skaala, 0., D. Dahle, K.E. Dorstad, and G. Naevdal. 1990. Interactions between natural and farmed fish populations: information from genetic markers. *J. Fish Biol.* 36: 449-460.

Solar, I.I., I.J. Baker, E.M. Donaldson, G.A. Hunter, and E.T. Stone. 1986. Coded wire tag recoveries from the first release of all-female and sterile groups of coho salmon *(O. kisutch)* into the marine environment. *Can. Data Rep. Fish. Aquat. Sci.* No. 609. 29 pp.

Suzuki, R. and Y. Fukuda. 1973. Sexual maturity of F1 hybrids among salmonid fishes. *Bull. Freshwater Fish. Res. Lab.* 23:57-74.

Teskeredzic, E., E.M. Donaldson, Z. Teskeredzic, E. McLean, and I.I. Solar. 1991. Comparison of heat and heat-electro shocks to induce triploidy in coho salmon *(Oncorhynchus kisutch)*. *Can. Tech. Rep. Fish. Aquat. Sci.* 1785, 7 pp.

Thorpe, J.E., C. Talbot, and M.S. Miles. 1987a. Irradiation of Atlantic salmon eggs to overcome early maturity when selecting for high growth rate, in: "Selection, Hybridization., and Genetic Engineering in Aquaculture," Vol. I, Tiews, K., ed., p. 361-374. Heenemann Verlags mbH, Berlin.

Thorpe, J.E., R.S. Wright, C. Talbot, and M.S. Miles. 1987b. Secondary sexual characteristics developed by 60-Co sterilised Atlantic salmon. Proc. Third Int. Symp. on the Reproductive Physiology of Fish, St. John's, Nfld. Aug. 2-7, 1987. p. 139.

Thorgaard, G.H. 1983. Chromosome set manipulation and sex control in fish, in: "Fish Physiology, Vol. IX, Reproduction, Part B, Behavior and Fertility Control," Hoar, W.S., Randall, D.J., and Donaldson, E.M., eds., pp. 405-434. Academic Press, New York.

Thorgaard, G.H. 1986. Ploidy manipulation and performance. *Aquaculture* 57: 57-64.

Tsoi, R.M. 1969. The effect of nitrosomethyl urea and dimethylsulphate on sperm of the rainbow trout *(Salmo irideus* Gibb.) and the peled *(Coregonus peled Gmel.)*. *Dokl. Biol. Sci.* USSR 189: 849-852.

Ueda, T., M. Kobayashi, and R. Sato. 1986. Triploid rainbow trouts induced by polyethylene glycol. *Proc. Japan Acad.* 62B: 161-164.

Ueda, T., R. Sato, and J. Kobayashi. 1988. Triploid rainbow trout induced by high-ph • high calcium. *Nippon Suisan Gakkaishi*. 54: 2045.

Utter, F.M., O.W. Johnson, G.H. Thorgaard., and Rabinovitch, P.S. 1983. Measurement and potential applications of induced triploidy in Pacific salmon. *Aquaculture* 35: 125-135.

Watson, W. 1755. An account of Mr. Samuel Tull's method of castrating fish. *Phil. Trans. Roy. Soc.* (Lond.). 48(2): 870-874.

Wunder, W. 1977. Experimentelle untersuchungen zur frage der regeneration der maennlichen und weiblichen keimdruesen (gonaden) der regenbozen forelle *(Salmo irideus* Gibb.). *Zool. Anz.* 198: 245-262.

Yamazaki, F. 1983. Sex control and manipulation in fish. *Aquaculture* 33: 329-354.

Zhang, P., M. Hayat, C. Joyce, L.I. Gonzalez-Villasenor, C.M. Lin, R.A. Dunham, T.T. Chen, and D.A. Powers. 1990. Gene transfer, expression and inheritance of pRSV-rainbow trout-GH cDNA in the common carp, *Cyprinus carpio* (Linnaeus). *Mol. Reprod. Dev.* 25: 3-13.

Germplasm Repositories for Plants

RAYMOND L. CLARK

1. Organization and Operation

From the simple "system" of pre-1898, the program for plant germplasm preservation in the United States evolved into the regional station network of the 1940s and added a higher level of security with the establishment of the National Seed Storage Laboratory in the 1950s, and the inclusion of clonally propagated material at ten repositories established in the 1980s.

The major components of the present National Plant Germplasm System (NPGS) are:

The National Germplasm Resources Laboratory (NGRL) in the Plant Sciences Institute at the Beltsville Agricultural Research Center, Beltsville, Maryland. The NGRL includes the Plant Introduction Office, National Plant Germplasm Quarantine Center, the Plant Exploration Office, and the Germplasm Resources Information Network (GRIN).

The curator sites for working collections are located throughout the mainland U.S., Puerto Rico, and Hawaii. The four regional plant introduction (PI) stations have responsibility for, collectively, over 5,000 species and 150,000 individual accessions.

The National Seed Storage Laboratory (NSSL), Fort Collins, Colorado, is the long-term backup storage facility for all plant germplasm in the NPGS. There are more than 240,000 samples now stored at NSSL in aluminum foil laminated, sealed containers. Seeds are stored at -18C or, if data are available to ensure its safe use, in liquid nitrogen. Requests for seeds from NSSL are referred to the curator sites that maintain the working collections for the species involved.

2. Acquisition of Germplasm

Domestic germplasm, though scarce in our country, is often collected by scientists working on breeding programs to improve native species. This material may be sent directly to the curator site or to the Plant Introduction Office in Beltsville, Maryland. In other cases, domestic exploration trips are planned and funded by USDA-ARS to obtain new material of specific native crops.

To be properly logged into the GRIN database, such material should be accompanied by certain data relating to: the name of the species, the collector's name, collector's sample number, the location from which the sample was obtained, the date, the number of plants in the population and the number sampled, habitat information, latitude, longitude, and elevation

Plant Germplasm Introduction and Testing Laboratory, U.S. Department of Agriculture, Pullman, Washington 99164-4236, U.S.A.

Genetic Conservation of Salmonid Fishes, Edited by J.G. Cloud and G.H. Thorgaard, Plenum Press, New York, 1993

of the collection site, as well as any unusual plant traits observed. In addition to this wild or native plant material, advanced breeding lines, germplasm populations, and cultivars that are registered in the journal, *Crop Science*, are now automatically entered into the NPGS and assigned a P.I. number.

Foreign germplasm acquisitions are the more important source of new materials for the NPGS. As with domestic acquisitions, these can come from donations or from plant collecting trips. Scientists who wish to initiate — and perhaps participate in — a collecting trip, either domestic or foreign, must follow an established protocol to obtain funding from the NPGS. Each year several such trips are planned, funded, and carried out, resulting in several hundred new germplasm accessions being available for researchers. Material from these collecting trips enters the country through the Quarantine Center. The Plant Introduction Office then sends the germplasm to the curator responsible for that species.

3. Maintenance

When germplasm is received by a curator, there is a logging in protocol that documents the event. As soon as possible the curator stores the material under conditions designed to prolong its viability. In the case of seeds of most temperate species this means cool, dry storage. Vegetatively propagated germplasm may require cool, moist conditions or immediate planting or grafting to suitable rootstocks. Viability of most seeds is easily maintained for several years — and often for 20 to 30 — if good quality seeds are stored at 4C and at 35% or less relative humidity.

Proper storage is only the first part of an effective maintenance program. The second, and much more difficult, part is regeneration. For species normally stored and distributed as seeds there are several factors to consider: (1) what is the method of pollination — self or outcrossing; (2) if outcrossing, what is the agent for moving the pollen from one flower to another; (3) what is the degree of heterozygosity in the population; (4) is this species adapted to the regeneration site; (5) if this species normally is self-pollinated, what effect would pollinator activity have on the amount of outcrossing; and (6) are suitable cultural practices available and used. The regeneration process is fraught with pitfalls and it is successful only if all people involved do each part of the operation accurately and in a timely manner. Protocols have been worked out for many of our species that will ensure, when carefully followed, a successful regeneration that is representative, genetically, of the parent population.

Vegetatively propagated germplasm presents a much different set of problems for maintenance and storage. Most of this material will not produce seeds — or it is a specific genotype (clone) or an important cultivar and must be maintained without going through a sexual cycle.

4. Evaluation

What do we really have in our plant germplasm repositories? One of the most revealing findings about our plant germplasm resources was brought to light when GRIN began generating reports on evaluation data — and we found it to be full of gaps. These gaps pointed out two important faults of most previous evaluation work: (1) it was incomplete and (2) it was not systematic. As a result, Crop Advisory Committees were formed to try to help the NPGS overcome these deficiencies.

There are two general types of evaluation data, that which relates to environmentally neutral traits and that which relates to traits strongly affected by environment. The former are such things as flower color, leaf arrangement and venation, plant pubescence, seed color, root branching, and number of petals. Such data can be collected during regeneration at the curator site. The latter are such traits as maturity, yield, disease reactions, response to fertilizers, and seedling vigor. To obtain reliable data on these traits, special, replicated tests are required, thus involving separate plantings from those used for regeneration of the germplasm.

5. Distribution and Utilization

When a curator has properly maintained, evaluated, and stored the germplasm, it is available for distribution to scientists upon request. The Pullman P.I. Station (Pullman, Washington, U.S.A.) has sent out nearly 20,000 seed packets per year in the past five years. About one-fourth of these packets have gone to foreign scientists. Distribution is free and involves 50 to 200 seeds per sample. For clonal materials, budwood or cuttings are the usual form of distribution but meristem tissue cultures are becoming more important. Pollen may also become a more important part of germplasm exchange, especially between countries. There are relatively few pathogens transmitted through pollen so its use decreases the quarantine problems associated with seed or vegetative materials.

At the end of each growing season, we send out requests to germplasm recipients for data on the performance of our material. We also request acknowledgement of the use of P.I. germplasm in publicly released material, such as parental lines, germplasm populations, and cultivars.

Almost all present day U.S. crops are based on introduced plant germplasm. Many of the earlier introductions were lost or inadequately documented in relation to their use in our agricultural system. A few notable exceptions provide some idea of the importance of utilization of introduced germplasm.

A single gene from P.I. 217407, a popcorn, was found to provide outstanding resistance to northern corn leaf blight. This Ht gene was soon incorporated into all corn belt hybrids and resulted in more than half a million dollars in increased production per year in the 1940s and 1950s.

Similarly, a single gene for Fusarium wilt resistance in tomato was found in P.I. 79532, the wild currant tomato from Peru. Essentially all modern tomato varieties carry this gene and it continues to provide protection for the nearly billion dollar tomato industry. Another wild relative of tomato, P.I. 128657, also from Peru, provided the first gene for resistance to root knot nematodes.

The wheat industry in the United States is dependent on disease resistant accessions such as P.I. 178383. This accession, collected by Harlan in Turkey, showed very little promise of importance from an agronomic standpoint. It was not until it was tested against some of the major wheat diseases that its value was revealed. It has good resistance to four races of stripe rust, 35 races of common bunt, and usable tolerance to flag smut and snow mold. This is a good example of the value of maintaining germplasm even when many of its plant traits are poor.

There are many other specific examples of the value of individual accessions in providing genes for improved varieties or in protecting major crops against the dangers of a narrow genetic base. In recent years, many molecular biologists have used plant germplasm collections as a source of specific genes to study gene action, to aid in gene mapping, and insertion into other species to study gene activation in foreign genomes.

Thus, in spite of earlier predictions that the development of molecular biology would lessen the need for germplasm collections, it seems to be increasing the utility of this material.

6. Advisory Groups

Just as with the federal government, the NPGS has a complex series of checks and balances to assure a well-thought-out approach to germplasm management. It is a system that continues to evolve in a direction that seems to be leading to a more effective program.

There are now five different advisory committees, or groups of committees, that serve the NPGS.

The National Plant Genetics Resources Board, appointed by the Secretary of Agriculture, is charged with giving advice to him and the State Agriculture Experiment Station Directors on needs and priorities of the national program. Board members are chosen based on their expertise and to represent federal, state, and private sectors.

The National Plant Germplasm Committee includes the Administrative Advisors of the four P.I. Stations, selected Agriculture Research Service (ARS) Area Directors and representatives from National Program Staff, Cooperative State Research Service, and private companies. It provides coordination for the research and service of groups involved with plant germplasm, developing policy and establishing research priorities.

The Plant Germplasm Operations Committee includes the Director of the Germplasm Resources Laboratory, the Director of the NSSL, and the leaders of the various curator sites. It provides coordination of day-to-day operations, reviews exploration proposals, and identifies problems and needs of the system.

The Germplasm Matrix Team is a relatively new group chaired by the National Program Staff's Leader for plant germplasm. The other members of the team are NPS people with vested interest in germplasm matters. This team makes specific recommendations to the ARS Administrator relative to the funding of plant explorations and policy and operational matters.

The Technical Advisory Committees (separate ones for each P.I. station and clonal unit) provide technical, budgetary, and policy recommendations for site personnel and recommend budget requests to the Experiment Station Directors for RRF funds. They also provide annual updates of germplasm use by researchers in their states.

The Crop Advisory Committees have recently become a key group of advisors to the system as a whole. There are now 40 of these committees made up of crop experts from USDA, State Agriculture Experiment Stations, and private concerns. An attempt is made in constituting these groups to include as many scientific disciplines as possible, even though the majority usually are plant breeders or geneticists.

The Crop Advisory Committees were originally conceived because of the need, once a national germplasm database was established, to standardize evaluation data. The committees, in cooperation with the crop curators, prepared standardized plant descriptor lists, described each descriptor state, and recommended methods of obtaining reliable data.

7. International Relationships

The Consultative Group on International Agricultural Research (CGIAR), is the oversight group for the International Board for Plant Genetic Resources (IBPGR), headquartered in Rome. The IBPGR sponsors and helps advise and coordinate the activities of the International Agricultural Research Centers (IARCs) throughout the world. Their objectives are similar to our NPGS and we do collaborate in several areas, especially in plant exploration and exchange. The NSSL also serves as the official backup site for many of the IBPGR-sponsored germplasm collections in other countries.

The Food and Agricultural Organization (FAO) of the United Nations, also with headquarters in Rome, is involved in plant germplasm work on an international scale through its parent organization. The U.S. NPGS has no formal working relationships with FAO since our government interacts with the United Nations through the State Department.

The Office of International Cooperation and Development (OICD) is in the U.S. Department of Agriculture and often sponsors visits and exchanges with foreign scientists that allow additional interaction between our NPGS and similar foreign groups. The Department of State's Agency for International Development (USAID) also helps coordinate and fund collaborative work between scientists in the U.S. and abroad. Many of these programs produce benefits in information and germplasm exchange not possible through other avenues.

8. Summary

The program I have described is a many-faceted, evolving system. It has taken nearly a century to develop. Many of the biological techniques necessary for efficient preservation and utilization of germplasm are in a much more advanced state. Cryopreservation, which I just briefly mentioned for preservation of plant seeds, has enormous potential for other types of germplasm as well.

References

Burgess, Sam (ed.) 1971. The national program for conservation of crop germ plasm. (A Progress Report on Federal/State Cooperation.) Univ. of Georgia, June 1971. 73 p.

Frankel, O.H. and E. Bennett (eds.) 1970. Genetic resources in plants — their exploration and conservation. IBP Handbook No. 11, F.A. Davis, Philadelphia. 554 p.

Janick, Jules (ed.) 1989. The national plant germplasm system of the United States. *Plant Breeding Reviews* Vol. 7, Timber Press, Portland, OR. 230 p.

Advances in the Cryopreservation of Embryos and Prospects for Application to the Conservation of Salmonid Fishes

W. F. RALL

1. Introduction

Considerable controversy has developed within the conservation community as to the precise magnitude of the biological diversity crisis and the appropriate tactics needed to prevent mass extinctions. Most biologists agree that: (1) an increasing number of species are facing extinction as a result of native habitat loss or fragmentation (Wilson, 1988) and (2) nearly all habitat destruction results from the direct or indirect action of man (Ehrlich and Ehrlich, 1981). Uncertainties about the number of species at risk and the relationship between habitat destruction and species extinction (Mann, 1991) prevent qualitative estimates of what Ehrlich and Wilson (1991) describe as "the epidemic of extinctions now under way." Such uncertainty complicates the formulation of effective public policy directed at preserving biological diversity.

Survival of a species or population in the wild is thought to depend on a secure native habitat that is of sufficient size to support a population meeting certain genetic and demographic criteria (Soulé, 1987). Most of the important requirements are related to the properties and characteristics of the population as a whole, including size, life-history characteristics and the nature of the gene pool. The latter, especially genetic variations within populations or communities of individuals (i.e. polymorphism), represents the key to effective conservation. The ultimate goal of all conservation programs is to preserve genetic variation within the targeted population(s) of a species. The extent of genetic variation determines if the population will maintain sufficient fitness and flexibility to undergo further evolutionary changes (Frankel and Soulé, 1987). Much of the controversy associated with the field of conservation biology is related to the best method to maintain "genetic fitness" for future generations.

Conservation efforts often are classified into: (1) *in situ* programs that protect and manage animal populations within their natural, native habitat; and (2) *ex situ* programs that remove individuals, gametes or embryos from wild populations for controlled breeding and management in captivity. In general, habitat protection is acknowledged as the most efficient approach for conserving bio- and genetic-diversity. For some species, however, *in situ* conservation alone can not be relied upon to ensure the long-term viability of species at risk (Conway, 1988). This is especially true when habitat has undergone extensive change or is likely to be adversely affected by biological, social or political factors. *Ex situ* approaches

National Zoological Park, Smithsonian Institution, Washington, DC 20008-2598, U.S.A.

Genetic Conservation of Salmonid Fishes, Edited by J.G. Cloud
and G.H. Thorgaard, Plenum Press, New York, 1993

require a higher degree of technological and managerial input, but usually provide better long-term security for maintaining biodiversity. Soulé (1991) proposed an actuarial approach to determine the appropriate mixture of *in situ* and *ex situ* conservation tactics. According to this view, the effectiveness of *in situ* tactics is determined by the "half-life" of native habitats and protected areas (or reserves). Where the expected lifetime of such areas do not meet the demographic requirements for security, Soulé argues that *ex situ* tactics can be used to reach conservation objectives.

In many respects, the current status of some populations of North American and European salmonids may serve as a useful model of the need for dynamic application of Soulé's approach. Native habitats of most threatened salmonid populations have been extensively modified by fisheries, hydroelectric power facilities and/or environmental (e.g., acid rain) policies. The likelihood of continued habitat modification and the critical nature of population declines dictate that *ex situ* management approaches provide the best (or perhaps only) chance for some species. *Ex situ* management offers the only hope for some populations until the current crises can be resolved and sufficient genetic diversity can be preserved for future reintroduction into restored native habitat. Once viable populations are restored in secure reserves, less intensive *in situ* programs can be used to manage the population.

This paper describes the most secure *ex situ* method for preserving the genetic diversity of a population, namely, the cryopreservation, storage and use of animal germ plasm in the form of embryos. The current status of embryo cryopreservation is described, including the theory and practice of the two approaches currently used to cryopreserve embryos and practical obstacles to applying embryo cryopreservation to salmonid fishes.

2. Current Status of Embryo Cryopreservation

The cryopreservation and storage of animal germ plasm (spermatozoa, embryos and oocytes) at low temperatures offers unique opportunities for *ex situ* conservation programs (Rall, 1991a; Ballou, 1992). The ability to place germ plasm into a state of suspended animation for any desired period provides a powerful technique to preserve the genetic diversity in a current population. Once preserved, germ plasm banks can be used to "infuse" genetic diversity into future generations at any time. Germ plasm, preserved in the form of embryos, offers advantages for conservation. The most important of these is that the entire genomic constitution is known at the time of cryopreservation. The basic requirements for an effective embryo banking program are: (1) an action plan that integrates the goals and strategies of the *in situ* and *ex situ* components of the conservation program, (2) suitable reproductive and cryopreservation biotechniques for collecting, preserving, storing and using embryos, and (3) systematic sampling schemes for ensuring that most of the genetic diversity in the population is preserved.

The successful cryopreservation and banking of animal germ plasm evolved from Polge et al.'s discovery in 1949 that glycerol protects spermatozoa from the harmful effects of freezing and thawing. Since that time, a wide variety of biological cells have exhibited high survival following controlled freezing and storage at temperatures below -150°C (see Mazur, 1984). Although the first reports of the successful cryopreservation of embryos appeared in the late 1960s, it was Whittingham et al.'s report of normal offspring from cryopreserved mouse embryos in 1972 that marked the breakthrough in embryo cryopreservation. In the twenty years since that report, basic and applied research has yielded similar success with 13 other species of mammals and one species each of insect and marine rotifer (Table 1).

The ability to cryopreserve embryos and store them at low temperatures for any desired period of time has led to important applications. Three examples of practical uses are: (1) Banks

Table 1. Reports of Successful Embryo Cryopreservation Leading to Normal Live Offspring.

	Species	First Report
Mammalian:		
	Mouse	Whittingham et al., 1972
	Cattle	Wilmut & Rowson, 1973
	Rabbit	Bank & Maurer, 1974
	Sheep	Willadsen et al., 1976
	Rat	Whittingham, 1975
	Goat	Bilton & Moore, 1976
	Horse	Yamamoto et al., 1982
	Eland	Kraemer et al., 1983
	Human	Trounson et al., 1983
	Baboon	Pope et al., 1984
	Marmoset	Summers et al., 1986
	Cynomolgus monkey	Balmaceda et al., 1987
	Cat	Dresser et al., 1988
	Pig	Hayashi et al., 1989
Insect:		
	Fruit fly (*Drosophila melanogaster*)	Steponkus et al., 1990
Sea Urchin:		
	Hemicentrotus pulcherrimus	Asahina and Takahashi, 1978
Marine rotifer:		
	Brachionus plicatilia	Okamoto et al. 1987

of embryos from laboratory mice and rats to ensure the continued availability of rare genotypes and mutants for research. The Jackson Laboratory in Bar Harbor, Maine, maintains 272 strains of genetically-unique, but infrequently used, mice as cryopreserved embryos stored in liquid nitrogen (L.E. Mobraaten, personal communication). (2) Embryo cryopreservation has become an integral component of the cattle embryo transfer industry and is used to transport cattle germ plasm internationally. And (3) embryo cryopreservation is increasingly used as an adjunct of *in vitro* fertilization techniques in the treatment of human infertility.

In the past 20 years, two approaches have been developed to preserve embryos at low temperatures. The first, termed "controlled slow freezing," evolved from Polge et al.'s 1949 discovery of the cryoprotective properties of glycerol and subsequent research during the 1950s and 1960s with microorganisms and mammalian tissue-culture cells. The second approach, termed "vitrification", was originally proposed by Luyet (1937; 1940), but practical application of this approach was achieved 45 years later (Rall and Fahy, 1985; Rall, 1987).

3. Controlled Slow-Freezing Procedures

Controlled slow-freezing procedures are characterized by: (1) the addition of molar concentrations of glycerol or another cryoprotectant to the cell suspension; and (2) the use of a controlled rate of freezing to the storage temperature. The basic steps required to preserve embryos by this approach are listed in Table 2. Cryopreservation invariably results in stresses that lead to a small decrease at least in embryo viability. Therefore, the first step is to ensure that only high quality embryos at the proper developmental stage are processed further. The next step is to suspend the selected embryos in buffered saline containing a cryoprotective solute. For example, mouse embryos usually are collected at the 8-cell stage and then are suspended in buffered saline containing 1.5 molar glycerol or DMSO (see Wood et al., 1987, for detailed protocol). Once the cryoprotectant has partially permeated the blastomeres, the suspension is then cooled and frozen to temperatures below about -120°C. The appropriate cooling and freezing conditions depend upon the specific embryo properties and must be controlled within close tolerances. In our mouse embryo example, it takes approximately 15 min for glycerol to fully permeate into the cells at 20°C. Then, the mouse embryo suspensions are cooled to -7°C and seeded with ice crystals to ensure that ice forms at near equilibrium conditions. Finally, the seeded suspensions are cooled slowly at 0.5°C/min to - 40°C before transfer into liquid nitrogen (-196°C).

The long-term viability of frozen suspensions is assured only when the storage temperature remains below -130°C. Submersion in the liquid phase of a liquid nitrogen refrigerator is the most practical storage method. At the end of the storage period, the suspensions are warmed and thawed using controlled conditions. For most embryo suspensions, the container is transferred directly into warm water (20 to 37°C) until the suspension thaws. Next, the cryoprotectant is removed from the suspension. The large size and low permeability of most embryos to cryoprotectants usually increases the risk of osmotic shock during this step. Mouse embryos are diluted by one of two methods (Leibo, 1984). In the first, the concentration of cryoprotectant in the suspension is reduced in a series of 4 to 6 equimolar steps by dilution with isotonic saline at 5 min intervals (stepwise dilution). In the second method, embryos are placed in a saline containing sucrose (0.5 to 1 molar) and held until the cryoprotectant leaves the cells (sucrose dilution). This process requires about 10 minutes at 20°C for mouse embryos. Then, the embryos are rehydrated in isotonic saline. Once the cryoprotectant has been removed, the embryos are returned to normal physiological conditions. The development of a successful controlled freezing protocol for embryos (or any other cells) is complicated by many interacting factors (reviewed by Mazur, 1984) that can be divided into two groups. The first is related to the specific intrinsic properties of the embryos in question. These include the permeability characteristics of the cells to water and cryoprotectants, the size of the blastomeres, the presence of any special physiological considerations (e.g. a sensitivity to cold shock or chilling) and any heterogeneity in the properties of individual blastomeres or embryos in the suspension. These cellular properties can vary enormously for different types of cells (e.g. spermatozoa versus embryos) and the same cell type isolated from different species. The second group of factors is related to the procedural steps or conditions that have been selected to prepare and cryopreserve the suspension. Examples include the type of freezing container, the type and concentration of cryoprotectant, and the rates of cooling and warming. Successful cryopreservation is influenced strongly by interactions between these diverse factors. As a result, the steps and conditions of controlled freezing for each type of embryo or cell must be adjusted to minimize the many diverse sources of injury.

The interaction of cellular permeability with other cryobiological factors is one of the most important considerations for developing effective freezing protocols (Mazur, 1970). First, the permeability of the cell membrane to water interacts with the rate of cooling (see third step of controlled freezing, Table 2) and determines the rate at which water leaves the cytoplasm by exosmosis during freezing (Mazur, 1963). Optimum survivals are achieved when the rate of cooling allows loss of most of the cell water. Cells cooled too rapidly or too slowly usually exhibit lower survival. Rapid cooling fails to provide sufficient time for the cells to dehydrate and avoid the detrimental affects associated with intracellular freezing. In contrast, slow cooling results in excessive dehydration and subjects cells to "solution-effects" injury (Mazur, 1984).

Another interaction between the permeability properties and procedural factors occurs during the second step of freezing (Table 2). The permeability properties of the cell membrane

Table 2. Steps of Embryo Cryopreservation by Controlled-Rate Freezing.

1.	Collect and assess embryo quality.
2.	Equilibrate embryos in a solution containing molar concentrations of a cryoprotective solute (e.g., glycerol, DMSO).
3.	Freeze embryo suspension using controlled cooling to temperatures below -130°C.
4.	Low temperature storage at -196°C.
5.	Warm and thaw embryo suspension using controlled conditions.
6.	Remove cryoprotective solute from embryo suspension.
7.	Return embryos to normal physiological conditions.

to cryoprotectants interact with the conditions (especially time and temperature) selected for equilibration to determine the extent of cryoprotectant permeation into the cytoplasm (Mazur and Miller, 1976). The amount of cryoprotectant permeating into cells before freezing plays an important role in determining the extent of cryoprotection (Taylor et al., 1974). The third interaction occurs when embryos are diluted out of the cryoprotectant solution (see step 6, Table 2). The amount of cryoprotectant in the cell interacts with the dilution procedure to determine the extent of osmotic swelling when the suspending solution is diluted (Levin and Miller, 1981). The most important consequence of these and other interactions is that each step of the freezing process must be optimized to yield high post-thaw survival.

Successful cryopreservation protocols for embryos and cells usually produces a characteristic sequence of changes in the osmotic volume of the cells (Leibo, 1977; Rall, 1991b). The sequence of changes occurring during each step of controlled freezing (Table 2) is shown in Figure 1. No changes in cell volume would be expected during the first step (collection) provided that the embryos are suspended in an isotonic saline. However, during the second step, a transient shrink-swell change in the volume of cells occurs when exposed to a saline containing molar concentrations of a permeable cryoprotectant, such as glycerol. The initial shrinkage results from the exosmosis of water as the cell restores osmotic equilibrium with the hypertonic cryoprotectant solution. Then the cell gradually swells to its original volume as the cryoprotectant permeates the cytoplasm. The time required for a cell to shrink and return to its original volume depends on the permeability of the cell membrane to water and cryoprotectant (Mazur et al., 1974). The permeation properties of cells to cryoprotectants varies widely depending on the type of cell and the size and chemical nature of the molecules. For example, small molecules, such as methanol, usually permeate mammalian embryos very rapidly (<1 min, Rall et al., 1984), whereas large molecules, such as sucrose, do not permeate at all (Leibo, 1984).

During the third step, the cells progressively shrink when frozen by an optimized cooling procedure. This shrinkage is a consequence of a gradual increase in the suspending solution's osmolality when ice forms and gradually grows during cooling. Cells restore osmotic equilibrium with the freeze-concentrated suspending solution by exosmosis of water from the cytoplasm (Mazur, 1970). When the temperature decreases to about -120°C, the freeze-con-

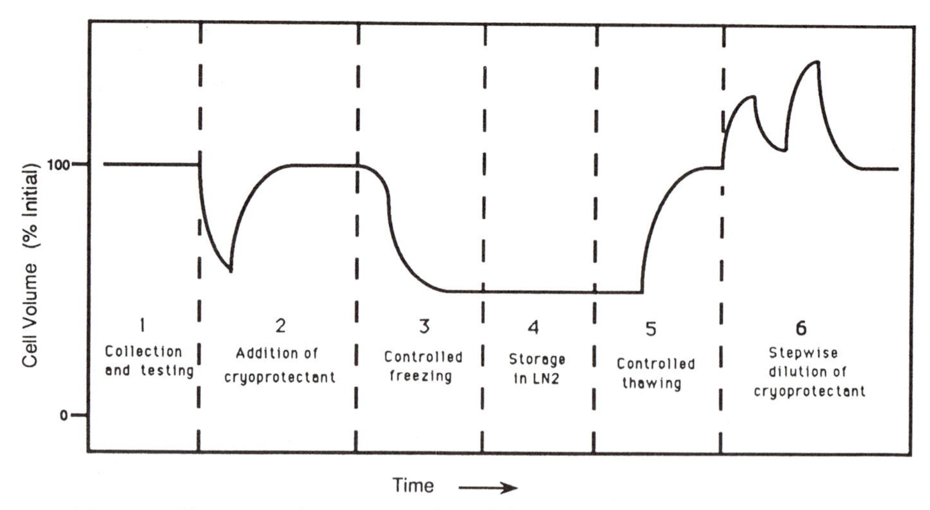

Figure 1. Diagrammatic representation of the cell volume changes during each step of successful controlled freezing. These characteristic changes provide a useful guide for optimizing each step of cryopreservation for any type of embryo or cell. See Table 2 and text for details. [Figure reprinted from Rall (1991b) with modifications.]

centrated residual liquid in the extracellular solution and the dehydrated cytoplasm solidify (vitrify) into a glass (Rall et al., 1984). An example of the osmotic behavior of mouse embryos during controlled freezing is shown in Figure 2.

During the fourth step of freezing (low-temperature storage), the cell volume does not change because the entire suspension has solidified completely. However, important osmotic changes occur when cell suspensions are thawed and the cryoprotectant is diluted from the suspension (steps 5 and 6, respectively). Controlled warming results in a gradual decrease in the osmolality of the suspending solution as extracellular ice melts. Cells undergo a gradual increase in volume as water flows into the cytoplasm to restore osmotic equilibrium. It is important that the rate of warming be sufficiently high to prevent further crystallization or recrystallization of ice (Mazur, 1984). In general, embryo suspensions are usually warmed much more rapidly than they are cooled. Rapid warming results in complete thawing of the suspension before significant embryo rehydration. Dilution of a thawed suspension with isotonic saline usually results in a transient osmotic swelling. The initial swelling results from the movement of water into the cells to restore osmotic equilibrium with the hypotonic suspending solution. Once equilibrium is restored, the cells gradually shrink as cryoprotectant leaves the cytoplasm. Excessive osmotic swelling is prevented by reducing the concentration of cryoprotectant in a series of small steps. The appropriate step size and interval between steps is determined by the permeability of the cell membrane to cryoprotectant and water (Schneider and Mazur, 1984).

An alternative dilution approach (not shown in Figure 1) is the so-called sucrose procedure (Leibo, 1984). Cells are placed into a saline containing a 0.25 to 1 molar concentration of an impermeable solute (usually sucrose). The absence of cryoprotectant in the suspending solution allows rapid efflux of cryoprotectant from the cytoplasm. The impermeable solute prevents (or reduces) the transient increase in cell volume because it increases the osmolality of the suspending solution. As the cryoprotectant leaves the cytoplasm, the cell progressively shrinks due to the hypertonic extracellular solution. Once the cryoprotectant leaves the cells, the suspending solution is replaced with isotonic saline.

4. Vitrification Approaches

The most important difference between vitrification and controlled rate freezing is the method used to dehydrate cells before low-temperature storage (Rall, 1987). Vitrification approaches produce osmotic dehydration by placing cells in a highly concentrated solution of cryoprotectant (>6 molar) *prior to cooling*. Then the entire cell suspension is transformed from the liquid state into a glassy solid by cooling to the storage temperature (-196°C). Vitrification offers considerable promise for simplifying and improving the cryopreservation of cells because potential injury associated with ice formation in the suspension is eliminated (Rall and Fahy, 1985). Vitrification has been applied to a wide variety of embryos (Table 3). Most of the reports have been for mammalian systems that have been successfully cryopreserved by controlled freezing methods. However, in one case, *Drosophila* embryos, vitrification has succeeded for cryopreservation, where controlled freezing has, so far, failed (Steponkus et al., 1990).

Despite differences in the method used to produce osmotic dehydration, the basic steps of vitrification (Table 4) are similar to those of controlled rate freezing (Table 1). In fact, the development of a successful vitrification protocol is complicated, in large part, by the same factors that influence controlled freezing methods (see above). Most of the differences are related to the need for a vitrification solution consisting of one or several cryoprotectants. The greatest challenge in developing a successful vitrification protocol is to formulate a vitrification solution that satisfies two requirements (Fahy et al., 1984; Rall, 1987). The first is related to the physical-chemical properties of the vitrification solution; it must be sufficiently concentrated to avoid crystallization during cooling and to vitrify into a glassy solid. The second is to match the choice of cryoprotectants with the intrinsic permeability and toxicity properties of the cells in question. Ideally, at least one of the cryoprotectants in the vitrification solution should permeate the cytoplasm, but the overall composition must not produce excessive osmotic stress or chemical toxicity (Rall, 1987). Inappropriate cryoprotectants are those that permeate very rapidly, or very slowly, or that produce toxic injury at high concentration. A vitrification solution that is appropriate for an embryo of one species and developmental stage may be inappropriate for embryos of other species or developmental stages, primarily because of differences in cellular permeability characteristics.

Table 3. Reports of the Successful Vitrification of Embryos.

Species	Reference
Mammalian embryos:	
Mouse*	Rall and Fahy, 1985
Cattle*	Massip et al., 1986
Hamster oocytes	Critser et al., 1986
Rat*	Kono et al., 1988
Rabbit*	Smorag et al., 1989
Goat*	Yuswiati and Holtz, 1990
Sheep*	Schiewe et al., 1991
Cat	Rall et al., unpubl.
Plant somatic embryos:	
*Asparagus officinalis**	Uragami et al., 1989
Insect embryos:	
*Drosophila melanogaster**	Steponkus et al., 1990

*Normal live offspring reported from vitrified embryos.

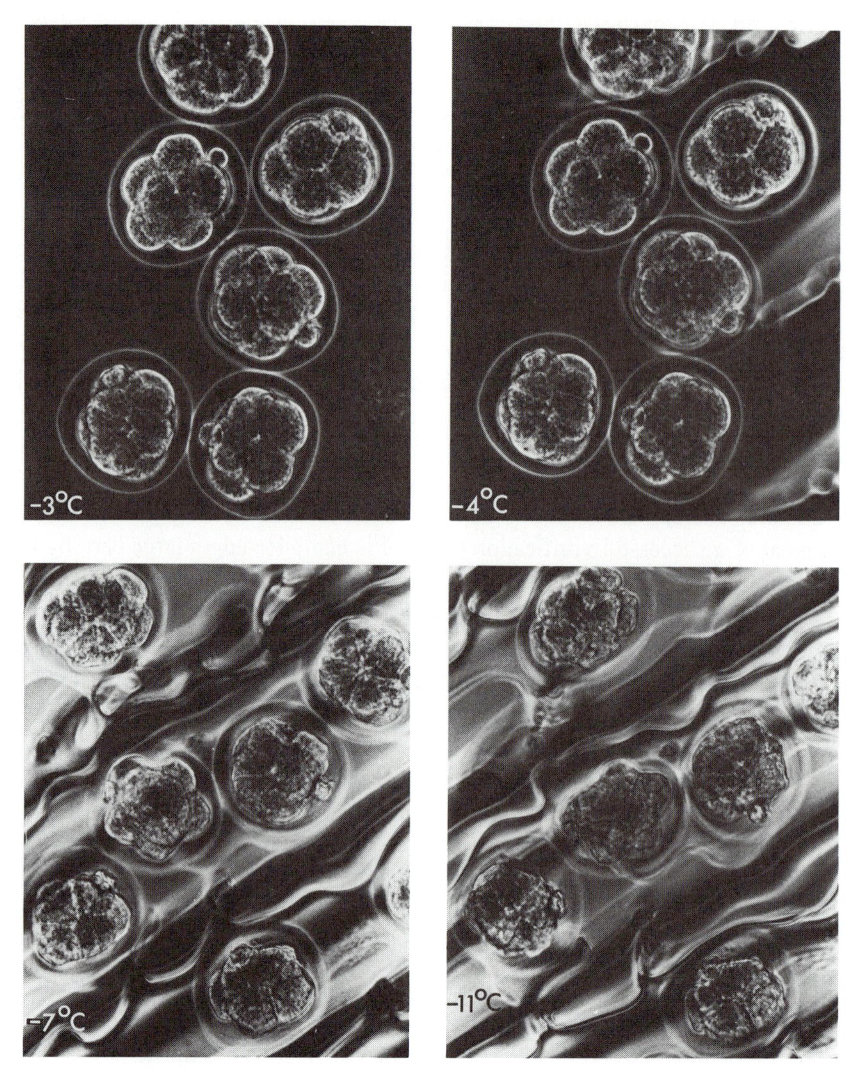

Figure 2. Microscopical appearance of six 8-cell mouse embryos during controlled slow freezing. Briefly, embryos were equilibrated in 1.5 molar glycerol in saline until the glycerol fully permeated into the cells. Embryos were then placed onto the stage of a cryomicroscope and frozen at 0.5°C/min to -31°C and photographs were taken at the indicated temperatures. During controlled slow freezing ice began to grow around the embryos at -4°C. The embryos gradually shrank to approximately half their original diameter as the suspending solution gradually crystallized. The embryos were then cooled rapidly to about -147°C. Each embryo is approximately 75 μm in diameter. [Figure reprinted from Rall and Polge (1984) with modifications.]

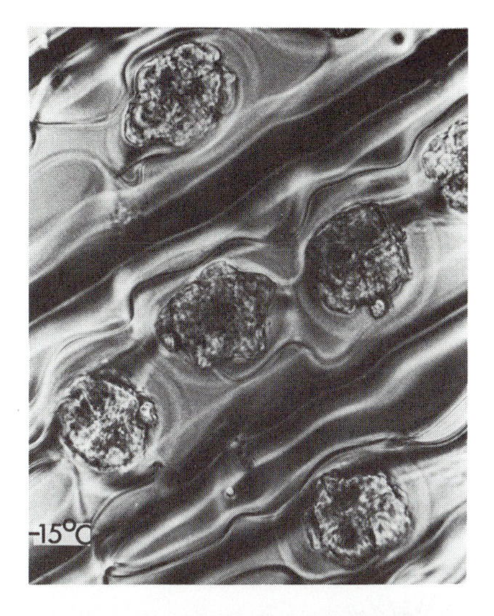
-15°C

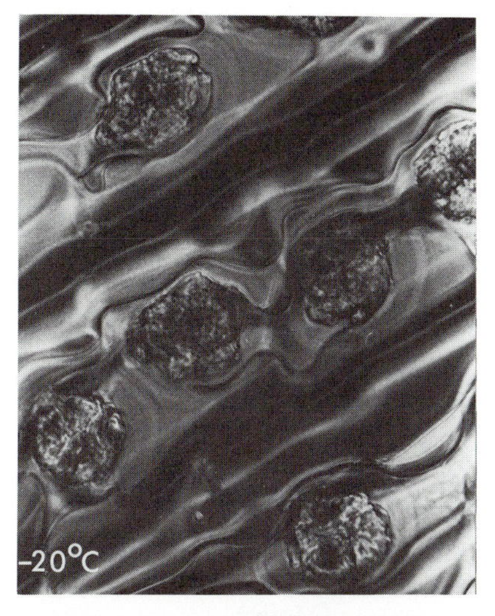
-20°C

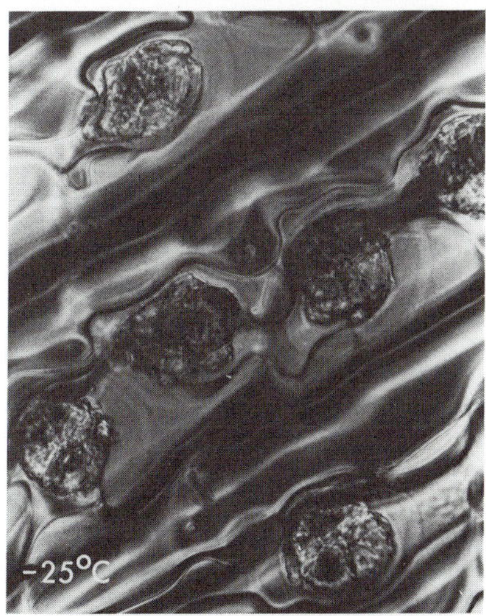
-25°C

-31°C

-147°C

Table 4. Steps of Embryo Vitrification.

1. Collect and assess embryo quality.
2. Equilibrate and dehydrate embryos in a concentrated, but nontoxic, solution of cryoprotectants (Vitrification Solution).
3. Vitrify embryo suspension by cooling to temperatures below -130°C.
4. Low temperature storage (-196°C).
5. Warming and softening of glassy solution into the liquid state.
6. Immediate removal of cryoprotective solutes from embryo suspension.
7. Return embryos to normal physiological conditions.

Table 5. Examples of Vitrification Solutions for Embryo Cryopreservation.

	Cryoprotectant		
Solution Name	Permeating (Molar)	Nonpermeating (% Wt./Vol.)	Reference
VS1	DMSO (2.62 M) Acetamide (2.62 M) Propylene glycol (1.3 M)	PEG (6%)	Rall, 1987
VS2	Propylene glycol (5.5 M)	PEG or BSA (6%)	Rall, 1987
VS3	Glycerol (6.5 M)	PEG or BSA (6%)	Rall, 1987
Massip's VS	Glycerol (2.2 M) Propylene glycol (3.2 M)	BSA (0.4%)	Massip et al., 1986
Kasai's EFS	Ethylene glycol (7.2 M)	Ficoll (18%) Sucrose (0.3 M)	Kasai et al., 1990

Each solution also contains an isotonic saline.
Key: DMSO = dimethyl sulfoxide; PEG = polyethylene glycol (8000 MW); BSA = bovine serum albumin; Ficoll = polymer of sucrose (70,000 MW).
See Rall (1987), Massip et al. (1986) and Kasai et al. (1990) for details.

4.1. Vitrification Solutions

Vitrification solutions for embryos (Table 5) have three common features. First, each contains a mixture of low and high molecular weight cryoprotectants. The partial permeation of low molecular weight solutes into the cells protects cells from the potentially harmful effects of cellular dehydration and ensures that the cytoplasm vitrifies during cooling. High molecular-weight polymers protect cells by stabilizing cell membranes and increasing the ability of the solution to vitrify (Fahy et al., 1984; Rall, 1987). Second, the total concentration of solutes in each solution is high. This ensures that the solution vitrifies during cooling to the storage temperature and avoids crystallization (devitrification) during subsequent warming. The concentration required to prevent crystallization depends on the rates of cooling and warming (Fahy et al., 1984). The first three solutions listed in Table 5 were formulated to vitrify when cooled at rates of 20°C/min and avoid devitrification when warmed at rates greater than 100°C/min (Rall, 1987). The remaining solutions require much higher rates of cooling and warming to avoid crystallization (about 1,000 and 2,000°C/min, respectively). And third, each vitrification solution contains an isotonic saline component to provide normal levels of extracellular electrolytes.

4.2. Osmotic Consequences of Vitrification

Successful cryopreservation by vitrification requires careful control of the conditions used to equilibrate embryos in the vitrification solution. The appropriate choice of conditions depends upon the permeability properties of the cells to water and the cryoprotectants in the solution, the toxic sensitivities of the cells, and the effect of temperature on permeability and toxicity. The ultimate aim of the equilibration procedure is to produce a concentrated, dehydrated cytoplasm while controlling the extent of cryoprotectant permeation, osmotic stresses and chemical toxicity.

Optimized vitrification procedures result in a characteristic sequence of changes in the osmotic volume of cells (Rall, 1991b). The sequence of changes in cell volume occurring during each step of the vitrification procedure (Fig. 3) closely parallel those during controlled-rate freezing (Fig. 1). The most important differences occur during the equilibration and cooling steps. Figure 3 illustrates the use of a three-step equilibration procedure to achieve the necessary

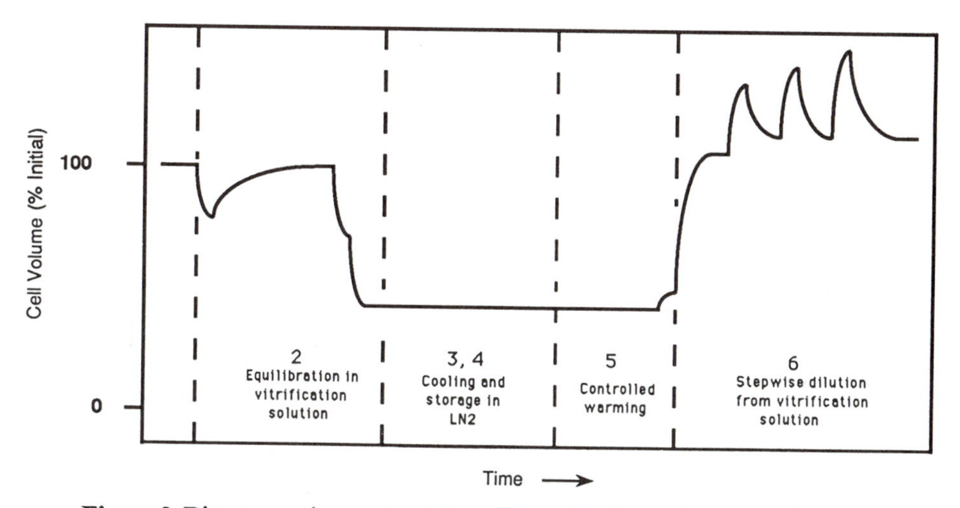

Figure 3. Diagrammatic representation of the cell volume changes during each step of successful vitrification. These characteristic changes provide a useful guide for optimizing each step of vitrification for any type of embryo or cell. See Table 4 and text for details. [Figure reprinted from Rall (1991b) with modifications.]

degree of intracellular dehydration for vitrification. The first step is identical to that used for freezing; embryos are transferred into a saline containing about 1.5 to 2 molar concentration of the same cryoprotectants in the final vitrification solution. The embryo suspension is then held at room temperature until the cryoprotectants permeate into the cells. This results in a shrink-swell change in cell volume. Then the embryos are transferred into the final vitrification solution in two short steps to limit further permeation of cryoprotectants and produce osmotic dehydration. The appropriate conditions for equilibration depend on the permeability of the cells to the cryoprotectants in the vitrification solution. For example, 8-cell mouse embryos require about 20 min for 1.6 molar glycerol to fully permeate into the cells at 22°C. Then, dehydration of the cytoplasm is accomplished by one minute exposures to 4.2 molar glycerol and vitrification solution VS3a at 22°C.

Once the cytoplasm has dehydrated, the entire cell suspension is vitrified by cooling to temperatures below -130°C. Therefore, no change in cell volume occurs during cooling and storage in liquid nitrogen. Vitrified suspensions usually require rapid warming to prevent crystallization from the physical-chemical process called devitrification (Rall, 1987). There is no change in cell volume during warming until about 0°C. Once thawed, the cryoprotectants in the suspension must be diluted immediately to prevent further permeation and reduce the likelihood of toxicity (Rall and Fahy, 1985). Figure 3 shows the effects of stepwise dilution of the suspension immediately after warming. As in the case of controlled-rate freezing (Fig. 1), embryos undergo a transient osmotic swelling when the concentration of cryoprotectants is reduced. The reasons for cellular swelling during stepwise dilution are the same as those described earlier during controlled-rate freezing. Cells diluted from vitrification solutions often require more steps to prevent excessive swelling due to the presence of higher amounts of cryoprotectants in their cytoplasm (Rall, 1987). The use of sucrose dilution procedures for vitrified embryos has been very effective (Rall, 1987; Massip et al., 1986; Schiewe et al., 1991).

5. Cryopreservation of Fruit Fly (*Drosophila melanogaster*) Embryos by Vitrification

A recent report on the vitrification of fruit fly embryos is an example of some of the inherent species-specific problems that can be encountered (Steponkus et al., 1990). The approaches used to overcome impediments to cryopreserving large-sized insect embryos may provide useful insights for attempts to cryopreserve salmonid embryos.

5.1. Features of Drosophila Embryos That Complicate Cryopreservation

Three intrinsic properties of *Drosophila* embryos complicate cryopreservation. First,

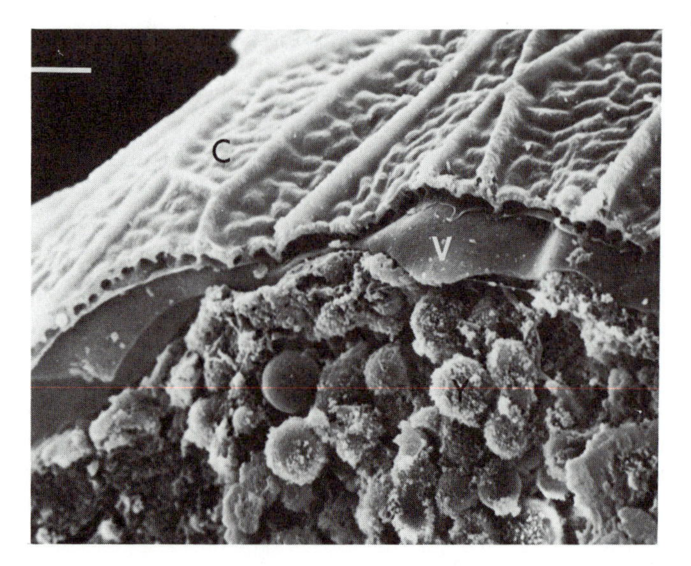

Figure 4. Scanning electron micrograph of freeze-fractured *Drosophila melanogaster* embryo. The major membrane systems, chorion (C), vitelline membrane (V) and intracellular yolk spheres (Y), are visible. The length of the scale bar is 5 μm. [Figure reprinted from Fullilove et al., 1978.]

they are large ovoids, approximately 0.5 mm long and 0.15 mm wide, with a flattened dorsal surface and slightly convex ventral surface. One consequence of this large size is a low surface area to volume ratio compared to mammalian embryos. This reduces the rate at which water and cryoprotectants can flow into and out of the embryo during the steps of cryopreservation (Mazur, 1984). For example, theoretical calculations suggest that *Drosophila* embryos require very low rates of controlled freezing (<0.5°C/min) to ensure adequate osmotic dehydration and avoid the deleterious effects of intra-cellular freezing (Lin et al., 1989).

Second, a complex membrane system surrounds and isolates embryos from their environment from fertilization to hatching 24 h later, (Fig. 4). The membranes of native embryos are impermeable to solutes but allow gas exchange and limited movement of water vapor. The outer membrane, called the eggcase or chorion, is a tough, opaque barrier that surrounds an inner membrane, called the vitelline membrane. The vitelline membrane is covered with a waxy layer which is thought to be the primary barrier to water and solute permeation. Native embryos tolerate exposure to solutions of formaldehyde, methanol and

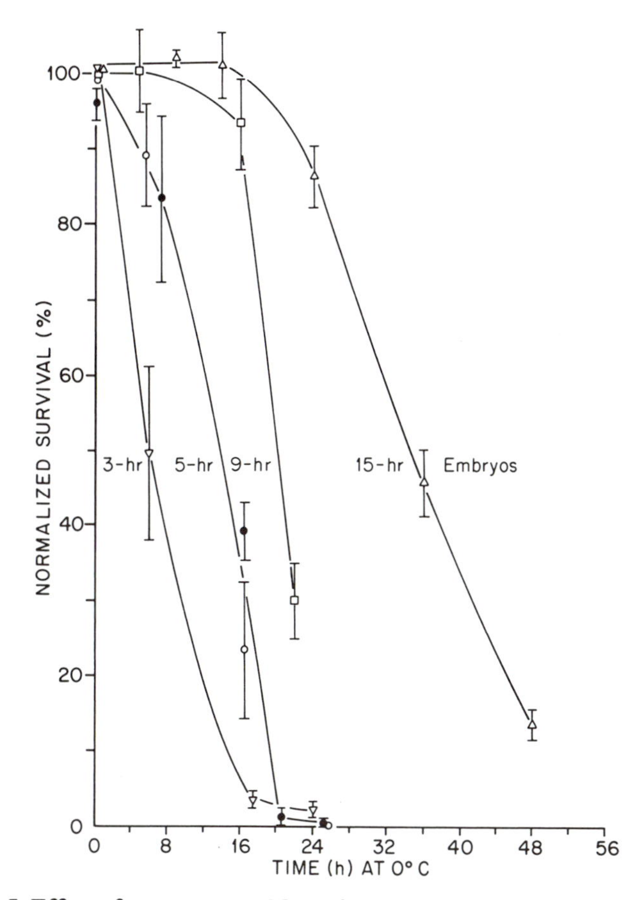

Figure 5. Effect of exposure to 0°C on the survival of 3, 5, 9 and 15 hour-old *Drosophila melanogaster* embryos. Embryos were cooled at 5°C/min from room temperature to 0°C. Survival is based on the percentage of treated embryos that develop to hatched larvae, and were normalized to room temperature controls, which averaged 81%. [Data reprinted from Mazur et al., 1992.]

chloroform with no ill effects (Limbourg and Zalokar, 1978). These permeability barriers must be removed before cryopreservation is possible.

Third, *Drosophila* embryos usually are killed when exposed to low environmental temperature (Myers et al., 1988). The sensitivity of *Drosophila* embryos to chilling injury depends upon their developmental stage. Mazur and colleagues (1992) report that the earlier the developmental stage, the more rapidly embryos die when held at 0°C (Fig. 5). Approximately half of 3, 5, 9 and 15 h old embryos die, respectively, after 6, 14, 20 and 34 h of exposure to 0°C. Although 15 h old embryos tolerate up to 18 h of exposure to 0°C with no ill effects, exposure to subzero temperatures greatly increases the rate of death (Fig. 6). For example, cooling to -30°C for less than 1 minute results in 100% mortality.

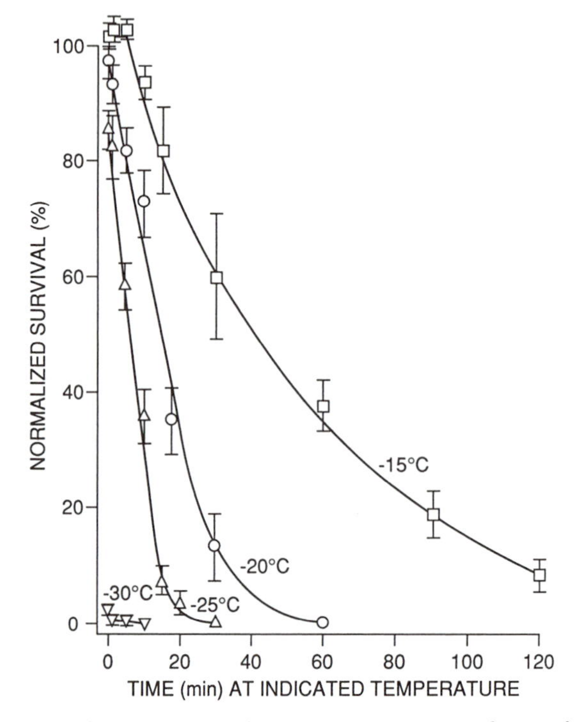

Figure 6. Effect of exposure to subzero temperatures on the survival of 12 to 15 hour-old *Drosophila melanogaster* embryos. Embryos were cooled rapidly to the indicated temperatures, held for the appropriate time and then warmed rapidly to room temperature. Survival is based on the percentage of treated embryos that develop to hatched larvae, and were normalized to room temperature controls, which averaged 82%. [Data reprinted from Mazur et al., 1992.]

5.2. Removal of Permeability Barriers

Perhaps the most important property of *Drosophila* embryos that limits current cryopreservation techniques is their permeability barriers. The chorion can be removed relatively easily (Hill, 1945), but removal of the waxy layers of the vitelline membrane requires controlled exposure to organic solvents (Widmer and Gehring, 1974; Limbourg and Zalokar, 1978).

Lynch et al. (1989) examined many factors related to the permeabilization of *Drosophila* embryos and developed an optimized procedure for cryopreservation. The steps of their procedure are:

1. Solubilize chorion by suspending embryos in 2.6% sodium hypochlorite (50% bleach) for 2 min and then rinse embryos with distilled water.
2. Extract waxy layers by sequential extraction with organic solvents:
   a. isopropanol (20 sec).
   b. n-hexane (30 sec).
   c. rinse with *Drosophila* Ringer's solution.
3. Suspend embryos in isotonic saline during subsequent steps.

Most embryos (80-90%) develop normally to hatched larvae when permeabilized by this procedure, and about 80-90% of the embryos are permeable to water and cryoprotectants. It is important to note that these two populations often do not coincide (i.e. some permeabilized embryos may not be viable and some viable embryos may not be permeabilized). Mazur et al. (1991) recently reported that the repeatability and efficacy of permeabilization can be improved by controlling the amount of alcohol carried from step 2a to 2b. Mazur et al.'s modifications of step 2 are: (a) rinse embryos with isopropanol for 20 seconds; (b) remove all adhering alcohol from embryos by air-drying; (c) rinse embryos in n-heptane containing 0.3% 1-butanol for 90 seconds; and (d) remove residual alkane/alcohol by air drying and rinsing with isotonic saline.

5.3. Cryopreservation Strategy

Initial attempts to cryopreserve permeabilized embryos by controlled freezing methods were unsuccessful (Leibo et al., 1988). Embryos did not survive cooling below -35°C at the low rates (0.2 to 0.5°C/min) required to ensure osmotic dehydration. The low rates of cooling resulted in long periods of exposure to subzero temperatures, and the lack of survival presumably resulted from the high sensitivity of *Drosophila* embryos to chilling injury (see Fig. 6). The inability to prevent this injury during controlled-rate freezing precluded this approach for cryopreservation. Vitrification approaches were examined in the hope that rapid cooling would preclude the kinetic processes associated with chilling injury (Steponkus et al., 1990).

Fortunately, chilling injury could be prevented by using ultra-rapid cooling (≥4,000°C/min) in combination with vitrification. After examining a range of developmental stages, cryoprotectants, equilibration procedures, cooling and warming conditions, vitrification solutions and dilution procedures, a successful vitrification procedure was identified (Steponkus et al., 1990). The specific conditions and procedures are listed in Table 6.

The ability of embryos to survive and develop to hatched larvae following various steps of the vitrification procedure is shown in Table 7. Control replicates indicate that permeabilization and equilibration in 2.125 molar ethylene glycol do not significantly reduce the rate of development of embryos to hatched larvae (88% and 80% survival, respectively). However, these steps in combination with dehydration in the vitrification solution resulted in a significant decline in survival (55%). An even smaller proportion of embryos (6.8%) developed to larvae after dehydration and cooling at 4,000°C/min in liquid nitrogen. Higher survivals were obtained when embryos were dehydrated and cooled rapidly in either liquid propane or partially-solidified liquid nitrogen (9.4% and 18.3%, respectively). The high level of variation in survival observed in these experiments probably reflects the difficulty in producing embryo suspensions with the same permeability characteristics before cryopreservation (Mazur et al., 1991).

The process used to develop an effective cryopreservation procedure for *Drosophila* embryos provides important lessons for other biological materials. First, innocuous chemical and/or physical modifications of membrane systems may eliminate intrinsic permeability barriers that limit cryoprotectant permeation or osmotic dehydration. Second, the entire life cycle of an animal must be examined to determine the most appropriate developmental stage for cryopreservation. In this case, 12 to 15 h *Drosophila* embryos were selected for two reasons: (a) embryos at this stage exhibited the lowest sensitivity to chilling injury; and (b) embryos begin to develop a new membrane system (cuticle) at about 16 h that limits permeation of cryoprotectants and water. The last important lesson is that vitrification, when used in

Table 6. Cryopreservation of *Drosophila melanogaster* Embryos by Vitrification.

1. Age of embryos: 13 to 15 hours.
2. Vitrification solution: 8.5 M ethylene glycol plus 6% (wt/vol) bovine serum albumin in *Drosophila* medium.
3. Permeabilization of embryos by method listed in text.
4. Stepwise equilibration in vitrification solution:
    a. Place embryos into 2.125 molar ethylene glycol plus 6% BSA in *Drosophila* medium for 20 min at 22°C.
    b. Place embryos in vitrification solution for 8 min at 0°C.
5. Ultra-rapid cooling of about 25 embryos on an electron microscope grid to -196°C (4,000 to 25,000°C/min).
6. Storage in liquid nitrogen (-196°C).
7. Rapid warming (1,000°C/min) and rapid dilution of cryoprotectant at 22°C.

Table 7. Survival of *Drosophila melanogaster* Embryos after Various Steps of the Vitrification Procedure.

Step*	Treatment Description	Survival** (Mean ± SD)
3	Permeabilization control	88.0% ± 5.2
4a	Equilibration in 2.125 molar control	80.0% ± 7.7
4b	Dehydration in vitrification solution control	55.0% ± 13.3
5	Vitrification in liquid nitrogen (-196°C)	6.8% ± 4.4
5	Vitrification in liquid propane (-196°C)	9.4% ± 7.5
5	Vitrification in liquid nitrogen slush (-204°C)	18.3% ± 8.9

*See Table VI and text for details.
**Values represent the means of at least 30 replicates of ≥100 embryos per replicate. (Data from Steponkus et al., 1990)

combination with ultra-rapid cooling, may provide unique opportunities to avoid the deleterious effects of chilling injury.

6. Factors Complicating the Application of Embryo Cryopreservation to Salmonid Fishes

Little progress has been made to date in applying embryo cryopreservation procedures to salmonids (see Stoss, 1983 for review). This probably reflects four specific features of salmonid fish embryos that present serious obstacles to cryopreservation. First, salmonids, like most teleosts, have a complex membrane system that plays an important role in controlling the immediate environment of the embryo (Groot and Alderdice, 1985). Immediately after shedding, salmonid eggs are flaccid and surrounded by a proteinaceous membrane, the chorion or egg case. When transferred into water, the egg undergoes a series of changes independent of fertilization, called water activation or hardening (Hayes, 1949). Water activation results in the release of cortical granules and colloidal materials from the cortical alveoli into the future perivitelline space. The perivitelline fluid is produced when water flows across the chorion in response to an osmotic pressure gradient between the external water and the colloids. Within about 3 h, net flow of water across the chorion ceases due to a counteracting hydrostatic pressure within the perivitelline space. The intra-embryonic pressure of salmonids after water hardening varies from 20-50 mm Hg (Eddy, 1974; Alderdice et al., 1984) and yields turgid embryos that resist crushing when covered by gravel (Hayes, 1949). The presence of a chorion ranging in salmonids from 28 to 61.6 μm thick (Groot and Alderdice, 1985), large amounts of colloidal material in the perivitelline fluid (42% solids by weight; Eddy, 1974), and membranes surrounding both the embryo and yolk constitute a complex system for developing effective procedures for cryoprotectant permeation and osmotic dehydration.

Second, the embryos of salmonid fishes are very large ellipsoids. Groot and Alderdice (1985) report that the equivalent spherical diameters of five species of Pacific salmon range from 5.96 mm for sockeye (*Oncorhynchus nerka*) to 8.67 mm for chinook (*O. tshawytscha*). When compared to *Drosophila* embryos, the diameter and volume of salmonid embryos are, respectively, approximately 20-fold and 10^3-10^4 fold greater. This large size results in a much lower surface area to volume ratio when compared to *Drosophila* or mammalian embryos. One consequence of such a low ratio is a reduction in the rate at which water and cryoprotectants can move into and out of the embryo during the steps of cryopreservation (Mazur, 1984).

A third complicating feature is the presence of a large quantity of yolk. Yolk provides all nutrients for embryonic development in a high density form (41% solids by weight; Blaxter, 1969) and comprises at least 90% of the dry weight of salmonid embryos prior to hatching (Marr, 1966). During the steps of cryopreservation, yolk probably acts as an independent compartment and responds osmotically in a manner analogous to the cellular cytoplasm. The development of a single effective protocol for cryoprotectant permeation and osmotic dehydration of the yolk and cell compartments may be difficult due to known large differences in their volume and water contents, and the likelihood that their membranes have different permeability characteristics.

A fourth potentially complicating factor is a sensitivity to chilling injury. At present it is not known if salmonid embryos are injured by subzero temperature exposure in an manner analogous to *Drosophila*. The only relevant data are for unfertilized eggs of *Salmo gairdneri* which reportedly undergo "normal" water hardening following super-cooling or freezing to -20°C (Erdahl and Graham, 1980; Harvey and Ashwood-Smith, 1982). These reports and the fact that salmonid embryos normally develop in cold water (2-15°C) suggests that chilling injury may not be a major problem.

6.1. Strategies for Cryopreservation

It should be emphasized that none of the intrinsic features listed above present insurmountable obstacles to the application of cryopreservation to salmonid embryos. However, considerable basic and applied research is required to investigate the cryobiological implications of each of these features. The first step is to establish a model teleost system to evaluate inherent properties of fish embryos and develop protocols. The characteristics of an ideal model system include: (1) small adult size; (2) short generation interval; (3) the ability to maintain fish and embryos *in vitro* throughout the entire life cycle; (4) the reproductive physiology, embryology and developmental genetics must be well-characterized; (5) eggs, embryos and spermatozoa must be available daily (i.e. nonseasonal breeder); and (6) appropriate reproductive biotechniques must be available (e.g.,oocyte, semen and embryo collection, embryo micromanipulation and survival surgery). Two candidate teleosts, zebrafish (*Brachydanio rerio*) and medaka (*Oryzias latipes*), exhibit many of these features (Laale, 1977; Yamamoto, 1975; Winfield and Nelson, 1991; Grady et al., 1991) and have been the subject of previous cryobiological studies (Harvey and Chamberlain, 1982; Harvey et al., 1983; Harvey, 1983; Arii et al., 1987).

The first specific study is to determine the permeability characteristics during development from eggs and fertilized zygotes to hatching. The permeation of water and a range of potential cryoprotective solutes should be examined, including small molecules (e.g. methanol) and larger solutes (e.g. sugars). Previous reports of water "exchange" in teleosts (Loeffler and Lovtrup, 1970; Harvey and Chamberlain, 1982) and cryoprotectant permeation (Harvey and Ashwood-Smith, 1982; Harvey et al., 1983) greatly underestimated these properties. That is because the osmotically inactive volumes of perivitelline fluid, yolk and cytoplasm were not taken into account in estimating the intraembryonic volume available for water and solutes. As noted above, the solids content of yolk is very high and cytoplasm is somewhat lower (respectively, 41 and 15% by volume). Therefore, the "unavailable volume" should be determined by measuring the Boyle van't Hoff relationship (Leibo, 1980) for each compartment of the embryo.

Another important area for study concerns the membrane systems surrounding the embryo. It is likely that success will require modification or removal of the chorion and perivitelline fluid to speed the permeation of cryoprotectants and water (Harvey, 1983). Current studies of chimera formation, transgenesis and other genetic/embryological manipulations in teleosts may provide useful procedures and approaches. Those studies often require the insertion of micropipets or other instruments through the chorion (Maclean et al., 1987). Alternative approaches, such as disassociation of membranes with proteolytic enzymes or chemicals, have potential for simplifying procedures (Iwamatsu, 1983). A less attractive alternative to the cryopreservation of whole embryos is to isolate embryonic cells from developing embryos for cryopreservation and eventual thawing and transfer into recipient embryos. Under ideal conditions the resulting chimeric embryos will yield chimeric adults that produce viable gametes from the transferred cells. If genetic or developmental barriers prevent the recipient embryo from forming viable gametes (e.g. triploids, Utter et al., 1983; or perhaps homozygous clones, Streisinger et al., 1981), the transplanted cells may serve as the sole source of gametes.

A third area of study is related to potential chilling injury. Studies should be designed to determine the inherent sensitivity of teleost embryos to low temperatures. Studies involving mammalian embryos suggest that a high sensitivity to chilling injury is associated with large amounts of intraembryonic lipids (Polge et al., 1974). Therefore, the sensitivity of the yolk and cell compartments to low temperature should be examined.

7. Conclusions

In conclusion, the development of an effective cryopreservation procedure for teleost embryos has great potential for facilitating the conservation of endangered populations of salmonids. The formation of salmonid embryo banking programs to preserve unique genetic diversity within populations can offer benefits as an integral part of *ex situ* and *in situ* conservation efforts. A concerted effort to systematically examine the feasibility of applying cryobiological procedures to fishes is needed to provide the necessary information for making this possible.

ACKNOWLEDGEMENTS: I thank Dr. D. E. Wildt for comments on the manuscript, Dr. F. R. Turner for prints of Figure 4, and Dr. P. Mazur for supplying Figures 5 and 6. Research supported by grants from NIH (NIGMS R01-GM37575; NIA 1Y01AG10164-01) and Friends of the National Zoo.

References

Alderdice, D.F., J.O.T. Jensen, and F.P.J. Velsen. 1984. Measurement of hydrostatic pressure in salmonid eggs. *Can. J. Zool.* 62:1977-1987.

Arii, N., K. Namai, F. Gomi, and T. Nakazawa. 1987. Cryoprotection of medaka embryos during development. *Zool. Sci.* 4:813-818.

Asahina, E. and T. Takahashi. 1978. Freezing tolerance in embryos and spermatozoa of the sea urchin. *Cryobiology* 15:122-127.

Ballou, J. D. 1992. Potential contribution of cryopreserved germ plasm to the preservation of genetic diversity and conservation of endangered species in captivity. *Cryobiology* 29:19-25.

Blaxter, J.H.S. 1969. Development: Eggs and larvae, in: "Fish Physiology," W. S. Hoar and D. J. Randall, eds., vol. 3, pp. 178-252, Academic Press, New York.

Conway, W. 1988. Can technology aid species preservation?, in: "Biodiversity," E. O. Wilson, ed., pp. 263-268, National Academy Press, Washington, DC.

Critser, J., B.W. Arneson, D.V. Aaker, and G.D. Ball. 1986. Cryopreservation of hamster oocytes: Effects of vitrification or freezing on human sperm penetration of zona-free hamster oocytes. *Fertil. Steril.* 46:277-284.

Eddy, F.B. 1974. Osmotic properties of the perivitelline fluid and some properties of the chorion of Atlantic salmon eggs (*Salmo salar*). *J. Zool. Lond.* 174:237-243.

Ehrlich, P.R. and A.H. Ehrlich. 1981. "Extinction: The Causes and Consequences of the Disappearance of Species," Random Press, New York.

Ehrlich, P.R. and E.O. Wilson. 1991. Biodiversity studies: Science and policy. *Science* 253:758-762.

Erdahl, D.A. and E.F. Graham. 1980. Preservation of gametes of freshwater fish, in: "Proc. 9th International Congress on Animal Reproduction and Artificial Insemination, RT-H-2," pp. 317-326.

Fahy, G.M., D.R. MacFarlane, C.A. Angell, and H.T. Meryman. 1984. Vitrification as an approach to cryopreservation. *Cryobiology* 21:407-426.

Frankel, O.H. and M.E. Soulé. 1981. "Conservation and Evolution." Cambridge University Press, Cambridge.

Fullilove, S.L., A.G. Jacobson, and F.R. Turner. 1978. Embryonic development: Descriptive, in: *"The Genetics and Biology of Drosophila,"* M. Ashburner and T. R. F. Wright, eds., vol. 2C, p. 151, Academic Press, New York.

Grady, A.W., I.E. Greer, and R.M. McLaughlin. 1991. Laboratory management and husbandry of the Japanese medaka. *Lab Anim.* 20:22-28.

Groot, E.P. and D.F. Alderdice. 1985. Fine structure of the external egg membrane of five species of Pacific salmon and steelhead trout. *Can. J. Zool.* 63:552-566.

Hayashi, S., K. Kobayashi, J. Mizuno, K. Saitoh, and S. Hirano. 1989. Birth of piglets from frozen embryos. *Vet. Record* 125:43-44.

Harvey, B. 1983. Cooling embryonic cells, isolated blastoderms and intact embryos of the zebra fish *Brachydanio rerio* to -196°C. *Cryobiology* 20:440-447.

Harvey, B. and M.J. Ashwood-Smith. 1982. Cryoprotectant penetration and supercooling in the eggs of salmonid fishes. *Cryobiology* 19:29-40.

Harvey, B. and J.B. Chamberlain. 1982. Water permeability in the developing embryo of the zebrafish, *Brachydanio rerio. Can. J. Zool.* 60:268-270.

Harvey, B., R.N. Kelley, and M.J. Ashwood-Smith. 1983. Permeability of intact and dechorionated zebra fish embryos to glycerol and dimethyl sulfoxide. *Cryobiology* 20:432-439.

Hayes, F.R. 1949. The growth, general chemistry and temperature relations of salmonid eggs. *Q. Rev. Biol.* 24:281-308.

Hill, D.L. 1945. Chemical removal of the chorion from *Drosophila* eggs. *Drosoph. Inform. Serv.* 19:62.

Iwamatsu, T. 1983. New techniques for dechorionation and observations on the development of the naked egg in *Oryzias latipes. J. Exp. Zool.* 228:83-89.

Kono, T., O. Suzuki, and Y. Tsunoda. 1988. Cryopreservation of rat blastocysts by vitrification. *Cryobiology* 25:170-173.

Laale, H.W. 1977. The biology and use of the zebra fish *Brachydanio rerio* in fisheries research. *J. Fish Biol.* 10:121-173.

Leibo, S.P. 1977. Fundamental cryobiology of mouse ova and embryos, in: 'The Freezing of Mammalian Embryos," K. Elliott and J. Whelan, eds., pp. 69-92, Elsevier, Amsterdam.

Leibo, S.P. 1980. Water permeability and its activation energy of fertilized and unfertilized mouse ova. *J. Membr. Biol.* 53:179-188.

Leibo, S.P. 1984. A one-step method for direct nonsurgical transfer of frozen-thawed bovine embryos. *Theriogenology* 21:767-790.

Leibo, S.P., S.P. Myers, and P.L. Steponkus. 1988. Survival of *Drosophila melanogaster* embryos cooled to subzero temperatures. *Cryobiology* 25:545-546.

Levin, R.L. and T.W. Miller. 1981. An optimum method for the introduction and removal of permeable cryoprotectants. *Cryobiology* 18:32-48.

Limbourg, B. and M. Zalokar. 1978. Permeabilization of *Drosophila* embryos. *Dev. Biol.* 35:382-387.

Lin, T.-T., R.E. Pitt, and P.L. Steponkus. 1989. Osmometric behavior of *Drosophila melanogaster* embryos. *Cryobiology* 26:453-471.

Loeffler, C.A. and S. Lovtrup. 1970. Water balance in the salmon egg. *J. Exp. Biol.* 52:291-298.

Luyet, B.J. 1937. The vitrification of organic colloids and of protoplasm. *Biodynamica* 1(39):1-14.

Luyet, B.J. and P.M. Gehenio. 1940. Life and Death at Low Temperatures, Biodynamica Press, Normandy, Missouri.

Lynch, D.V., T.-T. Lin, S.P. Myers, S.P. Leibo, R.J. MacIntyre, R.E. Pitt, and P.L. Steponkus. 1989. A two-step method for permeabilization of *Drosophila* embryos. *Cryobiology* 26:445-452.

Maclean, N., D. Penman, and Z. Zhu. 1987. Introduction of novel genes into fish. *Bio/Technol.* 5:257-261.

Mann, C.C. 1991. Extinction: Are ecologists crying wolf. *Science* 253:736-738.

Marr, D.H.A. 1966. Influence of temperature on the efficiency of growth of salmonid embryos. *Nature* 212:957-959.

Massip, A., P. van der Zwalmen, and F. Ectors. 1986. Pregnancies following transfer of cattle embryos preserved by vitrification. *Cryo-Letters* 7:270-273.

Mazur, P. 1963. Kinetics of water loss from cells at subzero temperatures and the likelihood of intracellular freezing. *J. Gen. Physiol.* 47:347-369.

Mazur, P. 1970. Cryobiology: The freezing of biological systems. *Science* 168:939-949.

Mazur, P. 1984. Freezing of living cells: Mechanisms and implications. *Amer. J. Physiol.* 247C:125-142.

Mazur, P., K.W. Cole, and A.P. Mahowald. 1991. Critical role of alcohol in the permeabilization of 12-hr *Drosophila* embryos by alkanes. *Cryobiology* 28:524.

Mazur, P., K.W. Cole, P.D. Schreuders, and A.P. Mahowald. 1991. Confirmation of the ability of permeabilized 12-hr *Drosophila* embryos to survive cooling to -200°C. *Cryobiology* 28:524-525.

Mazur, P., R.H. Miller, and S.P. Leibo. 1974. Survival of frozen-thawed bovine red cells as a function of the permeation of glycerol and sucrose. *J. Membr. Biology* 15:137-158.

Mazur, P. and R.H. Miller. 1976. Survival of frozen-thawed human red cells as a function of the permeation of glycerol and sucrose. *Cryobiology* 13:523-536.

Mazur, P., U. Schneider, and A.P. Mahowald. 1992. Characteristics and kinetics of subzero chilling injury in *Drosophila* embryos. *Cryobiology* 29:39-68.

Myers, S.P., D.V. Lynch, D.C. Knipple, S.P. Leibo, and P.L. Steponkus. 1988. Low-temperature sensitivity of *Drosophila melanogaster* embryos. *Cryobiology* 25:544-545.

Okamoto, S., M. Tanaka, H. Kurokura, and S. Kasahara. 1987. Cryopreservation of parthenogenic eggs of the rotifer *Brachionus plicatilia. Nippon Suisan Gakkaishi* 53:2093.

Polge, C., A.U. Smith, and A.S. Parkes. 1949. Revival of spermatozoa after vitrification and dehydration at low temperatures. *Nature* 164:666.

Polge, C., I. Wilmut, and L.E.A. Rowson. 1974. The low temperature preservation of cow, sheep, and pig embryos. *Cryobiology* 11:560.

Rall, W.F. 1987. Factors affecting the survival of vitrified mouse embryos. *Cryobiology* 24:387-402.

Rall, W.F. 1991a. Guidelines for establishing animal genetic resource banks: Biological materials, management, and facility considerations, in: "Proceedings of the Wild Cattle Symposium," D. L. Armstrong and T. S. Gross, eds., pp. 96-106, Henry Doorly Zoo, Omaha, Nebraska.

Rall, W.F. 1991b. Prospects for the cryopreservation of mammalian spermatozoa by vitrification, in: 'Reproduction in Domestic Animals, Supplement 1, Proceedings of the 2nd

International Conference on Boar Semen Cryopreservation," L. A. Johnson and D. Roth, eds., pp. 65-80, Paul Parey Scientific Publishers, Hamburg.

Rall, W.F. and G.M. Fahy. 1985. Ice-free cryopreservation of mouse embryos at -196°C by vitrification. *Nature* 313:573-575.

Rall, W.F. and C. Polge. 1984. Effect of warming rate on mouse embryos frozen and thawed in glyceol. *J. Reprod. Fertil.* 70:285-292.

Rall, W.F., D.S. Reid, and C. Polge. 1984. Analysis of slow-warming injury of mouse embryos by cryomicroscopical and physio-chemical methods. *Cryobiology* 21:106-121.

Schiewe, M.C., W.F. Rall, L.D. Stuart, and D.E. Wildt. 1991. Analysis of cryoprotectant, cooling rate and in situ dilution using conventional freezing or vitrification for cryopreserving sheep embryos. *Theriogenology* 36:279-283.

Schneider, U. and P. Mazur. 1984. Osmotic consequences of cryoprotectant permeability and its relation to the survival of frozen-thawed embryos. *Theriogenology* 21:68-79.

Smorag, V., B. Gajda, B. Wieczorek, and J. Jura. 1989. Stage-dependence viability of vitrified rabbit embryos. *Theriogenology* 31:1227-1231.

Soulé, M.E. 1987. "Viable Populations for Conservation," Cambridge University Press, Cambridge, U.K.

Soulé, M.E. 1991. Conservation: Tactics for a constant crisis. *Science* 253:744-750.

Steponkus, P.L., S.P. Myers, D.V. Lynch, L. Gardner, V. Bronshteyn, S.P. Leibo, W.F. Rall, R.E. Pitt, T.-T. Lin, and R.J. MacIntyre. 1990. Cryopreservation of *Drosophila melanogaster* embryos. *Nature* 345:170-172.

Stoss, J. 1983. Fish gamete preservation and spermatozoan physiology, in: "Fish Physiology," W.S. Hoar, D.J. Randall, and E.M. Donaldson, eds., vol. 9B, pp. 305-350, Academic Press, New York.

Streisinger, G., C. Walker, N. Dover, D. Knauber, and F. Singer. 1981. Production of clones of homozygous diploid zebra fish (*Brachydanio rerio*). *Nature* 291:293-296.

Taylor, R., G.D.J. Adams, C.F.B. Boardman, and R.G. Wallis. 1974. Cryoprotective-permeant vs nonpermeant additives. *Cryobiology* 11:430-438.

Uragami, A., A. Sakai, N. Nagai, and T. Takahashi. 1989. Survival of cultured cells and somatic embryos of *Asparagus officinalis* cryopreserved by vitrification. *Plant Cell Rpts.* 8:418-421.

Utter, F.M., O.W. Johnson, G.H. Thorgaard, and P.S. Rabinovitch. 1983. Measurement and potential applications of induced triploidy in Pacific salmon. *Aquaculture* 33:329-354.

Whittingham, D.G., S.P. Leibo, and P. Mazur. 1972. Survival of mouse embryos frozen to -196 and -269°C. *Science* 178:411-414.

Widmer, B. and W.J. Gehring. 1974. A method for permeabilization of *Drosophila* eggs. *Drosoph. Inform. Serv.* 51:149.

Wilson, E.O. 1988. The current status of biological diversity, in: "Biodiversity," E. O. Wilson, ed., pp. 3-18, National Academy Press, Washington, DC.

Winfield, I.J. and J.S. Nelson. 1991. "Cyprinid Fishes: Systematics, Biology and Exploitation," Chapman and Hall, London.

Wood, M.J., D.G. Whittingham, and W.F. Rall. 1987. The low temperature preservation of mouse oocytes and embryos, in: "Mammalian Development," M. Monk, ed., pp. 255-280, IRL Press, Practical Approach Series, Oxford.

Yamamoto, T. 1975. "Medaka (Killifish) Biology and Strains," Keigaku Publ. Co., Tokyo.

Yuswiati, E. and W. Holtz. 1990. Successful transfer of vitrified goat embryos. *Theriogenology* 34:629-631.

Genetic Resource Banks and Reproductive Technology for Wildlife Conservation

DAVID E. WILDT, ULYSSES S. SEAL* and WILLIAM F. RALL

1. Introduction

Reproduction is fundamental to the continued existence and success of a species. Therefore, reproductive physiology plays a critical role in the emerging field of conservation biology and the preservation of bio- and genetic-diversity. Cryobiology, or low temperature biology, dictates our ability to successfully store animal germ-plasm, tissues and DNA which will be vital to managing both species and genetic variation. Within each of these disciplines, scientists can wield an arsenal of techniques sometimes categorized under the rather broad terms of "assisted reproduction" or "reproductive biotechnology". For more than two decades, there has been much speculation and debate about the potential uses of artificial insemination (AI), embryo transfer (ET), in vitro fertilization (IVF) and "frozen zoos" for species conservation. But, preservation of a species requires routine and efficient production of offspring, and most of the pregnancies generated in various wildlife species using assisted reproduction have been one-time events (Wildt et al., 1992a,b). Also, many failed attempts of artificial breeding go unreported (Wildt et al., 1986).

Although reproductive biotechnology, including the use of frozen gametes and embryos, has not as yet contributed to practical conservation, great strides have been made, enough to suggest that within the next decade artificial breeding will find management application, at least for some wildlife species. If the ultimate goal is to maintain the planet's biological and genetic diversity, then we contend that organized genetic resource banks containing sperm, oocytes, embryos, tissue and DNA will be an integral component of other ongoing conservation efforts. The benefits of such a strategy are enormous, and a primary objective of this chapter is to discuss this potential. We also provide evidence of real scientific progress in terms of what works (and has not worked) in applying these biotechniques to wildlife species. Even if all of these technical advancements were ignored, biologists have benefited immeasurably from their mistakes. Those of us interested in this field now are much more in awe of the overall challenge and much more confident about the strategies needed to allow reproductive biotechnology to contribute to conservation. This also is an exciting time from an organizational perspective. As will be described, systematic networks and formalized programs are in place,

NOAHS Center, National Zoological Park, Smithsonian Institution, Washington, DC 20008-2598, and *Captive Breeding Specialist Group, World Conservation Union, 12101 Johnny Cake Ridge Road, Apple Valley, MN 55124
Telephone for correspondence - (202) 673-4793, Telefax - (202) 673-4733

Genetic Conservation of Salmonid Fishes, Edited by J.G. Cloud and G.H. Thorgaard, Plenum Press, New York, 1993

and national and international cooperation is the norm for dealing with captive and wild species in crisis.

2. Developing Strategies for Conservation (Cross-Institutional Cooperation and Getting Organized)

The viability of wildlife depends, in part, upon the level of genetic variation within species, populations and individuals. Captive wildlife populations are especially vulnerable to additional losses in genetic diversity. Prior to the 1980s, zoos paid little attention to which individual animals were breeding. As a result, many populations became genetically stagnant, a finding that became alarmingly apparent to the zoo community in 1979. Ralls et al., in that year, discovered an extremely high incidence of inbreeding in 29 zoo species which correlated directly with neonatal mortality. Subsequent studies confirmed that creeping genetic monomorphism was adversely affecting the health of many captive as well as free-living species. Coincidentally, the American Association of Zoological Parks and Aquariums developed the concept of Species Survival Plans (SSPs) and, more recently, Taxon Advisory Groups (TAGs) to assist in managing species, populations and individuals for the purpose of ensuring genetic health (Hutchins and Wiese, 1991). The SSP and TAG programs have been bolstered by the Captive Breeding Specialist Group (CBSG) working under the auspices of the International Union for Conservation of Nature and Natural Resources (IUCN or World Conservation Union). The CBSG's mission is global, and its goals are to: (l) organize a global network of people and resources; (2) collect, analyze and distribute information; (3) develop global captive breeding programs; and (4) integrate management programs for captive and wild populations (Seal, 1991). The success of SSPs, TAGs and the CBSG largely relies upon the voluntary efforts of a coherent assembly of experts that includes reproductive specialists. We contend that the development of successful genetic resource banks will depend upon the ability of these specific groups to: (l) interact and formulate sound plans based upon proven technology; and (2) translate research findings into an applied benefit, that is, the conservation of species and genetic diversity. Because of its global mission, the CBSG has formulated guidelines for establishing, operating and reviewing genetic resource banking programs for species conservation (Rall et al., 1991).

3. A Call for Genetic Resource Banking

For more than 12 years, there has been a consistent demand for establishing genetic resource banks for wildlife species.

"New agencies should be established, or existing agencies charged, with the preservation of particular germ plasm resources. Funds should be provided to support these agencies and to train the personnel necessary for the maintenance of these essential resources. What is done for domestic species (e.g., AI, sperm and blastocyst freezing, and implantation) should be done for all species reproducing in captivity." (Conservation of Germ Plasm Resources: An Imperative, National Research Council, Report of Committee on Germ Plasm Resources, National Academy of Sciences, Washington, DC, 1978.)

"Establishment of a program in the U.S. to coordinate the management of animal germ plasm resources would be in the national interest." (Animal Germplasm Preservation and Utilization in Agriculture, Council for Agricultural Science and Technology, Report No. 101, Ames, Iowa, September 1984.)

"Preservation of germ plasm requires that institutions should be developed and/or strengthened for collection, maintenance and dissemination of genetic resources." (U.S. Strategy on the Conservation of Biological Diversity, Inter-Agency Task Force Report to Congress, U.S. Agency for International Development, Washington, DC, 1985.)

"A program of research could be administered through the National Science Foundation and channel funds to both basic studies on the reproductive biology and cryobiology of wild animals and to applied studies on the control of reproduction, AI and ET. Another approach could be establishing a few centers for study of the reproductive biology of wild animals. These centers could serve as focuses for programs of basic and applied research. They should be sufficiently well-funded to allow broad programs of research on-site as well as extramural research with cooperating institutions. These centers could likewise serve as repositories for frozen gametes and embryos from endangered populations". (Technologies to Maintain Biological Diversity, U.S. Congress, Office of Technology Assessment, Report No. OTA-F-330, U.S. Government Printing Office, March, 1987.)

"High priority studies include... gamete and developmental biology and the collection, evaluation and long-term storage of genetic material including gametes and embryos. Successful cryobiology will have a major impact on conserving genetic diversity. A resource of frozen semen and embryos could be used interactively with living populations to periodically infuse genetic material from captive or wild animal stocks or to instill captive populations with thawed genes from previous generations. The cryopreservation and storage of haploid gametes and diploid embryos and cell cultures at low temperatures (-196°C) offer unique opportunities for facilitating the propagation of wild species and ensuring the conservation of genetic diversity." (Research Priorities for Single Species Conservation Biology, a workshop sponsored by the National Science Foundation and the National Zoological Park, Washington, DC, Wildt and Seal, editors, 1989.)

The potential of such resources already has been demonstrated by the successful preservation of germ plasm from important domestic food animals, companion animals and crop plants. Private enterprise as well as some governmental actions have been initiated to collect, protect and use these agricultural resources. For example, the commercial distribution of frozen cattle semen has been in place for more than 30 years. Recently, similar commercial efforts have spread to the routine use of frozen cattle embryos, frozen rodent embryos (for biomedical research) and frozen sperm from purebred dogs. Private and federal institutions have begun to systematically store biological material from specific animal and plant genotypes. As three examples: (l) the Jackson Laboratory and the National Institutes of Health maintain large storage facilities for frozen embryos collected from hundreds of genotypes of mice used as animal models for biomedical research; (2) the United States Department of Agriculture maintains a repository for plant seeds from crop species; and (3) the American Type Culture Collection acquires, preserves and distributes characterized strains of bacteria, fungi, protozoa, algae, viruses, cell/tissue cultures and the creations of recombinant DNA technology.

4. Justification and Potential Utility of Genetic Resource Banks

In light of the publicity directed at dwindling habitat, the loss of species and genetic diversity and the potential of artificial breeding technology, it is remarkable that no organized effort exists, either in the U.S. or elsewhere, to sample, evaluate, cryopreserve, catalog, maintain and use germ plasm from wild animal and plant species. Genetic resource banking can act to facilitate conservation. For example:

1. A Genetic Resource Bank would not be merely a static warehouse of biological material but would serve a vital, interactive role between living populations of captive and free-living species. These interactions are required to prevent undesirable selection pressures in captivity, preserve new diversity resulting from natural evolutionary processes and allow small or fragmented populations to receive "infusions" of genetic diversity from cryopreserved germ plasm.
2. The combination of frozen gametes and embryos and reproductive techniques such as AI, ET and IVF offers unique opportunities for improving the efficiency of captive breeding programs (Wildt, 1989). Cryopreservation of germ plasm extends the generation interval of a species indefinitely. The genetic diversity of the founder animal does not die with the animal, but remains viable and available for future generations.
3. Germ plasm banking has the effect of reducing the number of animals needed to ensure that high levels of genetic diversity are retained within a population (Ballou, 1992). This reduces the capital and operating costs of zoo breeding programs and provides space for other species at risk for extinction.
4. Other benefits include the incorporation of germ plasm from wild stocks into captive breeding programs without removing animals from the wild, and the insurance banking offers against the loss of diversity or entire species from epidemics, natural disasters and social/political upheaval.
5. Transporting frozen sperm or embryos eliminates the considerable risks and costs associated with the transport or exchange of live animals.
6. The judicious interspecies use of germ plasm (e.g., hybridization of rare species of wild cattle with common cattle) may provide avenues for improving the genetics, food production and general agri-economy of developing countries.
7. Technologies developed for animal and plant germ plasm can be used to expand the genetic bank to include other biologicals including tissues, blood products and DNA. These new elements also can provide a service repository allowing more wide-spread access to rare specimens.

5. Utility of Genetic Resource Banks for Both Habitat and Single Species Conservationists

Conservation biologists constantly debate the relative merits of saving habitats versus species (Wildt, 1990). Habitat proponents suggest that representative ecosystems can be identified and permanently isolated from human interference. This approach focuses upon the long range problem and protects the ecosystem and the many species within it. The problem, of course, is in choosing which habitats to preserve since many species inevitably will be excluded. Additionally, those species most susceptible to extinction (i.e., large-sized predators at the top of the food chain) often require extraordinarily large home ranges. Therefore, it may be too expensive or socially disruptive to provide sufficiently-sized natural reserves for such animals. The contrasting view is to preserve single species, an approach exemplified by captive breeding programs. In this scenario, animals are maintained in a semi-controlled environment, and modern science is charged with identifying and manipulating the factors influencing reproductive success. In theory and given modern management techniques, rare species should thrive in captivity. But single species conservationists encounter many of the same problems that plague programs designed to save native habitats. The sheer cost of establishing "naturalistic" captive environments and prioritizing which species deserve the most attention are major concerns. These proponents also are faced with the paradoxical threat of unbridled success, that is, that captive breeding may increase animal numbers to the point that all captive habitat is saturated and then overwhelmed.

There is no doubt that habitats and single species should be preserved simultaneously and that the cryopreservation of biological materials can play a major role in both types of conservation. For example, free-living animals produce surplus germ plasm that can be recovered safely, cryopreserved and used to infuse captive populations with genetic vigor while eliminating the need to remove more animals from the wild. Likewise, cryobiology offers a resolution to the problem of limited captive breeding space. It simply is more cost-effective to preserve genetic material from populations, genotypes and individuals at low temperatures, eventually re-deriving these species when needed.

6. Overview on State-of-the-Art Reproductive Biotechnology Including Cryopreservation of Germ Plasm and Embryos

In humans and domestic animals, successes with germ plasm cryopreservation, AI, ET and IVF were the result of considerable basic research that was translated into significant improvements in reproductive performance. In reality, these advances have sparked a revolution in combating human infertility and improving livestock production (Wildt, 1989). For wildlife species (including vertebrates and invertebrates), these techniques could be useful for enhancing propagation and sustaining current levels of bio- and genetic diversity. This speculation is bolstered by considering the surge of interest and the development of effective and practical strategies for improving captive breeding of rare animal species over the past 10 years. Witness the formulation of species survival/action plans and the evolution of the population viability analysis which allows wildlife managers (those concerned with both *in situ* and *ex situ* conservation) to work together to (1) assess species/population robustness and (2) generate recovery plans for the future based on scientific fact (Seal and Foose, 1983; Foose, 1987). This progress, combined with the actual release of captive-born animals (e.g., golden lion tamarins, red wolves, bison, Arabian oryx; Kleiman, 1989; Kleiman et al., 1991) into wild habitats have revealed one common theme, there is an all-out need for more quality research (Wildt, 1989).

Logic dictates that considerable emphasis be placed on sustaining existing biodiversity using cryopreservation technology since further delay only accentuates ongoing and relentless losses in species and genetic variability. The technical details that translate frozen gametes and embryos into normal offspring have yet to be developed for many wildlife species. Nevertheless, current technology and in vitro test procedures are sufficient to ensure that most frozen germ plasm is biologically competent and eventually useful (Wildt, 1989). Another argument for developing genetic resource banks first is that this approach naturally spawns and mandates related research in other reproductive, biomedical and veterinary fields.

Reproductive biology is a complex discipline, and the actual applied use of a genetic resource bank will inevitably require interactive research among many disciplines and the development of new technologies. These, in turn, will generate massive data sets that will help scientists fully comprehend the fundamental biology of rare species, almost all of which have never been studied. For example, the "simple" collection and cryopreservation of Siberian tiger sperm will require interaction among: (1) population biologists (to help determine the kind and number of animals to sample to maximize genetic representation); (2) the Species Survival Plan Coordinator and SSP Propagation Committee (to choose individual animals to use); (3) veterinarians (responsible for developing optimal anesthesia); (4) biotechnical engineers (responsible for developing equipment to collect sperm); (5) gamete biologists (talented in sperm collection, processing, and evaluation); (6) cryobiologists (skilled in the freezing, packaging and storing of sperm); and (7) registrars of a global/regional databank (responsible for maintaining computerized records and catalogs detailing the pedigree of the sperm donor and the location, amount and distribution of the germ plasm). The actual use of the frozen sperm will stimulate even further research into an array of exciting areas (e.g., establishing the length of the estrous cycle; studying the impact of seasonality on reproductive performance; identifying and synchronizing estrus; predicting ovulation; developing techniques for AI, IVF and ET; diagnosing pregnancy and impending parturition). Therefore, genetic resource banks will serve as the incentive for developing and expanding other biological disciplines ranging from fundamental reproductive biology to genetic management of rare populations to applied aspects of veterinary medicine to global monitoring/computerized information systems.

6.1. Potential Benefits of AI, IVF or ET to Wildlife Conservation

The benefits of a genetic resource bank can be realized only if AI, IVF, and ET methods are available for the species of interest. AI is valuable for ensuring reproduction between behaviorally incompatible pairs, eliminating the risks of animal transport and providing an avenue for infusing genes from healthy wild stocks into insular or fragmented wild populations or genetically-stagnant, captive populations.

From an applied perspective, IVF has the potential of resolving many of the serious problems routinely encountered in modern captive-breeding programs (e.g., achieving representation from animals failing to reproduce because of inopportunity, management restrictions, behavioral peculiarities, physical handicaps, stress susceptibility or political boundaries) (Wildt, 1990). IVF also is highly attractive because it requires neither detection of overt estrus nor direct interaction between the male and female. In the context of a frozen germ plasm resource, an IVF program could be used to infuse gametes collected from free-living animals into captive populations. There also have been recent advances in related technology that allows recovering and maturing early stage ovarian oocytes in vitro, an approach that could be useful for salvaging genetic material from rare animals that die abruptly. Theoretically, with cryostored germ plasm, IVF could be combined with gamete maturation to produce embryos and then offspring from deceased parents. Pilot studies already have been successful with laboratory rodents and farm livestock (see reviews, Johnston et al., 1989, 1991).

For wildlife species, ET offers the possibility of accelerating the number of offspring produced and using gametes from females incapable of reproducing because of age or physical/medical handicaps. In conjunction with embryo freezing, ET could help preserve the combined genetic component of an individual in suspended (frozen) animation, thereby offering an approach for reintroducing available genetic material into later generations. The potential of ET increases if embryos can survive interspecies ET, that is, develop and be born to surrogate mothers of a more common, closely-related species.

6.2. Current Realities to the Use of Cryopreserved Sperm Plasm and AI, IVF and ET in Mammalian Species

Although the possibilities of using frozen wildlife germ plasm and embryos in concert with AI, IVF or ET are staggering, most successes have been limited to farm livestock and laboratory animals. Table 1 lists the 31 species in which AI with frozen-thawed sperm has resulted in live-born offspring. IVF has been successful in a total of 16 species (including a hybrid macaque study), and, in almost all cases, the young born was conceived using fresh (non-frozen) gametes (Table 2). Table 3 includes the 15 species in which frozen-thawed and transferred embryos have resulted in live-born offspring. Considering that more than 4,000 mammalian species inhabit our planet, it is obvious that the potential of reproductive biotechnology has been tested in an infinitesimal fraction of wild taxa.

The lack of application is the direct result of insufficient or, in most cases, nonexistent funding for both basic and applied wildlife research. Amazingly, no centralized organization has evolved to allow zoos to participate in a genetic resource conservation program (like the

Table 1. Mammalian species in which offspring have been produced by AI and frozen-thawed sperm.

Domesticated		Non-Domesticated	
Cattle	Water buffalo	Fox	Bison
Sheep	Domestic ferret	Wolf	Siberian ferret
Horse	Human	Addax	Giant panda
Pig		Blackbuck	Gaur
Rabbit		White-tailed deer	Reindeer
Dog		Fallow deer	Black-footed ferret
Goat		Chimpanzee	Cynomolgus monkey
Cat		Bighorn sheep	Chital deer
Mouse		Red deer	Eld's deer
			Leopard cat

Table 2. Mammalian species in which offspring have been produced by IVF followed by embryo transfer.

Domesticated		Non-Domesticated	
mouse	cattle	baboon	tiger
rabbit	pig	rhesus monkey	marmoset
rat	sheep	cynomolgus monkey	Indian desert cat
human	cat	hybrid pigtail x lion-tailed macaque	
hamster			

Table 3. Mammalian species in which offspring have been produced using frozen-thawed and transferred embryos.

Domesticated		Non-Domesticated
mouse	pig	eland
rat	cat	baboon
rabbit	goat	cynomolgus monkey
cow	horse	marmoset
sheep	human	red deer

Species Survival Plans). Even so, there is a general consensus that reproductive biotechnology could be used to better preserve genetic diversity and assist in captive propagation given that species reproductive norms were known (Wildt, 1989). In this respect, this same strategy was used to make the use of frozen sperm and embryos feasible in farm livestock. For example, the conventional use of AI and embryo transfer in domestic cattle became routine only after years of research into gamete biology and fundamental reproductive processes. This substantial progress could only be made after millions of research dollars were provided by federal, commercial and private sources.

6.3. Encouraging Progress with Wildlife Species

Basic research strategies, similar to those used in livestock, laboratory animals and humans, are possible and have been applied to wildlife species (primarily on a limited or pilot basis) at a few pioneering institutions. Advances in wildlife and zoo veterinary medicine now permit routine and safe animal anesthesia that allows blood sampling (for endocrine monitoring; Seal et al., 1985; Wildt et al., 1988; Brown et al., 1991a,b,c), electroejaculation (for semen collection; Howard et al., 1986), uterine catheterization (for nonsurgical embryo collection/transfer; Schiewe et al., 1991), laparoscopy (for oocyte recovery and intrauterine insemination; Miller et al., 1990; Donoghue et al., 1990; Wildt et al., 1992b) and ultrasound (for ovulation and pregnancy diagnosis; Donoghue et al., 1990). Other novel and exciting techniques have been developed to facilitate the eventual practical use of genetic material. The hormonal status of many wildlife species now can be tracked by measuring hormonal metabolites in voided urine or feces (Monfort et al., 1990, 1991a,b; Lasley and Kirkpatrick, 1991; Wasser et al., 1991). This innovative approach, which allows frequent sampling from individuals and eliminates the stresses associated with blood sampling under anesthesia, has been used to document seasonality, the estrous cycle, time of ovulation and even predict pregnancy and impending birth. Simple, cost-effective enzyme-linked immunosorbent assays (ELISA) can be used to predict critical events (like ovulation from a urine sample) and offer exciting means of assessing endocrine function under field conditions. New approaches, originally developed for assessing human fertility potential, also are finding application in wildlife species. "Heterologous" IVF systems are available whereby hamster or domestic cat oocytes can be used to test the penetrating or fertilizing capability of sperm collected from other species (Howard and Wildt, 1990). These in vitro assays of sperm viability will be important in testing the biological competence of frozen-thawed, wildlife sperm. Lastly, reproductive biotechnology is beginning to contribute significantly to nonmammalian and invertebrate species. Some progress has been made in the long-term storage of fish gametes (especially salmon and rainbow trout sperm; Stoss, 1983; Schmidt-Baulain and Holtz, 1989; Wheeler and Thorgaard, 1991) and mollusc (Gallardo et al., 1988) embryos. Insects also are benefiting as illustrated by continued progress in the freeze storage of, for example, honey bee spermatozoa and the recent successful cryopreservation of *Drosophila* embryos (Steponkus et al., 1990).

As demonstrated in Tables 1 to 3, there is similar tantalizing evidence in other wildlife species. The live births of any wildlife species as a result of biotechnology and the use of fresh or frozen germ plasm are laudable. However, certain events are particularly worthy of note as they either: l) demonstrate a biological "first" for wildlife species; 2) illustrate the ability to apply techniques developed for farm livestock to wild counterparts; or 3) focus upon a particularly difficult taxon that has received little or no research attention. Such milestones could include the birth of live offspring following:

- intraspecies embryo transfer in a baboon (first successful ET in a wildlife species: Kraemer et al., 1976)
- IVF in a baboon followed by embryo transfer (first successful IVF in a wildlife species: Clayton and Kuehl, 1984)
- AI of a puma using fresh sperm (Moore et al., 1981)
- interspecies embryo transfer in the gaur (to Holstein cow; Stover and Evans, 1984), eland (to cow: Dresser et al., 1982), bongo (to eland; Dresser et al., 1985a,b), zebra (to domestic horse: Summers et al., 1987a) and Przewalski's horse (to domestic horse: Summers et al., 1987a)
- intraspecies embryo transfer in the eland (Kramer et al., 1983), oryx (Pope et al., 1991), bongo (Dresser et al., 1985a,b) and suni antelope (Loskutoff et al., 1991)
- AI of a giant panda using frozen-thawed sperm (Moore et al., 1984)
- intraspecies transfer of frozen-thawed marmoset embryos (Summers et al., 1987b)
- AI of blackbuck with fresh or frozen-thawed sperm (Holt et al., 1988)
- IVF of Indian desert cat oocytes followed by embryo transfer to the domestic cat (Pope et al., 1989)
- AI of gaur with frozen-thawed sperm (Junior et al., 1990)
- IVF of tiger oocytes followed by embryo transfer to a surrogate tiger (Donoghue et al., 1990
- AI of black-footed ferrets with frozen-thawed sperm (Wildt et al., 1992b)
- AI of Eld's deer with frozen-thawed sperm (Wildt et al., 1992b)
- AI of leopard cats and cheetah using fresh sperm (Wildt et al., 1992b)

Although cryopreserved genetic material played a role in only six of these events, there is no biological reason why frozen gametes and embryos could not be used successfully to extend these early accomplishments.

Regardless of how a genetic resource is developed for a given species and which technique is used, some sperm and embryos always fail to survive freezing and thawing. Although much effort has been directed at evaluating various cryoprotectants and cooling rates for semen and embryos, the ease (or difficulty) of achieving post-thaw survival is, in part, dictated by species or genotype within species. In some cases, techniques that work well with one or more species fail or are only partially effective in a taxonomically-related counterpart (Wildt et al., 1992a,b). Perhaps most significant are findings that genotypes within a species can respond differently to a standardized protocol (Schmidt et al., 1985). For example, when 4- to 8-cell mouse embryos from 27 different genotypes were frozen and thawed using a regimented procedure, survival ranged from 27.4 to 75.2% (Schmidt et al., 1987).

Even given such observations, it is likely that only modest research will be needed for some species, and progress will benefit through the use of existing knowledge and common laboratory animal models (Wildt et al., 1986). There is, however, an almost complete lack of information on sperm and embryos for a vast number of species, suggesting that much more basic research is needed. Also of concern is the potential introduction and spread of diseases to domestic livestock and other wild stocks that could occur with the international transport of poorly monitored gametes and embryos (Schiewe, 1991). These factors must be taken into consideration during the first steps of formulating a genetic resource bank.

7. Recommendations for Establishing Genetic Resource Banks

Logic allows us to conclude that there is (l) an immediate need to maintain as much of the earth's bio- and genetic diversity as presently exists and (2) that genetic resource banks in combination with assisted reproductive technology have enormous potential. We also assert that significant progress has been made in beginning to adapt human/livestock technology to wildlife species. Certainly, compared to only a few years ago, we have increased our understanding of fundamental reproductive processes exponentially for a variety of wildlife species. Furthermore, although these strategies have not yet had a major conservation effect, live offspring from endangered species have been produced, despite an almost total lack of appropriate federal funding. Therefore, it is exactly the right time to begin considering the formal development of such resources for conservation. The interactive intramural and extramural features of a genetic bank infrastructure certainly are open to debate, but, in our opinion, the following characteristics are essential.

1. First, progress will be accelerated by supporting existing institutions. Organizations with professionals and technicians with combined skills in the fields of (domestic and nondomestic species) reproductive physiology, cryobiology, low temperature biology (cold storage), gamete function, embryology, AI, IVF, ET, endocrinology, veterinary medicine (especially anesthesia), disease transmission via germ plasm, population biology, genetic management, field studies and/or computer programming, are highly worthy of consideration.
2. The research record of the staff should reflect an appreciation for the importance of basic research as a prerequisite to practical collection and subsequent utilization of the genetic materials.
3. The institutions should have earned independent recognition and/or be closely affiliated with other organizations of national and international repute and have demonstrated a record of leadership in conservation issues, scientific research and training.
4. The institutions should have international contacts and existing collaborations with conservation organizations, research institutions and governmental/private wildlife authorities world-wide. Such relationships are a prerequisite to coordinating, organizing and gaining access to special or rare wildlife populations for genetic material recovery, storage and distribution. Special collaborative talents will be required to ensure that species, populations and individuals in crisis receive first priority attention.

The issue of number and size of genetic resource banks needs to be addressed. It may be most appropriate to establish single "continental" or national centers supported by multiple regional banks. The latter will be crucial to securing appropriate samples of genetic material unique to specific geographic locations. Regional centers could support the mission of the national bank by providing half the frozen sample to the central repository while maintaining half locally for "insurance-safety" purposes. National and regional banks also could be encouraged to conduct research based upon innovations as well as advances made in human and domestic animal research.

Most importantly, there must be considerable emphasis placed upon both basic and applied research by the various banking institutions. Areas of high priority include:

- Defining the effect of species on the efficiency of freezing sperm, ova and embryos (a strategy that will determine how well existing technology can be applied immediately to rare species);
- Identifying the best cryoprotectants and cooling and thawing processes for a particular species of interest;
- Developing accurate laboratory approaches for testing the viability of thawed germ plasm, embryos, tissues and DNA;

- Conducting detailed, longitudinal studies on the many species-specific factors that influence the production of live offspring from thawed material;
- Improving, testing and processing procedures to ensure that the transport of germ plasm does not contribute to disease transmission;
- Gamete fingerprinting for individual identification, relatedness and diversity status.

8. Summary

Certainly, two additional critical questions to be addressed in this decade will be (1) which species should be targeted and (2) what will be the source of funding for these very costly ventures? For now, those species or taxa managed under an SPP/TAG authority or those designated by the CBSG as "critical" should receive initial attention. The fact that an organized effort exists argues that there is general agreement that the species is experiencing difficulty and deserves high priority. Unfortunately, species priority also will be dictated by the continued discovery of species or subspecies approaching extirpation. Federal research monies often magically appear when species numbers decrease to levels which almost ensure extinction. The goal should be not to rely upon last ditch scientific heroics, but to use the scientific method to understand species biology before extinction becomes imminent.

Certainly, establishing and maintaining genetic resource banks will be expensive, requiring long-term, financial commitments. Funding will needed for: (1) capital development; (2) the support of existing staff and the hiring of new talent; (3) purchasing equipment and supplies to permit the safe collection, processing, long-term storage and active use of animal materials; (4) transporting and supporting research teams charged with collecting and preserving material; and (5) conducting basic and applied research. There is no "National Institutes of Health (NIH) for Wildlife Species", although developing a federally-supported National Institutes of the Environment (NIE) presently is being debated. An NIE likely would provide extramural support for wildlife research, much like the NIH does for human and animal model studies. Also, hopefully the National Science Foundation will increase the amount of money in "Basic Research in Conservation and Restoration Biology", a funding program initiated in 1990. Nonetheless, this issue is so crucial, that it is essential that Congress and appropriate federal agencies finally attend to what is rapidly becoming an almost historic plea to conserve our wildlife heritage using these promising strategies.

References

Anonymous. 1978. Conservation of Germ Plasm Resources: An Imperative, National Research Council, Report of Committee on Germ Plasm Resources, National Academy of Sciences, Washington, DC.

Anonymous. 1984. Animal Germplasm Preservation and Utilization in Agriculture, Council for Agricultural Science and Technology, Report No. 101, Ames, IA.

Anonymous. 1985. U.S. Strategy on the Conservation of Biological Diversity, Inter-Agency Task Force Report to Congress, U.S. Agency for International Development, Washington, DC.

Anonymous. 1987. Technologies to Maintain Biological Diversity, U.S. Congress, Office of Technology Assessment, Report No. OTA-F-330, U.S. Government Printing Office, March.

Ballou, J.D. 1992. Potential contribution of cryopreserved germ plasm to the preservation of genetic diversity and conservation of endangered species in captivity. *Cryobiology* 29: 19-25.

Brown, J.L., D.E. Wildt, C.R. Raath, V. de Vos, J.G. Howard, D. Janssen, S. Citino and M. Bush. 1991a. Impact of season on seminal characteristics and endocrine status of adult free-ranging African buffalo *(Syncerus caffer)*. *J. Reprod. Fert.* 92:47-57.

Brown, J.L., D.E. Wildt, J.R. Raath, V. de Vos, J.G. Howard, D.L. Janssen, S.B. Citino and M. Bush. 1991b. Seasonal variation in LH, FSH and testosterone secretion and concentrations of testicular gonadotropin receptors in free-ranging impala *(Aepyceros melampus)*. *J. Reprod. Fert.* 93: 497-505.

Brown, J.L., M. Bush, C. Packer, A.E. Pusey, S.L. Monfort, S.J. O'Brien, D.L. Janssen and D.E. Wildt. 1991c. Developmental changes in pituitary-gonadal function in free-ranging lions *(Panthera leo)* of the Serengeti Plains and Ngorongoro Crater. *J. Reprod Fert.* 91:29-40.

Clayton, 0. and T. Kuehl. 1984. The first successful in vitro fertilization and embryo transfer in a nonhuman primate. *Theriogenology* 21:228 (abstr.).

Donoghue, A.M, L.A. Johnston, U.S. Seal, D.L. Armstrong, R.L. Tilson, P. Wolf, K. Petrini, L.G. Simmons, T. Gross and D.E. Wildt. 1990. In vitro fertilization and embryo development in vitro and in vivo in the tiger *(Panthera tigris)*. *Biol. Reprod.* 43:733-747.

Dresser, B.L., L. Kramer, C.E. Pope, R.D. Dahlhausen and C. Blauser. 1982. Superovulation of African eland *(Taurotragus oryx)* and interspecies embryo transfer to Holstein cattle. *Theriogenology* 17:86 (abstr.).

Dresser, B.L., C.E. Pope, L. Kramer, G. Kuehn, R.D. Dahlhausen, E.J. Maruska, B. Reece and W.D. Thomas. 1985a. Successful transcontinental and interspecies embryo transfer from bongo antelope *(Tragelaphus euryceros)* at the L.A. Zoo to eland *(Taurotragus oryx)* and bongo at the Cincinnati Zoo. *Proc. Am. Assoc. Zool. Prk. Aquar.*, pp. 166-168.

Dresser, B.L., C.E. Pope, L. Kramer, G. Kuehn, R.D. Dahlhausen, E.J. Maruska, B. Reece and W.D. Thomas. 1985b. Birth of bongo antelope *(Tragelaphus euryceros)* to eland antelope *(Taurotragus oryx)* and cryopreservation of bongo embryos. *Theriogenology* 23:190. (abstr.).

Foose, T. 1987. Species survival plans and overall management strategies, in: "Tigers of the World: The Biology, Biopolitics, Management and Conservation of an Endangered Species," R. Tilson and U.S. Seal, eds, pp. 304-316, Noyes Publications, Park Ridge.

Gallardo, C.S., M.R. Del Campo and L. Filun. 1988. Preliminary trials of the cryopreserving of marine mollusc embryos as illustrated with the marine mussel *Choromytilus chorus* from southern Chile. *Cryobiology* 25:565 (abstr.).

Holt, W.V., H.D.M. Moore, R.D. North, T.D. Hartman and J.K. Hodges. 1988. Hormonal and behavioural detection of oestrus in blackbuck, *Antilope cervicapra*, and successful artificial insemination with fresh and frozen semen. *J. Reprod. Fert.* 82:717-725.

Howard, J.G. and D.E. Wildt. 1990. Ejaculate-hormonal traits in the leopard cat *(Felis bengalensis)* and sperm function as measured by in vitro penetration of zona-free hamster ova and zona-intact domestic cat oocytes. *Mol. Reprod. Devel.* 26:163-174.

Howard, J.G., M. Bush and D.E. Wildt. 1986. Semen collection, analysis and cryopreservation in nondomestic mammals, in: "Current Therapy in Theriogenology," D. Morrow, ed., pp. 1047-1053, W.B. Saunders Co., Philadelphia.

Hutchins, M. and R.J. Wiese. 1991. Beyond genetic and demographic management: The future of the Species Survival Plan and related AAZPA conservation efforts. *Zoo Biol.* 10:285-292.

Johnston, L.A., A.M. Donoghue, S.J. O'Brien and D.E. Wildt. 1991. "Rescue" and maturation in vitro of follicular oocytes of nondomestic felid species. *Biol. Reprod.* 45:898-906.

Johnston, L.J, S.J. O'Brien and D.E. Wildt. 1989. In vitro maturation and fertilization of domestic cat follicular oocytes. *Gamete Res.* 24:343-356.

Junior, S.M., D.L. Armstrong, S.H. Hopkins, L.G. Simmons, M.C. Schiewe and T.S. Gross. 1990. Semen cryopreservation and the first successful artificial insemination of gaur *(Bos gaurus)*. *Theriogenology* 33:262 (abstr.).

Kleiman, D.G. 1989. Reintroduction of captive mammals for conservation: Guidelines for reintroducing endangered species into the wild. *Bioscience* 39:152-161.

Kleiman, D.G., B.B. Beck, J.M. Dietz and L.A. Dietz. 1991. Costs of a reintroduction and criteria for success: Accounting and accountability in the golden lion tamarin conservation program. *Symp. Zool. Soc. Lond.* 62:125-142.

Kraemer, D.C., G.T. Moore and M.A. Kramen. 1976. Baboon infant produced by embryo transfer. *Science* 192:1246-1247.

Kramer, L., B.L. Dresser, C.E. Pope, R.D. Dahlhausen and R.D. Baker. 1983. Nonsurgical transfer of frozen-thawed eland embryos. *Proc. Am. Assoc. Zoo Vet.*, pp. 104-105.

Lasley, B.L. and J.F. Kirkpatrick. 1991. Monitoring ovarian function in captive and free-ranging wildlife by means of urinary and fecal metabolites. *J. Zoo Wildl. Med.* 22:23-31.

Loskutoff, N.M., B.L. Raphael, B.A. Wolfe, L.A.N. French, R.B. Buice, J.G. Howard, M.C. Schiewe and D.C. Kraemer. 1991. Embryo transfer in small antelope. *Proc. Soc. Therio.* pp. 341-342.

Miller, A.M., M.E. Roelke, K.L. Goodrowe, J.G. Howard and D.E. Wildt. 1990. Oocyte recovery, maturation and fertilization in vitro in the puma *(Felis concolor)*. *J. Reprod. Fert.* 88:249-258.

Monfort, S.L., C. Martinet and D.E. Wildt. 1991b. Urinary steroid metabolite profiles in female Pere David's deer *(Elapharus davidinus)*. *J. Zoo Wildl. Med.* 22:78-85.

Monfort, S.L., C. Wemmer, T.H. Kepler, M. Bush, J.L. Brown and D.E. Wildt. 1990. Monitoring ovarian function and pregnancy in the Eld's deer *(Cervus eldi)* by evaluating urinary steroid metabolite excretion. *J. Reprod. Fert.* 88:271-281.

Monfort, S.L., N.P. Arthur and D.E. Wildt. 1991a. Monitoring ovarian function and pregnancy by evaluating excretion of urinary oestrogen conjugates in semi-free-ranging Przewalski's horses *(Equus przewalski)*. *J. Reprod. Fert.* 91:155-164.

Moore, H.D.M., M. Bush, M. Celma, A.L. Garcia, T.D. Hartman, J.P. Hearn, J.K. Hodges, D.M. Jones, J.A. Knight, L. Monsalve and D.E. Wildt. 1984. Artificial insemination in the giant panda *(Ailuropoda melanoleuca)*. *J. Zool. (Lond.)* 203:269-278.

Moore, H.D.M., R.C. Bonney and D.M. Jones. 1981. Induction of oestrus and successful artificial insemination in the cougar, *Felis concolor*. *Vet. Rec.* 108:282-283.

Pope, C.E., E.J. Gelwicks, M. Burton, R. Reece and B.L. Dresser. 1991. Nonsurgical embryo transfer in the scimitar-horned oryx *(Oryx dammah)*: Birth of a live offspring. *Zoo Biol.* 10:43-51.

Pope, C.E., E.J. Gelwicks, K.B. Wachs, G.L. Keller, E.J. Maruska and B.L. Dresser. 1989. Successful interspecies transfer of embryos from the Indian desert cat *(Felis silvestris ornata)* to the domestic cat *(Felis catus)* following in vitro fertilization. *Biol. Reprod.* 40 (Suppl.) 61 (abstr.).

Rall, W.F., J.D. Ballou and D.E. Wildt. 1991. Cryopreservation and banking of animal germ plasm for species conservation: An imperative for action by the Captive Breeding Specialist Group. *Proc. Ann. Cap. Breed. Special. Grp.*, Singapore.

Ralls, K., K. Brugger and J. Ballou. 1979. Inbreeding and juvenile mortality in small populations of ungulates. *Science* 206:1101-1103.

Schiewe, M.C. 1991. The science and significance of embryo cryopreservation. *J. Zoo Wildl. Med.* 22:6-22.

Schiewe, M.C., M. Bush, L.G. Phillips, S. Citino and D.E. Wildt. 1991. Comparative aspects of estrous synchronization, ovulation induction and embryo cryopreservation in the scimitar horned oryx, bongo, eland and greater kudu. *J. exp. Zool.* 58:75-88.

Schmidt, P.M., C.T. Hansen and D.E. Wildt. 1985. Viability of frozen-thawed mouse embryos is affected by genotype. *Biol. Reprod.* 32:238-246.

Schmidt, P.M., M.C. Schiewe and D.E. Wildt. 1987. The genotypic response of mouse embryos to multiple freezing variables. *Biol. Reprod.* 37:1121-1128.

Schmidt-Baulain, R. and W. Holtz. 1989. Deep freezing of rainbow trout *(Salmo gairdneri)* sperm at varying intervals after collection. *Theriogenology* 32:439-443.

Seal, U.S. 1991. Life after extinction, in: "Beyond Captive Breeding: Reintroducing Endangered Mammals to the Wild," J.H.W. Gipps, ed, pp. 39-55, Clarendon Press, Oxford.

Seal, U.S. and T. Foose. 1983. Development of a masterplan for captive propagation of Siberian tigers in North American zoos. *Zoo Biol.* 2:241-244.

Seal, U.S., E.D. Plotka, J.D. Smith, F.H. Wright, N.J. Reindl, R.S. Taylor and M.F. Seal. 1985. Immunoreactive luteinizing hormone, estradiol, progesterone, testosterone and androstenedione levels during the breeding season and anestrus in Siberian tigers. *Biol. Reprod.* 32:361-68.

Steponkus, P.L., S.P. Meyers, D.V. Lynch, L. Gardner, V. Bronshteyn, S.P. Leibo, W.F. Rall, R.E. Pitt, T.-T. Lin and R.J. MacIntyre. 1990. *Cryopreservation of Drosophila melanogaster embryos. Nature* 345:170-172.

Stoss., J. 1983. Fish gamete preservation and spermatozoa physiology, in: "Fish Physiology," W.J. Hoar, D.J. Randall and E.M. Donaldson, eds, pp. 305-350, Academic Press, New York.

Stover, J. and J. Evans. 1984. Interspecies embryo transfer from gaur *(Bos gaurus)* to domestic Holstein cattle *(Bos taurus)* at the New York Zoological Park. *X Intl. Cong. Anim. Reprod. Artif. Insem.* 2:243.

Summers, P.M., A.M. Shephard, J.K. Hodges, J. Kydd, M.S. Boyle and W.R. Allen. 1987a. Successful transfer of the embryos of Przewalski's horse *(Equus przewalskii)* and Grant's zebra (*E. burchelli*) to domestic mares *(E. caballus)*. *J. Reprod. Fert.* 80:13-20.

Summers, P.M., A.M. Shephard, C.T. Taylor and J.P. Hearn. 1987b. The effects of cryopreservation and transfer on embryonic development in the common marmoset monkey, *Callithrix jacchus*. *J. Reprod. Fert.* 79:224-250.

Wasser, S.K., S.L. Monfort and D.E. Wildt. 1991. Rapid extraction of faecal steroids for measuring reproductive cyclicity and early pregnancy in free-ranging, yellow baboons *(Papio cynocephalus cynocephalus)*. *J. Reprod. Fert.* 92:415-423.

Wheeler, P.A. and G.H. Thorgaard. 1991. Cryopreservation of rainbow trout semen in large straws. *Aquaculture* 93:95-100.

Wildt, D.E. 1989. Reproductive research in conservation biology: Priorities and avenues for support. *J. Zoo Wildl. Med.* 20:391-395.

Wildt, D.E. 1990. Potential applications of IVF technology for species conservation, in: "Fertilization in Mammals," B.D. Bavister., J. Cummins, E.R.S. Roldan, eds, pp. 349-364, Serono Symposium, U.S.A.

Wildt, D.E., and U.S. Seal. 1988. Editors for monograph, *Research Priorities for Single Species Conservation Biology*. National Zoological Park, Smithsonian Institution, Washington, 23 P.

Wildt, D.E., A.M. Donoghue, L.A. Johnston, P.M. Schmidt and J.G. Howard. 1992a. Species and genetic effects on the utility of biotechnology for conservation. *Symp. zool. Soc. London*. 64: 45-61.

Wildt, D.E., S.L. Monfort, A.M. Donoghue, L.A. Johnston and J.G. Howard. 1992b. Embryogenesis in conservation biology — or how to make an endangered species embryo. *Theriogenology* 37: 161-184.

Wildt, D.E., L.G. Phillips, L.G. Simmons, P.K. Chakraborty, J.L. Brown, J.G. Howard, A. Teare and M. Bush. 1988. A comparative analysis of ejaculate and hormonal characteristics of the captive male cheetah, tiger, leopard and puma. *Biol. Reprod.* 38:245-255.

Wildt, D.E., M.C. Schiewe, P.M. Schmidt, K.L. Goodrowe, J.G. Howard, L.G. Phillips, S.J. O'Brien and M. Bush. 1986. Developing animal model systems for embryo technologies in rare and endangered wildlife. *Theriogenology* 25:33-51.

Cryopreservation of Fish Spermatozoa

BRIAN HARVEY

1. Why Freeze Fish Spermatozoa?

Experimental methods for cryopreserving fish spermatozoa, that is, freezing to the temperature of liquid nitrogen for indefinite storage, have appeared in the scientific literature for over twenty years. The reasons for doing so are more or less transplanted from cattle farming. Cryopreservation is not only a useful management tool, it also permits conservation of endangered or otherwise valuable germplasm. As the practice becomes widespread, opportunities exist for banking and distributing gametes. Genetic conservation as an end is most frequently cited for salmonids and tilapias (see for example Harvey, 1990 and Pullin, 1988); benefits to hatchery and broodstock management are similar for most cultured species of fishes, while banking for profit, which has been so successful in cattle, will remain stalled for any species until reliable large-scale methods appear.

The above objectives and means of achieving them have been reviewed in varying degrees of detail. The most recent general review of the technology is by Stoss (1983). Munkittrick and Moccia (1984) describe applications and methods for salmonids; Harvey (1987; 1990) discusses gene banking as it relates to aquaculture and genetic conservation, and Harvey and Carolsfeld (1993) review methods for warmwater cultured fishes with emphasis on criteria for developing laboratory-scale techniques.

2. What Are the Technical Problems?

Recovering motile, fertile frozen-thawed fish spermatozoa is not a great cryobiological challenge. The cells are small, have no acrosome, and are available in enormous quantities in season for experimentation. Yet, as the multitude of very different published methods shows, there are real difficulties in developing a reliable method. Why is this? The answer probably lies in the enormous diversity of fish spermatozoan physiology; although all the species studied are external fertilizers, the behaviour of the adults at spawning and the conditions in which sperm and egg must meet are very different. At two extremes are salmonids spawning in fresh water, where sperm is deposited as close as possible to the eggs and swims for 30 seconds at most, and marine species like herring, where spermatozoa are released in a diffuse cloud over the spawning grounds and remain motile for hours. Not surprisingly, cryopreservation of salmonid and herring sperm presents very different challenges. In species where spermatozoa swim for long periods in nature, cryopreservation is generally easier. Tilapias are a good

MTL Biotech Ltd., PO Box 5760, Station B, Victoria, British Columbia, Canada V8R 6S8.

Genetic Conservation of Salmonid Fishes, Edited by J.G. Cloud
and G.H. Thorgaard. Plenum Press, New York, 1993

example; they can spawn in fresh to full seawater, and their spermatozoa will swim for hours in a saline solution. Several reliable methods have appeared for cryopreserving the spermatozoa of these fishes (Harvey, 1983; Harvey and Kelley, 1988; Chao et al., 1987; Rana and McAndrew, 1987). Salmonids are more difficult, and a feature of the literature on cryopreservation of salmonid spermatozoa is the attention given to prolonging motility both before and after freezing (Benau and Terner, 1980; Christen et al., 1987; Terner, 1986; Wheeler and Thorgaard, 1991).

Since initiation of fish sperm motility is essentially a dilution effect (marked by a sudden reduction in potassium concentration and/or osmolality) upon expulsion of semen into the surrounding water, the membrane damage that to some degree accompanies freezing and thawing in any cell type tends to affect the sperm cell's osmotic tolerance. The practical result is premature initiation of motility and consequent shortening of duration of motility after thawing. This effect is most clearly seen when spermatozoa that were immotile in the freezing diluent are partially motile after thawing but before any "intentional" activation, and the end result is an often drastic reduction in the available time for fertilization.

3. A Partial List of Fish Species Whose Spermatozoa Have Been Frozen

Salmonids top the list in terms of numbers, with methods recently appearing for most economically important species including rainbow trout (Wheeler and Thorgaard, 1991; Baynes and Scott, 1987; Schmidt-Baulain and Holtz, 1989) and Atlantic salmon (Alderson and Macneil, 1984). Tilapias (mostly *Oreochromis mossambicus*) are also well represented (Harvey, 1983; Harvey and Kelley, 1988; Chao et al., 1987; Rana and McAndrew, 1987). Cyprinids, including both common and Chinese carps, have been studied for more than a decade, with successful methods recently appearing (Koldras and Bienarz, 1987; Kurokura et al., 1984). Some economically important marine species for which spermatozoa have been frozen include Asian sea bass (Leung, 1987), Atlantic halibut (Bolla et al., 1987), milkfish (Chao and Liao, 1987), black porgy (Chao et al., 1986) and bluefin tuna (Doi et al., 1982); various catfishes (Marian and Krasznai, 1987; Van Vuren and Steyn, 1987), walleye (Moore, 1987), whitefish (Piironen, 1987) and striped bass (Kerby, 1983) have also been studied. Coser et al. (1984) describe methods for cryopreservation of the spermatozoa of several economically important South American freshwater species.

4. Features of Successful Methods

Most successful methods now employ egg yolk in a simple diluent containing dimethyl sulfoxide (DMSO) as cryoprotectant, cooling rates between 20 and 50° C/minute, and rapid warming and prompt addition of the thawed semen to eggs. The use of egg yolk is well described in Baynes and Scott (1987), and is a successful method of membrane stabilization used for decades in the cryopreservation of mammalian spermatozoa (Polge, 1980). The trend in diluent composition is toward simplification, with the complex diluents of a decade ago being replaced by mixtures as simple as distilled water, glucose and DMSO (Stoss, personal communication).

An absolute requirement for inhibiting motility during dilution and before freezing is routinely cited; yet in our own experience with chinook salmon spermatozoa, post-thaw motility is as high in an activating diluent as in a diluent that inhibits motility. This should not be too surprising, as a search of the literature yields instances of high fertility with low motility sperm (Stoss and Holtz, 1983) as well as poor fertility with highly active sperm. If there is a lesson in these experiences, it is that cryopreservation results with fish spermatozoa mean little without fertility tests. In species where spermatozoa are active for a very short time (including all the salmonids), sperm motility after thawing is undoubtedly a problem, and many published methods employ activator solutions or "thawing solutions", based at least in part on solutions that prolong the activity of fresh spermatozoa, to prolong post-thaw motility sufficiently for fertilization to occur (Erdahl et al., 1987; Levanduski and Cloud, 1988; Steyn et al., 1989; Munkittrick and Moccia, 1984). Activator solutions may contain sodium bicarbonate (particularly effective in the pellet method, below), or a phosphodiesterase inhibitor such as caffeine or theophylline in a simple saline (Wheeler and Thorgaard, 1991).

5. Practical Limitations of Cryopreservation Methods

A problem facing fish biologists is the relatively large volumes of spermatozoa that must be handled. In cattle, a single 0.25 ml straw is routinely used to inseminate one cow; the eggs of a single Pacific salmon, however, may need several millilitres of thawed milt. Unfortunately, one of the best methods for cryopreserving fish spermatozoa, a pellet method based on techniques developed for freezing bull sperm, is difficult to scale up to the volumes needed for routine hatchery use. The method, most extensively described by Stoss (1983) and recently applied to masu salmon (Yamano et al., 1990) probably works well on a small scale because freezing and thawing rates are ideal in small pellets, and addition of the thawed, activated sperm to eggs is extremely rapid. Storage and manipulation of the pellets is, however, impractical on a large scale.

Small scale methods are entirely adequate for some purposes. Genetic conservation can be achieved by freezing small volumes of milt; at the extreme end of the scale, a simple method for cryopreserving microliter quantities of zebra fish spermatozoa (Harvey et al., 1982) is routinely used for genetic studies at the University of Oregon.

6. Making Cryopreservation User-Friendly

One often-overlooked aspect of cryopreservation of fish spermatozoa is that, while it is feasible to have trained technicians freeze the milt when it is available, the best moment for thawing and fertilization may be less predictable. As in cattle farming, this means that farmers or hatchery technicians will have to do the thawing and insemination themselves. Thawing, however, is as critical to success as freezing, and insemination is particularly tricky in species where motility is short-lived. Most published methods for cryopreservation of fish spermatozoa are unrepeatable by untrained technicians (and frequently even by other researchers); for methods to be truly practical, thawing and insemination will have to be easy and foolproof.

The real challenge facing cryobiologists interested in freezing fish spermatozoa is not in recovering enough viable sperm cells to fertilize small numbers of eggs in the laboratory, but in making the technique practical in the field. This means ruthlessly discarding any procedures that are any more complicated than those already performed by hatchery technicians working with fresh gametes. There is no question that industry and government are ready to use the technology for routine management, breeding programs and conservation of genetic resources, but the techniques will not be ready for these people until the above criteria have been met.

References

Alderson, R., and A.J. Macneil. 1984. Preliminary investigations of cryopreservation of milt of Atlantic salmon *Salmo salar* and its application to commercial farming. *Aquaculture* 43: 351-354.

Baynes, S.M. and A.P. Scott. 1987. Cryopreservation of rainbow trout spermatozoa: influence of sperm quality, egg quality and extender composition on post-thaw fertility. *Aquaculture* 66: 53-67.

Benau, D. and C. Terner. 1980. Initiation, prolongation and reactivation of the motility of trout spermatozoa. *Gamete Research* 3: 247-257.

Bolla, S., I. Holmefjord, and T. Refstie. 1987. Cryogenic preservation of Atlantic halibut sperm. *Aquaculture* 65: 371-374.

Bolla, S., I. Holmefjord, and T. Refstie. 1987. Cryogenic preservation of Atlantic halibut sperm. *Aquaculture* 65: 371-374.

Chao, N.H. and I.C. Liao. 1987. Application of honey in cryopreservation of sperm of milkfish *Chanos chanos* and black porgy *Acanthopagrus schlegeli*. Proceedings of the Third International Symposium on Reproductive Physiology of Fish. St. John's, Newfoundland, August 2-7, 1987. pp 94-96.

Chao, N.H., W.C. Chao, K.C. Liu, and I.C. Liao. 1986. The biological properties of black porgy *Acanthpagrus schlegeli* sperm and its cryopreservation. Proceedings of the National Science Council, Part B: Life Sciences. Taipei, Taiwan.

Chao, N.H., W.C. Chao, K.C. Liu, and I.C. Liao. 1987. The properties of tilapia sperm and its cryopreservation. *J. Fish. Biol.* 30: 107-118.

Christen, R., J.L. Gatti, and R. Billard. 1987. Trout sperm motility. *Eur. J. Biochem.* 166: 667-671.

Coser, A.M., H. Godinho, and D. Ribeiro. 1984. Cryogenic preservation of spermatozoa from *Prochilodus scrofa* and *Salminus maxillosus*. *Aquaculture* 37: 387-390.

Doi, M., T. Hoshino, Y. Taki, and Y. Ogasawara. 1982. Activity of the sperm of the bluefin tuna *Thunnus thynnus* under fresh and preserved conditions. *Bull. Japan. Soc. Sci. Fish.* 48: 495-498.

Erdahl, A.W., J.G. Cloud, and E.F. Graham. 1987. Fertility of rainbow trout (*Salmo gairdneri*) gametes: gamete viability in artificial media. *Aquaculture* 60: 323-332.

Harvey, B. 1990. Germplasm conservation in aquaculture, in: "The Preservation and Valuation of Genetic Resources." G.H. Orians, G.M. Brown, W.E. Kunin and J. Swierzbinski, eds., pp. 32-38. University of Washington Press, Seattle.

Harvey, B. 1987. Gamete banking and applied genetics in aquaculture. Proceedings of the World Symposium on Selection, Hybridization and Genetic Engineering in Aquaculture. Bordeaux, France, May 27-30, 1986, pp. 257-264.

Harvey, B. 1983. Cryopreservation of *Sarotherodon mossambicus* spermatozoa. *Aquaculture* 32: 313-320.

Harvey, B. and J. Carolsfeld. 1993. Induced breeding in tropical fish culture. International Development Research Centre, Ottawa. In press.

Harvey, B. and R.N. Kelley. 1988. Practical methods for chilled and frozen storage of tilapia spermatozoa, in "Second International Symposium on Tilapias in Aquaculture." R.S.V. Pullin, T. Bhukaswan, K. Tonguthai and J.L. MacLean, eds., pp. 197-189. ICLARM Conference Proceedings, Manila.

Harvey, B, R.N. Kelley, and M.J. Ashwood-Smith. 1982. Cryopreservation of zebra fish spermatozoa using methanol. *Can. J. Zool.* 60: 1867-1870.

Kerby, J.H. 1983. Cryogenic preservation of sperm from striped bass. *Trans. Am. Fish. Soc.* 112: 86-91.

Koldras, M., and K. Bienarz. 1987. Cryopreservation of carp sperm. *Pol. Arch. Hydrobiol.* 34: 125-134.

Kurokura, H., R. Hirano, M. Tomita, and M. Iwahashi. 1984. Cryopreservation of carp sperm. *Aquaculture* 37: 267-273.

Levanduski, M.J., and J.G Cloud. 1988. Rainbow trout (*Salmo gairdneri*) semen: effect of non-motile sperm on fertility. *Aquaculture* 75: 171-179.

Leung, L. 1987. Cryopreservation of the spermatozoa of the barramundi, *Lates calcarifer*. *Aquaculture* 64: 243-247.

Marian, T., and Z.L. Krasnai. 1987. Cryopreservation of European catfish *Siluris glanis* sperm. Proc. World Symp. on Selection, Hybridization, and Genetic Engineering in Fishes. 1986. Vol 2.

Moore, A.A. 1987. Short-term storage and cryopreservation of walleye semen. *Prog. Fish-Cult.* 49: 40-43.

Munkittrick, K.E. and R.D. Moccia. 1984. Advances in the cryopreservation of salmonid semen and suitability for a production scale artificial fertilization program. *Theriogenology* 21: 645-659.

Piironen, J. 1987. Factors affecting fertilization rate with cryopreserved sperm of whitefish *Coregonus mukshun Pallas*. *Aquaculture* 66: 347-357.

Polge, C. 1980. Freezing of spermatozoa, in: "Low Temperature Preservation in Biology and Medicine." M.J. Ashwood-Smith and J. Farrant, eds., pp. 45-60. Pitman Medical Ltd.

Pullin, R.S.V. 1988. Tilapia Genetic Resources for Aquaculture. Proceedings of the Workshop on Tilapia Genetic Resources for Aquaculture, March 1987, Bangkok. ICLARM, Manila. 108 pp.

Rana, K.J., and B.J. McAndrew. 1987. The viability of cryopreserved tilapia spermatozoa. *Aquaculture* 76: 335-345.

Schmidt-Baulain, R., and W. Holtz. 1989. Deep-freezing of rainbow trout *Salmo gairdneri* sperm at varying intervals after collection. *Theriogenology* 32: 439-443.

Steyn, G., J. Van Vuren, and E. Grobler. 1989. A new sperm diluent for the artificial insemination of rainbow trout (*Salmo gairdneri*). *Aquaculture* 83: 367-374.

Stoss, J. 1990. Personal communication.

Stoss, J. 1983. Fish gamete preservation and spermatozoan physiology, in: "Fish Physiology" Vol.9, Part B. W.S. Hoar, D.J. Randall, and E.M. Donaldson, eds. pp. 305-350. Academic Press, New York.

Stoss, J. and W. Holtz. 1983. Successful storage of chilled rainbow trout spermatozoa for up to 34 days. *Aquaculture* 31: 269-274.

Terner, C. 1986. Evaluation of salmonid sperm motility for cryopreservation. *Prog. Fish-Cult.* 48: 230-232.

Van Vuren, J.H.J. and G.J. Steyn. 1987. Cryopreservation of *Clarias gariepinus* sperm and fertilization success. Proceedings of the Third International Symposium on Reproductive Physiology of Fish. St. John's, Newfoundland, August 2-7, 1987. p 103.

Wheeler, P.A. and G.H. Thorgaard. 1991. Cryopreservation of rainbow trout semen in large straws. *Aquaculture* 93:95-100.

Yamano, K., Kasahara, N., E. Yamaha, and F. Yamazaki. 1990. Cryopreservation of masu salmon sperm by the pellet method. *Bull. Fac. Fisheries*, Hokkaido University 41: 149-154.

The Norwegian Gene Bank Programme for Atlantic Salmon (*Salmo salar*)

DAGFINN GAUSEN

1. Introduction

Norway has about 400 rivers which supported anadromous Atlantic salmon (*Salmo salar*). In 79 of these rivers the salmon population is either extinct or threatened by extinction. Acid precipitation has been the main cause of salmon mortality in the southern part of the country and 25 stocks are now virtually extinct (Hesthagen and Hansen, 1991). The parasite *Gyrodactylus salaris* has caused extensive fish kills in 34 salmon rivers in the middle region of the country.

Farmed salmon that escapes from fish farming cages in the sea, represent a new threat to the natural genetic resources of the Atlantic salmon. Farmed salmon are derived genetically from several wild populations and are subjected to an intensive selective breeding programme. Crossbreeding between farmed and wild salmon in nature is expected to lead to the loss of genetically determined characteristics and adaptive traits in local wild populations. Nation wide investigations has revealed that escaped farmed salmon presently comprise a substantial proportion of the mature salmon present on the spawning grounds in autumn (Gausen and Moen, 1991). A potential for large-scale genetic introgression thus exists.

It was against this background that the Norwegian government embarked on the gene bank programme for wild Atlantic salmon. The gene bank programme was strongly recommended by the Advisory Committee of the National Sea Ranching Programme. The gene bank was included in the sea ranching programme and was made financially possible by grants from the Ministry of Environment. The sea ranching programme was officially begun by the Directorate of Nature Management in January 1986 and was planned to continue for a ten year period.

2. Objectives and Strategies

The main purpose of the gene bank is to contribute to a nationwide preservation of the genetic diversity and characteristics of natural salmon stocks. During the initial developmental phases, the gene bank was based exclusively on deep freezing sperm. The goal was to preserve genetic material from more than 100 stocks and from at least 50 individuals from each stock. The resultant frozen sperm is expected to be used exclusively for broodstock production

Directorate for Nature Management, Tungasletta 2, N-7005 Trondheim, Norway

Genetic Conservation of Salmonid Fishes, Edited by J.G. Cloud
and G.H. Thorgaard, Plenum Press, New York, 1993

because of limited volume of sperm and the high costs involved in collection. Additionally, some of the frozen sperm will be stored for future use.

In 1988, the Advisory Committee of the Sea Ranching Programme recommended the establishment of broodstock stations, so-called "living gene banks", for Atlantic salmon. The main purpose of this recommendation was to establish a living reservoir of genetic material which could be used for the reestablishment or enhancement of threatened stocks. The Committee recognized that cryopreservation of eggs or embryos is a better way to maintain genetic material in a stable form, but the necessary techniques are not yet developed to cryopreserve these units. The resultant stations facilitate the accumulation of genetic diversity through the use of both frozen sperm and the aquisition of several year classes of parent fish from individual rivers.

The Directorate for Nature Management has also considered living gene banks based partially or exclusively on the release and retrieval of tagged fish. However, such a strategy does not offer the high level of security needed to preserve invaluable genetic material. The main problem with this consideration is the large natural mortality and the high catch mortality of the salmon which can cause the loss of family groups. The resultant loss of family groups will inevitably lead to loss of genetic diversity. Because these mortalities can only be compensated by releasing large numbers of tagged fish and because of the high cost involved, this is not a realistic gene bank alternative when dealing with a large number of threatened stocks.

The Directorate considers living gene banks as a temporary measure to be used only in cases where salmon stocks are threatened by extinction or massive invasion of farmed salmon. The gene bank stations are planned for a ten year period or approximately two salmon generations. At present two stations are operational and a third is being planned. These stations will be able to keep a maximum of 40 stocks. Further expansion will have to be decided upon by the government.

The basic gene bank strategy for Atlantic salmon in Norway is shown in Figure 1. Within this strategy, strong emphasis is placed on measures to prevent transmission of fish disease organisms.

All parent fish are maintained in a health-control program and strict rules for how each operation (handling of the fish, transportation of roe and milt etc.) is to be carried out have been established. Only disinfected eggs can be exported from the station in order to minimize the risk of spreading diseases to rivers. All fish production for stocking is carried out at local fish culture stations and the entire production is based on fresh water since seawater is often contaminated with Furunculosis bacteria and other fish disease organisms.

Adult fish caught in individual rivers are kept in local fish tanks for a few months until they become sexually mature. The resultant frozen sperm is then transported to a central storage facility while fresh milt and roe are transported to a disinfection facility. After fertilization and disinfection the roe are transported to the regional gene bank station where roe from each female (fertilized with sperm from only one male) are kept in quarantine in separate hatching cylinders. Eggs are removed from the quarantine when diseases which are transmissible to the interior of the eggs are discovered.

Within the gene bank station each family group is kept in separate tanks until the fish can be marked. Thereafter the families are pooled and each stock is kept in separate tanks throughout its lifecycle. The station is divided into four sections: (1) hatchery and initial feeding stage, (2) parr (3) smolt, and (4) older fish. These divisions makes it possible to manage diseases which may erupt within the station and it also eases the management of family groups and stocks.

Each of the captive stocks will spend two generations in the station. The production is based on the following guidelines aimed at retaining their genetic diversity:

1. Maximum survival.
2. Long generation time.
3. Identification of family groups.
4. Equal size of family groups.
5. A minimum effective population size of 50 for each generation.
6. Surplus fish-production for safety.
7. Mating schemes including the use of frozen sperm.

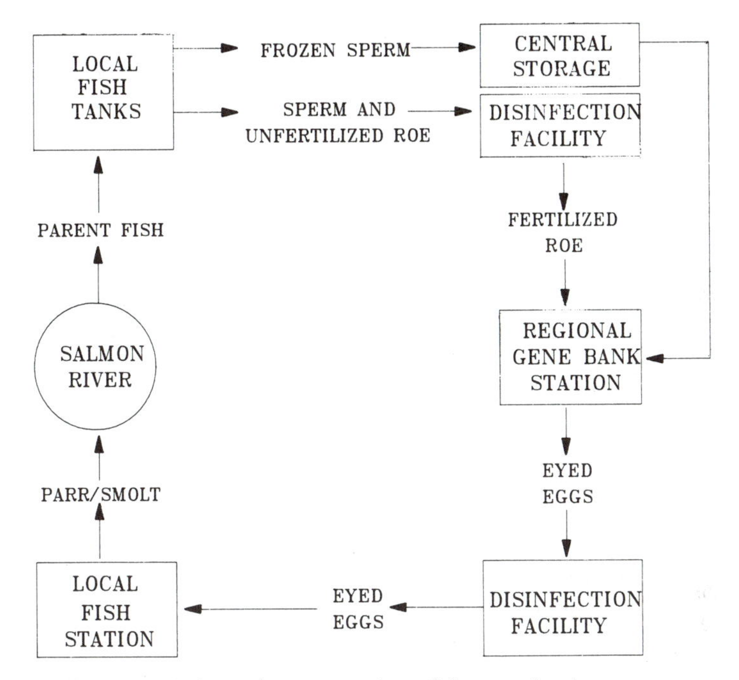

Figure 1. Schematic presentation of the gene bank strategy.

3. Method of Cryopreservation and Sampling Strategy

Deep freezing of sperm is a realistic gene bank alternative which facilitates gene storage over a nearly unlimited period of time (Stoss, 1983). The freezing method applied is the same as the one described by Stoss and Refstie (1983). Two millilitres of semen is taken from each fish; this volume is expected to fertilize 1000 eggs after thawing. The fertilizing ability of the frozen milt has been tested against controls using fresh milt, and based on the number of eyed eggs, was approximately 60% of that achieved when using fresh milt. In addition, biopsy and scale samples are taken from each fish for later analysis.

The population genetics of salmon stocks in Norway have not been mapped to any great extent. The sampling strategy of the gene bank programme is therefore based on general recommendations rather than knowledge about the genetic diversity. Marshall and Brown

(1975) considered optimum sampling strategies in plant populations when the genetic diversity is unknown and suggested a collection strategy that in most cases would obtain 95% of all alleles with a frequency greater than 0.05. This collection strategy is as follows:

1. 50-100 individuals should be collected from each location.
2. Sampling should be carried out at the maximum possible number of locations for a given time period.
3. Sampling locations should represent a wide ecological spectrum.

However, in the collection of plant material, one does not usually collect gametes from individual organisims, which is the case when collecting sperm from a male salmon. Allendorf and Phelps (1981) considered that a standard sample from a salmonid population should comprise 40-50 individuals.

The following sampling strategy has been employed by the gene bank programme:

1. Sperm from at least 50 individuals from each stock is frozen. Since the sampling cannot be carried out on identifiable stocks, each river is considered a sampling unit. Big tributaries are sampled separately. This choice of strategy is based on the empirical evidence that genetic differentiation along a single watercourse is minor compared to the between-river variation.
2. Sampling is carried out for a period of at least two years in each river to reduce the chances of gross overrepresentation of a single year-class.
3. Emphasis is placed on sampling from stocks representing a wide geographical and ecological range.
4. Stocks which are threatened by extinction are given priority over other stocks.
5. Stocks which are of particular scientific value, or valuable for fishing purposes, are also given priority.

4. Organization of the Gene Bank Programme

The gene bank programme is administered by the aquatic management division of the Directorate for Nature Management. Field work is planned in cooperation with the Environmental Protection Offices of the varying County Administrators. These offices provide advice about potential stocks for inclusion in the gene bank and arrange for local contacts.

The Advisory Committee of the Sea Ranching Programme has provided valuable scientific advice. A permanent advisory committee on genetic conservation of salmonid fishes will be instituted by the government in 1992.

Collection of sperm and eggs is carried out in close cooperation with local organizations which collect parent fish for use in the cultivation of individual rivers. These local groups include fishing and hunting associations, land owners associations, salmon boards and hydropower development boards. Staff from the Directorate, equipped with a mobile field laboratory, visits localities where parent fish are held. Sperm is frozen and eggs are collected at these locations. The permanent sperm bank is located at the breeding station for Norwegian Red Cattle (NRF) in Trondheim. Liquid nitrogen is provided by NRF.

The gene bank stations are owned and operated by private companies, hydropower boards and others. Contracts govern the relationship between the Directorate and the owners. All production costs and investments are financed by the Directorate. The Directorate decides upon all issues concerning the use of the stations facilities and also gives instructions to personnel.

5. Results and Experiences

The results of the cryopreservation programme is shown in Figure 2 and Table. 1. At present, the programme is halfway to reaching its current goal which is more than 100 stocks and 50 individuals from each stock. However a long term need for cryopreservation techniques is anticipated.

Figure 2. Locations of the 128 rivers included in the cryopreservation programme.

Table 1. Main Results of the Cryopreservation Programme.

Year	Individuals (N)	Stocks (N)
1986	364	46
1987	475	59
1988	410	49
1989	775	98
1990	774	90
Total	2798	128

The main costs of the cryopreservation programme lies in the field work. The Directorate has three field crews which operate in different parts of the country. The effectiveness of the field work is mainly restricted by the amount of time spent on transportation and execution of tasks other than fish stripping and cryopreservation procedures (preparation of equipment, disinfection, packing and unpacking of equipment, ordering of disposable supplies, and visits to local contacts). In 1990 the total cost of one sperm sample was about 800 NOK ($123).

The establishment of living gene banks for salmon is in its initial phases. So far the Directorate has put into operation two living gene banks for threatened salmon stocks (Fig. 3). At present the stations harbours 20 stocks but the collection of roe and milt continues. There will be about 30 stocks in these two stations when they are in full operation.

The total cost of a genebank station in full operation will amount to about 3 million NOK ($460,000) a year, or about $30,000 per stock.

Existing fish cultivation projects in individual rivers are of decisive significance for the success of the gene bank programme. The expenses would have been astronomical without the assistance of fish culture stations in catching and holding parent fish. The extent of fish cultivation activities therefore is a significant factor which may limit the number and selection

Figure 3. Situation of the regional gene bank stations (Haukvik and Eidfjord), and of the 19 salmon stocks (closed circles) and one trout stock (Salmo trutta; open circle) which are maintained at the stations.

of stocks in the gene bank in the future. However the Directorate's contribution and participation in such activities is increasing, especially in rivers with threatened stocks.

References

Allendorf, F.W. and S.R. Phelps. 1981. Isozymes and preservation of genetic variation in salmonid fishes, in: 'Fish Gene Pools," N. Ryman, ed., Ecological Bulletin, Stockholm.

Gausen, D. and V. Moen. 1991. Large-Scale Escapes of Farmed Atlantic Salmon (*Salmo salar*) into Norwegian Rivers Threaten Natural Populations. *Canadian Journal of Fisheries and Aquatic Sciences* 48:426-428.

Hesthagen, T. and L.P. Hansen. 1991. Estimates of the annual loss of Atlantic salmon, *Salmo salar* L., in Norway due to acidification. *Aquaculture and Fisheries Management* 22:85-91.

Marshall, D.R. and A.H.D. Brown. 1975. Optimum sampling strategies in genetic conservation, in: "Crop Genetic Resources for Today and Tomorrow," O. H. Frankel and J. G. Hawkes, eds., pp. 53-80, IBP 2, Cambridge Univ. Press, Cambridge.

Stoss, J. 1983. Fish gamete preservation and spermatozoan physiology, in: 'Fish Physiology, Vol 9B," W.S. Hoar, D.J. Randall, and E.M. Donaldson, eds., pp. 305-350, Academic Press, New York.

Stoss, J. and T. Refstie. 1983. Short-term storage and cryopreservation of milt from Atlantic salmon and sea trout. *Aquaculture* 30:229-236.

Reconstitution of Genetic Strains of Salmonids Using Biotechnical Approaches

GARY H. THORGAARD[1] and J. G. CLOUD[2]

1. Introduction

Genetic conservation of the existing salmonid stocks is an important goal in itself and as a component of programs designed to insure viable and sustainable fisheries under changing environmental conditions. These genetic resources may also become a valuable source of genes in support of the changing needs of the aquaculture industry.

Gene or genomic banking, achieved through the cryopreservation of sperm or germ cells (or cells that potentially develop into germ cells) is an important component of this conservation effort. There are at least three circumstances in which this technology would be useful. One is when a strain has become extinct in the wild but cryopreserved cells from the strain and suitable habitat within the native range are both available. A second is to recover the original genome of a strain without genetic alterations that have resulted from selection in hatcheries (Hynes et al., 1981; Allendorf and Ryman, 1987) or fisheries (Ricker, 1981). A third circumstance would be to recover genetic reference lines for research programs (Scheerer et al., 1991).

Although the advantages of these biotechnological approaches for preserving and recovering salmonid strains from cryopreserved cells have been recognized, concerns and objections to using this technology have been raised. One concern is that such approaches may detract from other efforts to conserve the strains, such as habitat protection. The rationale for this concern is that there are limited resources available for genetic conservation and that efforts associated with these approaches could reduce resources available for other activities. A second concern has been the opinion by some individuals that to rely on technological solutions to solve conservation problems is a mistake; these individuals (including some fishery biologists) believe that technology has created many of our conservation problems and cannot be relied on to help with the solutions. A third concern has been that fish strains preserved as cryopreserved cells at one point in time would not have been exposed to subsequent selection in the environment and consequently might be poorly adapted to future conditions. This concern is based on the premise that the conditions of the world are not static and that animals need to be adapting constantly to the changes in their environment; global warming is a good example of the type of change that might require ongoing adaptation in the environment.

[1]Departments of Zoology and Genetics and Cell Biology, Washington State University, Pullman, Washington 99164-4236, USA. [2]Department of Biological Sciences, University of Idaho, Moscow, Idaho 83843, USA

In spite of these concerns, biotechnological approaches to reconstituting salmonid strains have some clear advantages. Relative to the alternative of captive (hatchery) propagation, the advantages of a genomic bank include very low maintenance costs, high stability due to the lack of ongoing selection in captivity and a lower risk of loss due to system failures or disease. Since both captive propagation and cell cryopreservation have valid advantages for preserving and recovering strains, these approaches should be considered as complementary components of a complete program for strain conservation.

There is an immediate need for a genomic bank of cryopreserved cells. This is illustrated in our region by the recent extinction of the coho salmon and the dramatic depletion of the sockeye salmon in the Snake River, USA (Nehlsen et al., 1991). The primary cause for loss of the Snake River salmon has been construction of dams on the river's mainstem and the associated high mortality of downstream-migrating smolts. Efforts are being made to improve survival of the downstream migrants which, if successful, could lead to the reintroduction of coho in the Snake River basin. Native stocks could have been used in such recovery efforts if cells had been collected and cryopreserved from the locally-adapted stocks before their extinction.

The alternatives available for reconstitution of fish strains from cryopreserved cells include methods relying on the cryopreservation of (1) sperm or (2) embryonic cells.

2. Reconstitution from Cryopreserved Sperm

2.1. Repeated Backcrossing

Cryopreservation has been demonstrated to be a successful means of storing salmonid sperm for extended periods of time (Stoss, 1983; Baynes and Scott, 1987; Cloud et al., 1990; Wheeler and Thorgaard, 1991). Because only sperm can be successfully cryopreserved at this time, reestablishing an extinct population would be accomplished by fertilizing eggs of a closely related population with the stored sperm and utilizing the cryopreserved sperm to fertilize eggs of subsequent generations. Although several generations would be required to reconstitute a genetic strain by a program of repeated backcrossing, this approach is functional. There is already evidence that wild salmonid strains can be more readily reestablished if local sperm is used with foreign egg sources than if foreign strains are used (Bams, 1976).

2.2. Androgenesis

Since backcrossing takes a relatively long time to restore the equivalency of the original genome of an extinct population, androgenesis, the formation of embryos with all-paternal inheritance, is a potentially more rapid means of reconstituting a genetic strain from cryopreserved sperm. Androgenesis is induced by irradiating unfertilized eggs with gamma radiation and fertilizing the eggs with normal sperm (Romashov and Belyaeva, 1964; Purdom, 1969; Arai et al., 1979; Parsons and Thorgaard, 1985; Scheerer et al., 1986; May et al., 1988). Irradiation does not appear to excessively damage the eggs, as androgenetic diploids produced using sperm from tetraploid males show very good survival (Thorgaard et al., 1990). However, an optimal radiation dose does need to be identified for successful androgenesis, because doses above and below this level of radiation result in a decline in survival (Parsons and Thorgaard, 1984). In the absence of any additional treatment, the resulting haploid individuals die before or shortly after hatching.

Viable androgenetic diploids can be produced by inhibiting the first cleavage division and have been successfully reared to sexual maturity in spite of their complete homozygosity.

Successful treatments to block cleavage divisions of salmonid embryos have typically involved either heat or pressure regimes and need to be applied with precise timing. Production of androgenetic diploids by blocking cleavage has been relatively consistent, even though the yield has been low (typically below 10% to feeding). Offspring viability is apparently reduced both as a result of their homozygosity and the treatments applied to block the first cleavage division (Thorgaard et al., 1990). Another limitation of homozygous diploid androgenesis is the poor fertility of the homozygous diploid females which are produced (Scheerer et al., 1991). Additionally, examination of sex ratios in the progeny of androgenetic rainbow trout males has shown that XX as well as YY individuals can sometimes develop as males (Scheerer et al., 1991).

If homozygous diploid androgenesis is to be used for strain reconstitution, it will be necessary to cross among fertile homozygous androgenetic diploids to regenerate an outbred population. The low fertility of homozygous females and the additional time required to produce an outbred population are disadvantages of this approach.

Several alternative approaches are available for inducing androgenesis in fish. The production of outbred androgenetic diploids using sperm from tetraploid males (Thorgaard et al., 1990) has been demonstrated in rainbow trout. Alternatively, the production of outbred androgenetic individuals using fused sperm or a double-fertilization process may be possible, but these methods have not yet been fully demonstrated. Each of these alternatives has advantages and liabilities.

Production of outbred androgenetic diploids using sperm from tetraploid males is more demanding than homozygous diploid androgenesis because tetraploids must first be generated and producing viable tetraploids in salmonids has proven to be difficult (Chourrout, 1984; Chourrout et al., 1986; Myers et al., 1986). This approach requires that tetraploids be made for every strain which needs to be preserved while homozygous diploid androgenesis only initially requires that the sperm be cryopreserved. Another limitation of the tetraploid approach is that an excess of males is expected in the progeny (Chourrout et al., 1986); this can initially be overcome by hormonal sex reversal but would continue to be a problem in future generations. However, the viability of androgenetic offspring of tetraploids is expected to be higher than the viability of homozygous diploids because they are less inbred (Thorgaard et al., 1990).

Two potential androgenetic approaches which have not yet been demonstrated are the use of fused sperm or double fertilization to generate outbred androgenetic diploids. With the first approach, sperm would be fused by membrane or antibody treatments before being used to fertilize irradiated eggs. The success of androgenesis using sperm from tetraploid males provides encouragement for this fused sperm approach. The second approach would involve normal fertilization with one sperm followed by a second fertilization through a hole in the chorion; the micropyle normally acts as a block to fertilization with more than one sperm, but holes in the chorion or absence of a chorion can allow polyspermic fertilization (Yanagimachi, 1957; Sakai, 1961; Iwamatsu, 1983). An advantage that these methods could have over the tetraploid approach is that they would not require the extra generation for production of viable, fertile tetraploids. A potential advantage over the homozygous diploid approach would be that the offspring would be outbred and presumably more viable. However, the ease of applying androgenesis by sperm fusion or double fertilization is difficult to predict; neither has yet been definitively demonstrated. Sperm fusion has been reported using polyethylene glycol treatment of rainbow trout sperm (Ueda et al., 1986) but has not yet been combined with androgenesis. Polyspermic fertilization normally leads to pathological development (Iwamatsu, 1983) but if only two sperm fertilize the egg, normal development might proceed. Efforts to induce double fertilization by creating holes in the chorion before the second fertilization event were followed using pigmentation markers in a study in rainbow trout (Fields, 1991). Pigment mosaics, which could reflect incorporation of a second sperm nucleus in some but not all cells, were produced. A few pigmented triploids which could reflect complete contribution of the second nucleus in all cells were also produced. Further research on both sperm fusion and double fertilization as alternatives for androgenesis is needed.

3. Reconstitution Using Cryopreserved Embryonic Cells

Although cryopreservation of spermatozoa is an adequate technology with which to establish a genomic bank for salmonids, the time required to reconstitute a population from frozen sperm is relatively lengthy (as discussed earlier) and the maternally inherited mitochondrial genome of the population is lost. One obvious solution to overcome these problems is to cryopreserve embryos of the population; with this approach, both male and female outbred embryos would be available upon thawing. In fact this capability has been developed for a number of organisms. The successful cryopreservation of mammalian embryos was first reported by Whittingham et al. (1972) and Wilmut (1972) for the mouse and was quickly extended to a number of other mammalian species (Bank and Maurer, 1974; Whittingham, 1975; Willadsen et al., 1976; Willadsen et al., 1978). Although embryos from a wide variety of species (mainly mammals) have been successfully cryopreserved and a vast literature on the conditions and procedures used for mammalian embryo cryopreservation exists, attempts to freeze embryos of lower vertebrates have not been very successful (Mazur, 1979; also see Rall, this volume; Harvey, this volume). In those studies in which salmonid embryos have been used, only limited survival has been reported following relatively short exposures of the embryos to subfreezing temperatures (Zell, 1978; Erdahl and Graham, 1980; Stoss and Donaldson, 1983).

3.1. Fish Chimeras

The most successful protocol relative to fish embryos has been the cryopreservation of isolated blastomeres. Using zebrafish embryos, Harvey (1983) was the first to demonstrate that isolated cells from these embryos at half epiboly could successfully withstand freezing/thawing at liquid nitrogen temperatures with a high percentage of viable cells. Similar results have been obtained with rainbow trout blastomeres. Nilsson and Cloud (unpublished results) have frozen and stored isolated blastomeres from mid-blastulae of rainbow trout in liquid nitrogen; upon thawing approximately 39% of the cells were evaluated as being viable. The conclusion from the research conducted to date is that freezing salmonid embryos does not appear to be feasible with present technology but that isolated cells of the blastoderm are able to survive the freezing and thawing process.

In order for the freezing and storage of isolated blastomeres to be an important component of a genomic bank, the individual embryonic cells need to be reestablished into a germ-line. One method to accomplish this goal would be to transplant the isolated cells into a recipient embryo to form a resultant chimera. The assumption of this procedure is that the transplanted cells would participate in development and that a portion of the injected cells or their progeny would enter the germ-line. This technique of producing chimeras artificially was developed originally using mammalian embryos. From the pioneering work of Tarkowski (1961) and Mintz (1962), it was demonstrated unequivocally that mammalian chimeras could be produced experimentally by combining isolated blastomeres from different embryos. Later, Gardner (1968) developed an alternative method in which mammalian chimeras were produced by injecting cells from the inner cell mass of one embryo into the blastocoel of another. In the chimeras produced by either of these methods, it has been clearly shown that the newly introduced cells are able to enter the germ-line and give rise to functional gametes (McLaren, 1984).

The methodology of transplanting isolated blastomeres into recipient blastulae has been developed for rainbow trout (Nilsson and Cloud, 1989; 1992) and for the zebrafish (Lin et al., 1992). From an analysis of the distribution of labelled cells in different cell lines or tissues of embryos or hatched fry, it has been concluded that isolated cells from blastulae can reestablish themselves following transplantation into a recipient embryo and that they are pluripotential. Although there is presently no evidence that the transplanted cells can also reestablish themselves in the germ-line of salmonids, Lin et al. (1992) have demonstrated that the gametes of a number of their zebrafish chimeras were derived from the transplanted cells. This technology is only in the developmental phase; it is expected that fish chimeras will be produced from previously frozen blastomeres and that these blastomeres will contribute to the germ-line.

3.2. Nuclear Transplantation

Nuclear transplantation could also be developed for genetic conservation of salmonids. The successful transplantation of nuclei from blastomeres of a vertebrate embryo into enucleated, activated eggs was first reported by Briggs and King (1952) using an amphibian. Subsequently Tung et al. (1963) and Gasaryan et al. (1979) have reported the successful transplantation of nuclei from embryonic cells into enucleated eggs of *Carassius auratus* and *Misgurnus fossilius* respectively. Although the details concerning these investigations are limited, two results appear to be clear. Firstly, the nuclei of cells derived from mid-blastulae are totipotent, and secondly, with *Carassius*, fertile, adult fish have been produced by this technique (Research Groups, 1980).

Although this methodology is technically demanding, outbred fish could be produced in one generation by transferring nuclei from cryopreserved, embryonic cells into enucleated eggs of a nearby population of the same species.

4. Conclusions

As indicated in this and previous discussions, there is a need for an organized effort to collect and store germ plasm from our wild/native salmonid populations. At the very least, this activity could act as insurance against the demise of a population as other programs designed to restore or strengthen the populations are ongoing.

As we have presented, the technology is presently available to cryopreserve sperm and there is ongoing research to improve our capabilities to store and retrieve salmonid germ plasm.

References

Allendorf, F.W., and N. Ryman. 1987. Genetic management of hatchery stocks, in: "Population Genetics and Fishery Management." N. Ryman and F. Utter, eds. pp. 141-159. Washington Sea Grant Program, Seattle.

Arai, K., H. Onozato and F. Yamazaki. 1979. Artificial androgenesis induced with gamma irradiation in masu salmon (*Oncorhynchus masou*). *Bull Fac. Fish. Hokkaido Univ.* 30:181-186.

Bams, R.A. 1976. Survival and propensity for homing as affected by presence or absence of locally adapted genes in two transplanted populations of pink salmon (*Oncorhynchus gorbuscha*). *J. Fish. Res. Board Can.* 33: 2716-2725.

Bank, H. and R.R. Maurer. 1974. Survival of frozen rabbit embryos. *Exp. Cell Res.* 89:188-196.

Baynes, S.M. and A.P. Scott. 1987. Cryopreservation of rainbow trout spermatozoa: the influence of sperm quality, egg quality and extender composition on post-thaw fertility. *Aquaculture* 66:53-67.

Briggs, R. and T.J. King. 1952. Transplantation of living nuclei from blastula cells into enucleated frogs' egg. *Proc. Nat. Acad. Sci. USA.* 38:455-463.

Chourrout, D. 1984. Pressure-induced retention of second polar body and suppression of first cleavage in rainbow trout: production of all-triploids, all-tetraploids and heterozygous and homozygous diploid gynogenetics. *Aquaculture* 36:111-126.

Chourrout, D., B. Chevassus, F. Kreig, A. Happe, G. Burger, and P. Renard. 1986. Production of second generation triploid and tetrapolid rainbow trout by mating tetraploid males and diploid female-Potential of tetraploid fish. *Theor. Appl. Genet.* 72:193-206.

Cloud, J.G., W.H. Miller and M.J. Levanduski. 1990. Cryopreservation of sperm as a means to store salmonid germ plasm and to transfer genes from wild fish to hatchery populations. *Prog. Fish Cult. 52:51-53.*

Erdahl, D.A. and E. F. Graham. 1980. Preservation of gametes for freshwater fish. *Int. Congr. Anim. Reprod. Artif. Insem.* (Proceedings) pp. 317-326.

Fields, R.D. 1991. DNA fingerprinting and androgenesis in rainbow trout. Ph.D. Dissertation, Washington State University, Pullman, Washington. 122 pp.

Gardner, R.L. 1968. Mouse chimaeras obtained by the injection of cells into the blastocyst. *Nature* 220:596-597.

Gasaryan, K.G., N.M. Hung, A.A. Neyfakh and V.V. Invankov. 1979. Nuclear transplantation in teleosts *Misgurnus fossilis* L. *Nature* 280:585-587.

Harvey, B. 1983. Cooling of embryonic cells, isolated blastoderms, and intact embryos of the zebrafish *Brachydanio rerio* to -196C. *Cryobiology* 20:440-447.

Hynes, J.D., E.H. Brown, J., J.H. Helle, N. Ryman and D.A. Webster. 1981. Guidelines for the culture of fish stocks for resource management. *Can. J. Fish. Aquat. Sci.* 38:1867-1876.

Iwamatsu, T. 1983. A new technique for dechorionation and observations on the development of the naked egg in *Oryzias latipes*. *J. Exp. Zool.* 228-83-89.

Lin, S., W. Long, J. Chen, and N. Hopkins. 1992. Production of germ-line chimeras in zebrafish by cell transplants from genetically pigmented to albino embryos. *Proc. Natl. Acad. Sci. USA.* 89:4519-4523.

May, B., K.J. Henley, C.C. Krueger and S.P. Gloss. 1988. Androgenesis as a mechanism for chromosome set manipulation in brook trout (*Salvelinus fontinalis*). *Aquaculture* 75:57-70.

Mazur, P. 1979. Preservation of mammalian germ plasm by freezing, in: "Animal Models for Research on Contraception and Fertility." N.J. Alexander, ed., pp. 528-539. Harper and Row, Hagerstown.

McLaren, A. 1984. Germ cell lineages, in: "Chimeras in Developmental Biology," N. LeDouarin and A. McLaren, eds., pp.111-129, Academic Press, London.

Mintz, B. 1962. Formation of genotypically mosaic mouse embryos. *Am. Zool.* 2:432.

Myers, J.M., W.K. Hershberger and R.N. Kwamoto. 1986. The induction of tetraploidy in salmonids. *J. World Aquaculture Soc.* 17:1-7.

Naruse, K., H. Ijiri, A. Shima and N. Egami. 1985. The production of cloned fish in the medaka (*Oryzias latipes*). *J. Exp. Zool.* 236:335-341.
Nehlsen, W., J.E. Williams and J.A. Lichatowich. 1991. Pacific salmon at the crossroads: stocks at risk from California, Oregon, Idaho and Washington. *Fisheries* 16(2):4-21.
Nilsson, E.E. and J.G. Cloud. 1989. Production of chimeric embryos of trout (*Salmo gairdneri*) by introducing isolated blastomeres into receipient blastulae. *Biol. Reprod.* Suppl. 1, 40:90.
Nilsson, E.E. and J.G. Cloud. 1992. Rainbow trout chimeras produced by injection of blastomeres into receipient blastulae. *Proc. Natl. Acad. Sci.* USA, 89:9425-9428.
Parsons, J.E. and G.H. Thorgaard. 1984. Induced androgenesis in rainbow trout. *J. Exp. Zool.* 231:407-412.
Purdom, C.D. 1969. Radiation-induced gynogenesis and androgenesis in fish. *Heredity* 24:431-444.
Research Group of Cytogenetics, Research Group of Somatic Cell Genetics and Research Group of Nuclear Transplantation. 1980. Nuclear transplantation in teleosts. Hybrid fish from the nucleus of carp and the cytoplasm of crucian. *Scientia Sinica 23:517-523.*
Ricker, W.E. 1981. Changes in the average size and average age of Pacific salmon. *Can. J. Fish. Aquat. Sci.* 38:1636-1656.
Romashov, D.D. and V.N. Belyaeva. 1964. Cytology of radiation gynogenesis and androgenesis in loach (*Misgurnus fossilis*). Dokl. Akad. Nauk SSSR 157:964-967.
Sakai, Y.T. 1961. Method for removal of chorion and fertilization of naked egg in *Oryzias latipes*. *Embryologia* 5:357-368.
Scheerer, P.D., G.H. Thorgaard, F.W. Allendorf and K.L. Knudsen. 1986. Androgenetic rainbow trout produced from inbred and outbred sperm sources show similar survival. *Aquaculture* 57:289-298.
Scheerer, P.D., G.H. Thorgaard and F.W. Allendorf. 1991. Genetic analysis of androgenetic rainbow trout. *J. Exp. Zool.* 260:382-390.
Stoss, J. 1983. Fish gamete preservation and spermatozoan physiology, in: "Fish Physiology, Vol. 9B." W.S. Hoar, D.J. Randall, and E.M. Donaldson, eds., pp. 305-350. Academic Press, New York.
Stoss, J. and E.M. Donaldson. 1983. Studies on cryopreservation of eggs from rainbow trout (*Salmo gairdneri*) and coho salmon (*Oncorhynchus kisutch*). *Aquaculture* 31:51-65.
Streisinger, G., C. Walker, N. Downer, D. Kanuber and F. Singer. 1981. Production of clones of homozygous diploid zebrafish (*Brachydanio rerio*). *Nature* 291:293-296.
Tarkowski, A.K. 1961. Mouse chimaeras developed from fused eggs. *Nature* 190:857-860.
Thorgaard, G.H., P.D. Scheerer, W.K. Hershberger and J.M. Myers. 1990. Androgenetic rainbow trout produced using sperm from tetraploid males show improved survival. *Aquaculture* 85:215-221.
Tung, T.C., S.C. Wu, Y.Y.F. Tung, S.S. Yen, M. Tu and T.Y. Lu. 1963. Nuclear transplantation in fishes, *Scientia Sinica* 14:1244-1245.
Ueda, T., M. Kobayashi and R. Sato. 1986. Triploid rainbow trouts induced by polyethylene glycol. *Proc. Japan Acad.* B 62:161-164.
Wheeler, P.A. and G.H. Thorgaard. 1991. Cryopreservation of rainbow trout semen in large straws. *Aquaculture* 93:95-100.
Whittingham, D.G., S.P. Leibo and P. Mazur. 1972. Survival of mouse embryos frozen to -196C and -269C. *Science* 178:411-414.
Whittingham, D.G. 1975. Survival of rat embryos after freezing and thawing. *J. Reprod. Fertil.* 43:575-578.

Willadsen, S.M., C. Polge, L.E.A. Rowson and R.M. Moor. 1976. Deep freezing of sheep embryos. *J. Reprod. Fertil.* 46:151-154.

Willadsen, S.M., C. Polge and L.E.A. Rowson. 1978. The viability of deep-frozen cow embryos. *J. Reprod. Fert.* 52:391-393.

Wilmut, I. 1972. The effect of cooling rate, warming rate, cryoprotective agent and stage of development on survival of mouse embryos during freezing and thawing. *Life Sci.* 11:1071-1079.

Yanagimachi, R. 1957. Studies of fertilization in *Clupea pallasii*. VI. Fertilization of the egg deprived of the membrane. *Jap. J. Ichthyol* 6(3): 41-47.

Zell, S.R. 1978. Cryopreservation of gametes and embryos of salmonid fishes. *Ann. Biol. Anim. Biochem. Biophys.* 18:1089-1099.

Genetic Components in Life History Traits Contribute to Population Structure

A.J. GHARRETT and W.W. SMOKER

1. Introduction

The most important salmonid genetic resource is genetic variation. That variation is observed among populations throughout the range of Pacific salmon *(Oncorhynchus sp.)* as well as among the individuals within each population. Numerous electrophoretic studies of allozyme variation document the existence of both inter- and intra- populational genetic variation in Pacific salmon (e.g., Gharrett et al., 1987; Gharrett et al., 1988; Beacham et al., 1988). Much of the phenotypic variation observed also has a genetic basis (e.g., Ricker, 1972; Beacham et al., 1988); however, because the phenotype expressed is a result of both genetic and environmental influences, it is usually not straightforward to identify the genetic component. In addition, multiple genetic loci are often involved in expression of interesting and important life history traits such as morphology, development, and behavior.

Because of the difficulties involved in resolving genetic and environmental components from these quantitative traits, most fish population geneticists, for expediency, focus on the qualitative biochemical genetic variation often ignoring the quantitative genetic variation occurring within and among populations. That focus has provided and will continue to provide fisheries biologists with important biological and management information about salmonid fishes. However, by studying biochemical genetic variants, most of which are selectively neutral or nearly neutral, and ignoring life history traits, which undeniably affect fitness of individuals and populations, we are failing to study genetic variation that has evolutionary and conservation significance. Using allozyme variation, we have been observing only the most simplified models of population structure. Those models are important and necessary; but it is crucial that when they are used they are recognized as crude approximations of a complex situation and that the simplifying assumptions may not be justified.

For the past decade, we have been studying both the qualitative biochemical and quantitative genetic structure of a pink salmon (*O. gorbuscha*) population. Of particular interest has been the temporal and spatial timing of returning adults and of emigrant fry. We have identified this structure using a genetic marker bred into one subpopulation (Gharrett et al., in press) and have estimated the genetic component in return timing with a heritability study (Smoker et al., in press).

Fisheries and Ocean Sciences, University of Alaska Fairbanks, 11120 Glacier Highway, Juneau, Alaska 99801 USA

Genetic Conservation of Salmonid Fishes, Edited by J.G. Cloud and G.H. Thorgaard, Plenum Press, New York, 1993

2. Study Site and Species

Pink salmon have a rigid two-year life cycle (Gilbert, 1913; Davidson, 1934) which effectively segregates them into two reproductively isolated breeding units, those spawning in even years and those spawning in odd years (Aspinwall, 1974; Johnson, 1979; McGregor, 1982). Both even- and odd-year pink salmon return to Auke Creek, located near Juneau, Alaska, which flows approximately 350 m from Auke Lake to Auke Bay. Both emigrant fry and returning adults are counted through a weir located at tidewater. A small research hatchery operated by the National Marine Fisheries Service Auke Bay Laboratory is located just above the weir.

Five subpopulations of spawning adults have been identified in both even and odd years based on their time and place of spawning (Taylor, 1980; Gharrett et al., in press): early-run intertidal, early-run upstream, late-run intertidal, late-run upstream, and Lake Creek. Lake Creek is the primary tributary to Auke Lake and is small. Early fish return and spawn before September 1 and late fish return and spawn after September 1. There is usually a hiatus between the runs and the arrival of late-run fish is usually obvious because many of these fish still have some sea-bright scales. The weir is located at a natural demarcation (tidewater) between the intertidal and main stream segments of Auke Creek.

3. Timing of Adult Returns

In 1979 the late upstream run was genetically marked by altering the frequencies of electrophoretically detectable alleles that were present naturally at low frequencies (Lane et al., 1990). The marking was accomplished by spawning in the hatchery only late-run upstream adults possessing the *MDH-3,4*70* allele but not the *MDH-3,4*130* allele. When offspring from those fish returned in 1981 and interbred with naturally spawning late-run upstream fish, the frequency of the *MDH-3,4*70* allele increased from 0.05 to 0.24 and the frequency of the *MDH-3,4*130* allele decreased from 0.05 to 0.02 thereby marking that subpopulation. In subsequent generations, the frequencies of those marker alleles remained stable which indicates that there was little straying into that subpopulation and that the markers did not strongly affect fitness (Fig. 1).

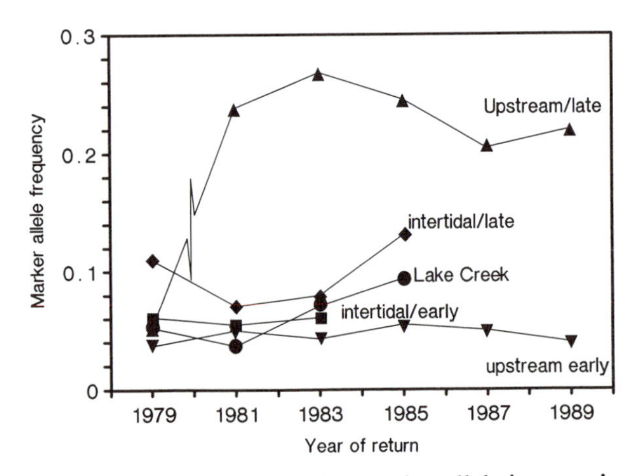

Figure 1. Frequency of the *MDH-3,4*70* marker allele in returning odd-year pink salmon in subpopulations of the Auke Lake drainage. The marker was bred into the late-upstream subpopulation in 1979 and has been monitored in this system since.

Because frequencies of the marker alleles did not change significantly in the other subpopulations (Fig. 1), it is also evident that few of the late-run upstream, genetically marked population interbred with fish from the other subpopulations. This is especially clear for the early-run subpopulations. These subpopulations have been monitored for five generations following the genetic marking. During this time, there has been no evidence of gene flow between early- and late-run fish, which suggests some degree of reproductive isolation.

Additional observations involving the genetic marker suggest even finer genetic structure exists, structure within subpopulations. In 1985, we began sampling returning adults daily at the weir to monitor the frequencies of early- and late-run markers for the upstream subpopulations. At first we were troubled because the MDH-3,4*70 frequency did not increase until after the late run began (Fig. 2). However, when we reviewed our marking procedure of 1979, we realized that we had not begun screening returning adults for spawners until the middle of the late run. This was done to reduce the possibility of inadvertently including early-run fish among the parents selected for genetic marking. As a result, we had selected for the later returning, late-run fish, and the inheritance of that return timing was still reflected in fish returning two generations later.

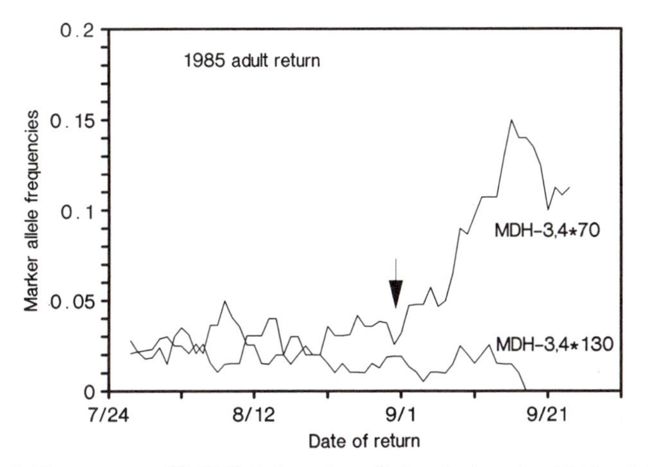

Figure 2. Frequency of *MDH-3,4* marker alleles during the 1985 adult return to the weir in Auke Creek. Approximately 10 fish were sampled daily. The curve was smoothed using running five day averages. The arrow indicates the arrival of late-run (silver bright) pink salmon at Auke Creek Weir.

From the genetic marking experiments, it is clear that there is a strong genetic component in the timing of returning adults. To estimate the genetic component of that timing, we conducted breeding experiments using a nested design, two dams per sire. Because we were aware that there were temporally (and probably genetically) distinct subpopulations, we performed two breeding experiments. One experiment bred 30 males and 60 females taken randomly from near the midpoint of the early run and the other experiment similarly made 60 families from fish taken randomly near the midpoint of the late run. All families of early-run ancestry returned during the early run; all families of late-run ancestry returned during the late run. The existence of genetic variability within subpopulations was also confirmed. Even after early-run and late-run effects were removed, heritabilities of return within a spawning run were about 0.4 and 0.2 for females and males respectively (Smoker et al., in press). We have also performed directional selection experiments that after one generation produced responses expected from a strong heritability for time of return (unpublished).

4. Timing of Juvenile Emigrations

We were also curious about the emigration timing of fry produced from early- and late-run fish. Specifically, we wanted to address the question of whether all fry emigrate simultaneously or do fry from late-run parents emigrate at a different time than fry from early run parents. To address that question, we sampled 10 to 20 fry each day during the emigration period of the marked, odd-year population which extended from March to June of even years. The samples were analyzed electrophoretically for the frequencies of *MDH-3,4* marker alleles. It is clear that the fry produced from the genetically marked, late-run subpopulation emigrate late in the seaward migration of pink salmon fry (Fig. 3). The pattern of adult return is similar to the pattern of fry emigration, which suggests that the two events have a genetic correlation.

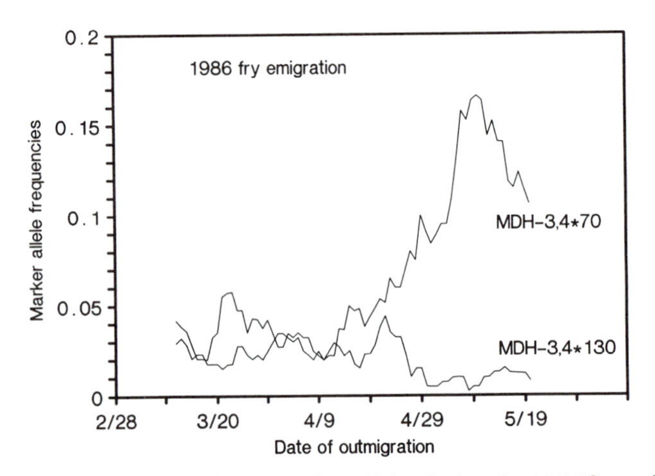

Figure 3. Frequency of *MDH-3.4* marker alleles during the 1986 fry emigration past the weir in Auke Creek. Approximately 10 fish were sampled daily. The curve was smoothed using running five day averages.

5. Discussion

Timing is clearly important to the survival of individuals and to the persistence of populations. If adults return too early or too late, the streams will often not be suitable for spawning or development of their embryos (Smoker et al., in press). If the fry emigrate too early or too late, they will miss the seasonally abundant food, essential for their early growth and survival. It is clear from the experiments we have reported here that there is a genetic component in timing. That idea is not new; many hatcheries have been selecting for run timing for a long time, either purposely or inadvertently. Selection of early spawning steelhead *(O. mykiss)* at the Skamania Hatchery on the Washougal River in Washington state changed the timing of the early run by two months in a 13 year timespan (Millenbach, 1973). In addition, Alexandersdottir (1987) suggests that timing of southeastern Alaskan pink salmon was altered as a result of directed fishing, and that alteration strongly affected the dynamics of pink salmon populations.

Although timing is an important genetically influenced trait, it is just one of many quantitative life history characters which are critical to the fitness of a population. Timing is particularly obvious to humans and is a good indicator of the importance of polygenic life history traits in general.

From the perspective of the importance of quantitative genetic characters to the long term success (fitness) of a population, it is actually the variability of these traits that is important. Most wild salmonid populations live in environments that vary over time, sometimes dramatically. Genetic variation provides the insurance that at least some of the individuals in the population will have a phenotype that can survive severe environmental perturbations that occur only occasionally.

The genetic variability underlying these life history characters is molded by the local environment. As a result, populations which are subjected to different environments have different genetic profiles. We have presented results which demonstrate that there is genetic control over migration timing. Moreover, because there is relatively little gene flow between early and late runs of pink salmon within the same stream, the two subpopulations are genetically different. The electrophoretic profiles of these temporally distinct subpopulations, however, are quite similar, as are those of other odd-year pink salmon runs near Auke Creek (McGregor, 1982).

The similarity of allozyme profiles suggests that some gene flow occurs among the many subpopulations. The genetic differences in timing indicates that genetic divergence resulted from selection for run timing and, by extrapolation, probably for other life history traits. These results are not contradictory. If the allozyme variants are neutral or nearly neutral and the life history traits are more subject to natural selection pressures, an equilibrium can be obtained that reflects the homogeneity of the neutral traits and the divergence of the non-neutral traits (Gillespie and Langley, 1976). However, the existence of two levels of divergence among populations indicates that allozyme data alone are insufficient to quantify genetic variability in populations. Moreover, populations spatially close to each other but occupying ecologically distinct habitats may resemble each other superficially (from allozyme frequency comparisons) but be genetically quite distinct at loci involved in expressing traits more sensitive to local selection regimes.

ACKNOWLEDGEMENTS. We appreciate the continued help and support of S.G. Taylor and the NMFS Auke Bay Laboratory for our pink salmon genetics research. J.A. Gharrett kindly reviewed the manuscript. This work was sponsored by Alaska Sea Grant College Program, cooperatively sponsored by NOAA, Office of Sea Grant and Extramural Programs, Department of Commerce, under Grant number NA90AA-D-SG066, project number R/02-17 and the University of Alaska with funds appropriated by the state. The U.S. government is authorized to produce and distribute reprints for governmental purposes notwithstanding any copyright notation appearing thereon.

References

Alexandersdottir, M. 1987. Life history of pink salmon *(Oncorhynchus gorbuscha)* and implications for management in southeastern Alaska. Ph.D. thesis. University of Washington, Seattle, WA. 148 p.

Aspinwall, N. 1974. Genetic analysis of North American populations of the pink salmon, *Oncorhynchus gorbuscha*, possible evidence for the neutral mutation-random drift hypothesis. *Evolution* 28:295-305.

Beacham, T.D., R.E. Withler, C.B. Murray, and L.W. Barner. 1988. Variation in body size, morphology, egg size, and biochemical genetics of pink salmon in British Columbia. *Transactions of the American Fisheries Society* 117:109-126.

Davidson, F.A. 1934. The homing instinct and age at maturity of pink salmon *(Oncorhynchus gorbuscha)*. *Bulletin of the U.S. Bureau of Fisheries* 48:27-39.

Gharrett, A.J., S. Lane, A.J. McGregor, and S.G. Taylor. in press. Use of a genetic marker to examine genetic interaction among subpopulations of pink salmon *(Oncorhynchus gorbuscha)*. *Canadian Journal of Fisheries and Ocean Sciences Special Publication.*

Gharrett, A.J., S.M. Shirley, and G.R. Tromble. 1987. Genetic relationships among Alaskan chinook salmon (*Oncorhynchus tshawytscha)* populations. *Canadian Journal of Fisheries and Aquatic Sciences.*

Gharrett, A.J., C. Smoot, A.J. McGregor, and P.B. Holmes. 1988. Genetic relationships of even-year northwestern Alaskan pink salmon. *Transactions of the American Fisheries Society* 117:536-545.

Gillespie, J. and C. Langley. 1976. Multilocus behavior in random environments. I. Random Levene models. *Genetics* 82:123-137.

Gilbert, C.H. 1913. Age at maturity of the Pacific coast salmon of the genus *Oncorhynchus*. *Bulletin of the U.S. Bureau of Fisheries*. 32:1-22.

Johnson, K. 1979. Genetic variation in populations of pink salmon *(Oncorhynchus gorbuscha)* from Kodiak, Alaska. M.S. thesis, University of Washington, Seattle, WA. 95 p.

Lane, S., A.J. McGregor, S.G. Taylor, and A.J. Gharrett. 1990. Genetic marking of an Alaskan pink salmon population, with an evaluation of the mark and the marking process. *American Fisheries Society Symposium* 7:395-406.

McGregor, A.J. 1982. A biochemical genetic analysis of pink salmon *(Oncorhynchus gorbuscha*) from selected streams in northern Southeast Alaska. M.S. thesis. University of Alaska-Juneau, Juneau, AK. 94 p.

Millenbach, C. 1973. Genetic selection of steelhead trout for management purposes. *International Atlantic Salmon Journal* 4:253-257.

Ricker, W.E. 1972. Hereditary and environmental factors affecting certain salmonid populations, in: Simon, R.C. and P.A. Larkin, eds., "The stock concept in Pacific salmon." H.R. MacMillan lectures in fisheries, University of British Columbia, Institute of Fisheries, Vancouver, B.C.

Smoker, W.W., A.J. Gharrett, and M.S. Stekoll. in press. Genetic variation in seasonal timing of anadromous migration in a population of pink salmon. *Canadian Journal of Fisheries and Aquatic Sciences Special Publication.*

Taylor, S.G. 1980. Marine survival of pink salmon fry from early and late spawners. *Transactions of the American Fisheries Society*. 109:79-82.

Status and Plight of the Searun Cutthroat Trout

PATRICK C. TROTTER,[1] PETER A. BISSON,[2] and BRIAN FRANSEN[2]

1. Introduction

The Endangered Species Committee of the American Fisheries Society recently identified all Washington, Oregon, and California populations of searun cutthroat trout, the anadromous form of the coastal cutthroat trout, *Oncorhynchus clarki clarki*, as being at some level of risk of extinction (Nehlsen et al., 1991). Here we focus on these at risk populations, listing reasons for their decline as set forth by the AFS Committee. We then review the life history, ecology, and genetic population structure of the subspecies, and from this identify gaps in our knowledge that will have to be filled if these populations are to be preserved.

2. Populations at Risk and Reasons for Declines

The status of anadromous populations of coastal cutthroat trout was surveyed in the continental U. S. by the AFS Endangered Species Committee (Nehlsen et al., 1991). As Figure 1 shows, California, Oregon and Washington comprise nearly one-third of the subspecies' historic range. The Committee reached the conclusion that all native naturally spawning populations within this area are at some level of risk, either on the threshold of endangered, on the threshold of threatened, or are species of special concern due to low numbers or special environmental sensitivities. Reasons given by the Committee for the declines in population numbers are:

1. Present or threatened destruction, modification, or curtailment of habitat or range due to logging in forests, urban and rural development, or mainstem passage. Logging as well as urban and rural development has been intense throughout the surveyed area in recent years.
2. Over-harvest in recreational fishing.
3. Negative interactions with hatchery stocks and/or introduced species. These include searun cutthroat trout of hatchery origin and other salmonids such as coho salmon and steelhead.

3. Historic Range and Life Cycle

The historic range of *O. c. clarki* (Fig. 2) extends from the Eel River, California to Gore Point, Kenai Peninsula, Alaska (Behnke, 1979, 1988), a range that corresponds remarkably closely with the Pacific coast rainforest belt defined by Waring and Franklin (1979).

[1]4926 26th Ave. S., Seattle, WA 98108, U.S.A. [2]Weyerhaeuser Co., Tacoma, WA 98477, U.S.A.

Genetic Conservation of Salmonid Fishes, Edited by J.G. Cloud
and G.H. Thorgaard, Plenum Press, New York, 1993

Searun cutthroat trout populations comprise one of four principal life history forms found within this range (Trotter, 1989), the others being:

- Fluvial-Adfluvial: river-dwelling populations that migrate to small tributaries for spawning and rearing.
- Lacustrine: lake-dwelling populations that migrate to tributaries for spawning and rearing.
- Fluvial: populations that dwell in headwater reaches; may disperse locally but do not migrate.

Anadromous populations favor streams with small to moderate drainage area (watersheds up to 130 km^2) with an abundance of low-gradient channels (Hartman and Gill, 1968).

Details of the searun cutthroat trout life cycle have been published by Pauley et al. (1989) and Trotter (1989), and are illustrated in Figure 3. In the following summary, we have highlighted several areas where vital information is lacking.

Spawning: Spawning takes place in early spring (late January, February, and early March in Oregon, Washington and California) in upper reaches of small tributaries (0.10.3 m^3/sec. summer low flows) in low-gradient riffles and the shallow downstream ends of pools.

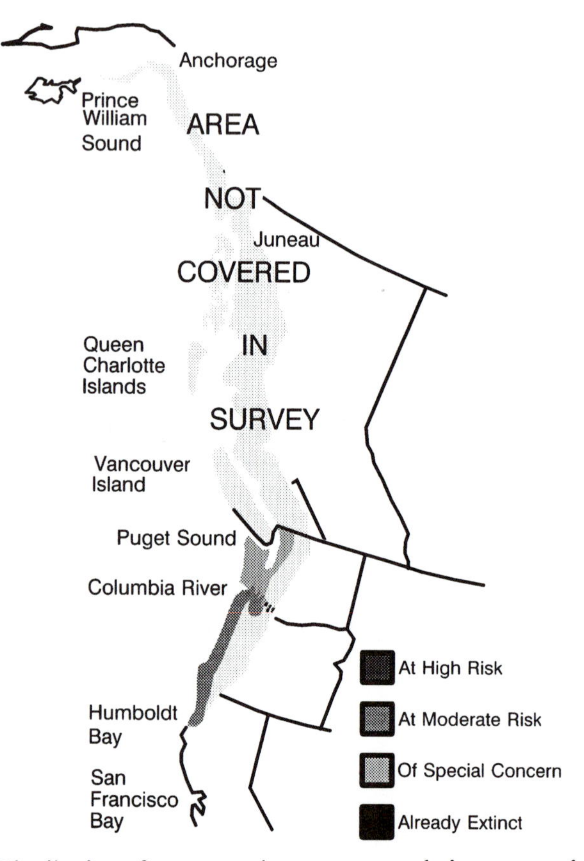

Figure 1. Distribution of searun cutthroat trout populations currently at risk of extenction in California, Oregon, and Washington (from Nehlsen et al., 1991).

Females typically spawn for the first time at age four. Searun cutthroat trout withstand the rigors of spawning rather well, and repeat spawning in subsequent years is common.

Egg incubation: Egg incubation requires 6 to 7 weeks and the alevins generally remain in the gravel about two weeks after hatching. Peak emergence usually occurs in mid-April but may be later in the northern portions of the range. Post-emergent fry move quickly into channel margins and backwaters for the first few weeks of their freeswimming lives.

Juvenile rearing in streams: Juvenile searun cutthroat trout prefer pools, but when other salmonids are present, the juvenile cutthroat trout move into low gradient riffles early in their first summer, then into pools as growth progresses. There is evidence of negative interactions with juvenile coho salmon and steelhead during this phase of searun cutthroat trout life history (Hartman and Gill, 1968; Glova, 1984, 1986; Bisson et al., 1988) which may limit searun cutthroat trout population size.

Populations exhibit variability during juvenile rearing. It has been suggested that this is an adaptation to the type of saltwater environment the young fish will be entering upon smoltification (Johnston, 1981). Where the fish enter relatively sheltered saltwater areas,

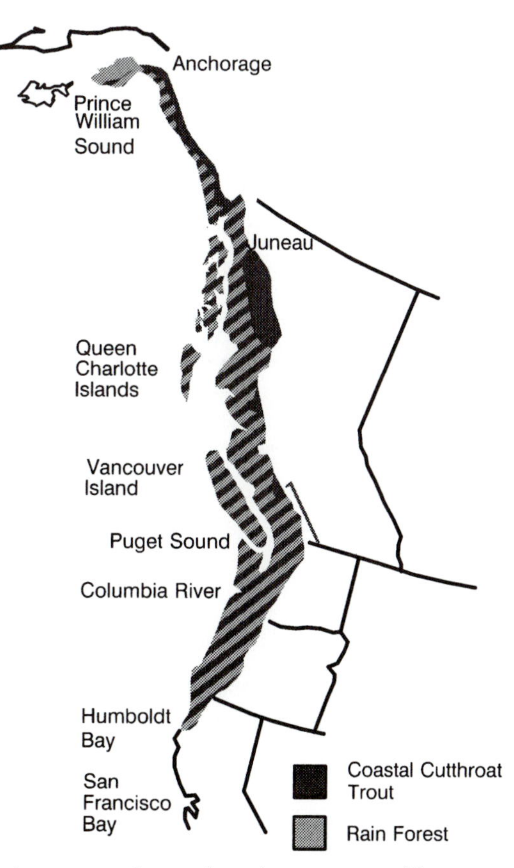

Figure 2. Native range of coastal cutthroat trout and its correspondence with the Pacific coast coniferous rainforest (from Behnke, 1979, 1988; and Waring and Franklin, 1979).

juveniles typically rear in the stream for two years. In areas where the young fish migrate directly into the open ocean, juvenile rearing extends for three years or more.

Smoltification: Peak outmigration occurs in mid-May in Washington, Oregon and California.

Saltwater Foraging: Fish stay close inshore, close to home streams, and limit their time in salt water to 5 to 8 months. Little else is known about saltwater movements and ecology.

Overwintering in Fresh Water: Searun cutthroat trout return from saltwater in late summer, fall, or winter. Not all of the returning fish will be ready to spawn the following spring, so the return to freshwater for these fish is truly an overwintering migration. Little if anything

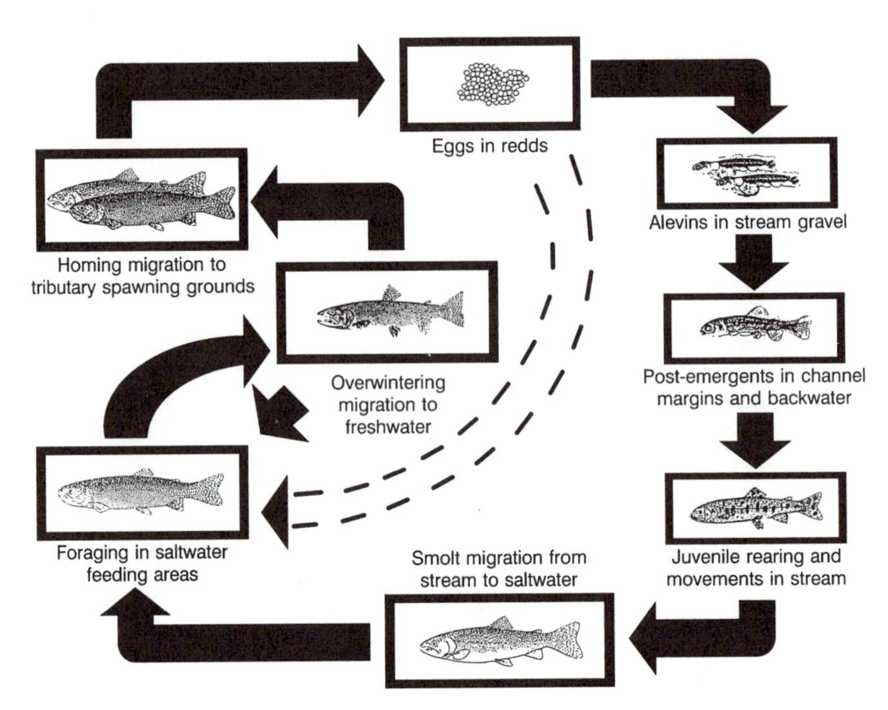

Figure 3. Life cycle stages of searun cutthroat trout.

is known about the requirements of the fish during the overwintering state. Fish that overwinter but do not spawn, along with any adults that have survived spawning, return to saltwater in the spring.

4. Amount and Distribution of Genetic Diversity

Information about the genetic composition of anadromous coastal cutthroat trout populations is available from only a small part of the subspecies range. Campton (1981) and Campton and Utter (1987) used protein electrophoresis to examine populations from northern and southern Hood Canal and north Puget Sound, Washington, and Tipping (1982) examined populations from the upper Cowlitz River, Washington (Fig. 4). Although the sampled populations were generally similar electrophoretically, variation was found to be apportioned among the four locales, suggesting that gene flow between these locales is restricted. Thus, populations from north Hood Canal, south Hood Canal, and north Puget Sound appear to represent distinct stocks, as does the Cowlitz River stock.

If the pattern of genetic variation found in these studies is reflected across the entire subspecies range, then the total subspecies gene pool could be composed of literally hundreds of genetically distinct breeding units.

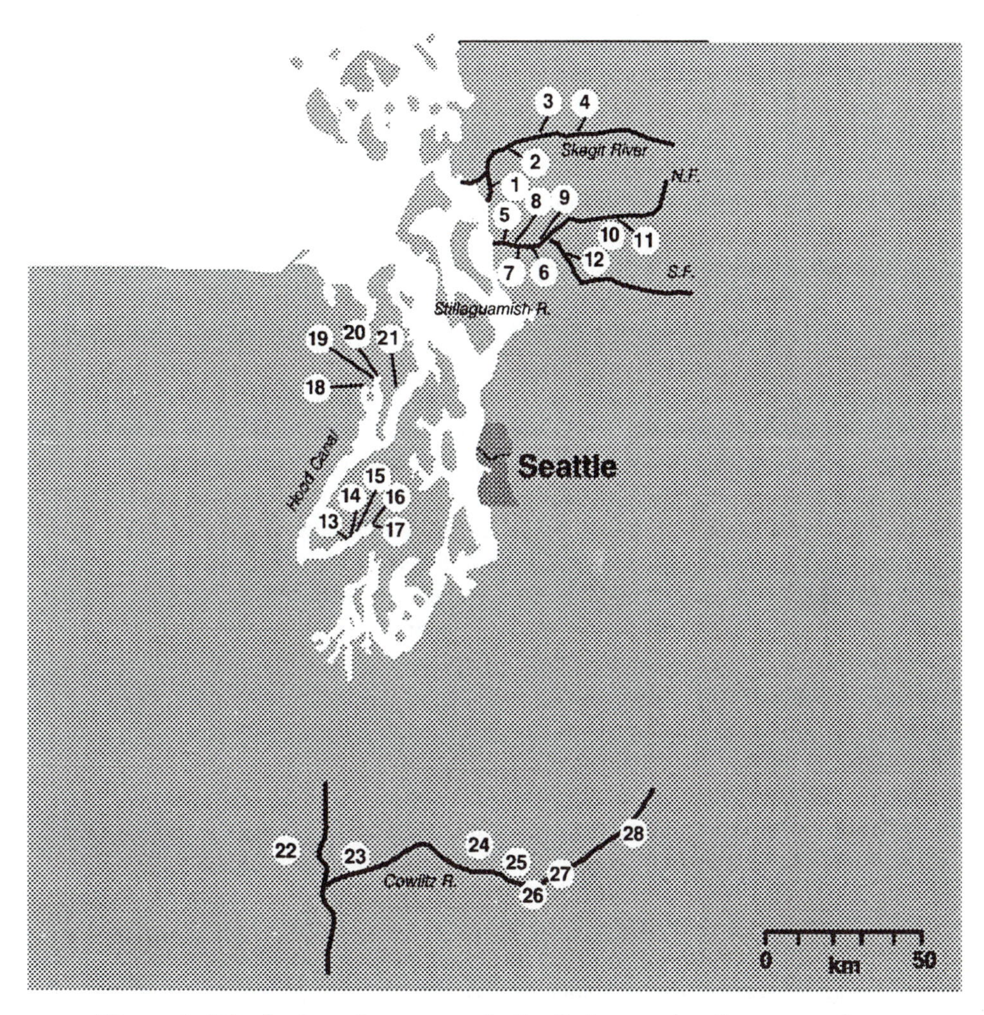

Figure 4. Distribution of some genetically distinct stocks of searun cutthroat trout in western Washington. Populations from sites 1-12 represent the northern Puget Sound stock; populations from sites 13-17 represent the southern Hood Canal stock; populations from sites 18-21 represent the northern Hood Canal stock; and populations from sites 22-28 represent the Cowlitz River stock (from Campton, 1981; Tipping, 1982; and Campton and Utter, 1987).

5. Key Research Needs

We suggest that further study is needed in the following key areas if populations at risk are to be preserved.

5.1. Freshwater Life History and Environmental Requirements

Although no other species of Pacific salmon appears to be as closely associated with the Pacific coastal rainforest as coastal cutthroat trout, there is still much to be learned about which environmental factors limit the production of anadromous populations. Cutthroat trout are known to heavily utilize stream habitat created by large coniferous woody debris (Bustard and Narver, 1975; June, 1981), and populations are known to be depressed by land use practices that result in losses of large woody debris and associated habitat (Lestelle and Cederholm, 1984; Hall et al., 1987). Yet other studies have documented shortterm increases in cutthroat trout populations after partial or complete removal of riparian trees (Murphy and Hall, 1981; Murphy et al., 1981; Bisson and Sedell, 1984). These increases have been shown to be related to increased food availability associated with elevated invertebrate production following forest canopy removal. Apparently complex interactions exist between the quality of physical habitat in streams and the abundance of prey, and our understanding of these interactions has still not reached the point that permits accurate forecasting of the outcome of environmental disturbances in forested watersheds on cutthroat trout populations (Hicks et al., 1991).

5.2. Limitation of Cutthroat Trout Populations Due to Interactions with Coho Salmon and Steelhead at the Juvenile Life History Stage

Studies carried out in British Columbia both in aquaria and in streams, have shown that juvenile coho salmon displace juvenile cutthroat trout from pools (Glova, 1984, 1986, 1987), and steelhead juveniles dominate juvenile cutthroat trout in riffles (Hartman and Gill, 1968). Juvenile coho and juvenile steelhead appear to be innately more aggressive than juvenile cutthroat trout, but there are also morphological differences between juveniles of the three species which confer performance advantage to juvenile coho over juvenile cutthroat trout in pools, and to juvenile steelhead over juvenile cutthroat trout in riffles (Bisson et al., 1988).

Since both riffles and pools are used by juvenile cutthroat trout, especially during that critical first summer of freshwater rearing, we suggest that competition for available habitat may limit cutthroat trout population size in streams where these species occur sympatrically. Measurements of juvenile salmonid biomass and density made over several years in a variety of habitat types in southwest Washington streams support this suggestion (Fig. 5).

When the pressures of competition for rearing space are removed, juvenile cutthroat trout densities may rebound. In one southwest Washington stream where juvenile coho are normally prevalent, a poor return of adult coho in the fall of 1989, followed by a scouring flood event that occurred in the winter of 1989-1990 after the few returning adult coho had spawned, nearly eliminated the 1990 coho year class. In the absence of young-of-the-year coho, juvenile cutthroat trout densities increased dramatically (Fig. 6).

This evidence is largely circumstantial, however. What is needed is a well-planned field study of juvenile interactions to confirm or reject the hypothesis that populations of anadromous cutthroat trout are controlled by exclusionary interactions at this life history stage. The results of this research would have an important management implication, particularly for salmon enhancement and supplementation programs. It has been the practice of fisheries managers in all three coastal states to stock hatchery coho or steelhead fry in rearing tributaries without actual knowledge of the numbers of natural fry present, or the potentially negative impacts of these releases on natural populations of other species such as coastal cutthroat trout. These practices would have to be curtailed if they were shown to be having severely adverse impacts on at-risk anadromous cutthroat trout populations.

5.3. Saltwater Movements and Ecology

This is an area that has never been systematically studied. Pearcy et al. (1990) recently reported on the distribution and biology of juvenile cutthroat trout in coastal waters off Oregon

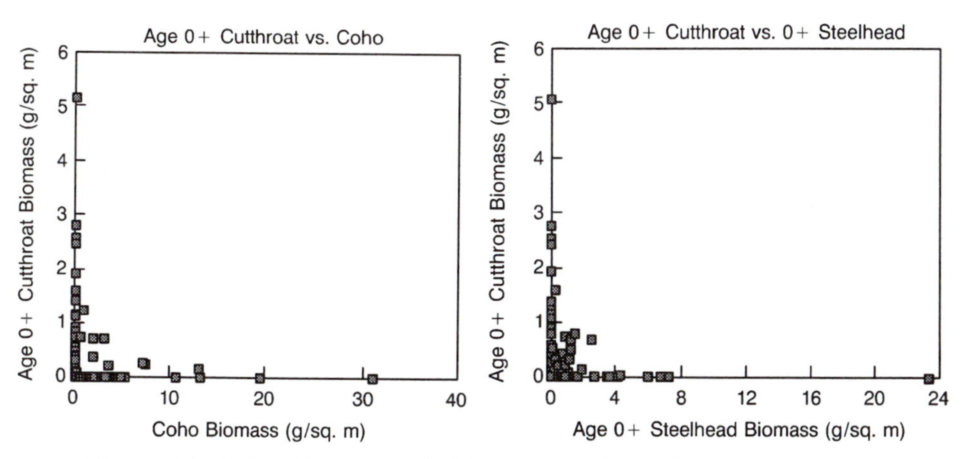

Figure 5. Relationship between the biomass of underyearling coastal cutthroat trout and underyearling coho salmon (left graph) and steelhead (right graph) in small streams in western Washington (data from B.R. Fransen and P.A. Bisson, unpublished).

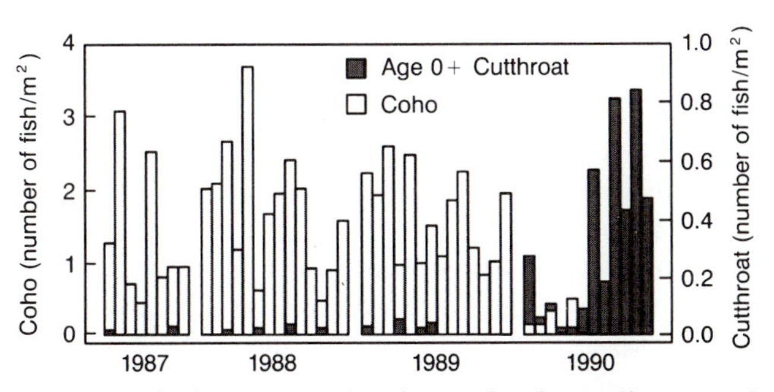

Figure 6. Increase in the summer abundance of underyearling coastal cutthroat trout in Huckleberry Creek, a tributary of the Deschutes River, Washington, after the absence of successful coho salmon spawning during the 1989-90 winter. Each bar represents the density of fish in an individual pool or riffle.

and Washington, but theirs was a compilation of data taken incidentally while pursuing other objectives.

5.4. Overwintering Ecology

Overwintering ecology is another area that has rarely been studied. The overwintering ecology of juvenile cutthroat presmolts has been investigated (Bustard and Narver, 1975; Sedell et al., 1984, Hartman and Brown, 1987) but not fish which have returned to overwinter after a period in saltwater.

5.5. Genetic Variation among Stocks

In addition to life history and ecological considerations, management plans for preserving *O. c. clarki* populations require detailed knowledge about how genetic variation is apportioned across the subspecies range, i.e., how many genetically significant breeding units exist, and how are they distributed geographically. This information is critical to establish population recovery programs. Emphasis should be placed on methods for obtaining genetic information that do not require sacrificing specimens, particularly in populations seriously at risk of extinction.

6. Summary

Wild native stocks of searun cutthroat trout are declining throughout the area surveyed by the AFS Endangered Species Committee. Information needed about the ecology, life history, and genetic population structure of anadromous *O. c. clarki* in order to form rational plans for preserving the declining stocks includes:

- A better understanding of Pacific coast rainforest ecosystem components that influence cutthroat trout production.
- Potential sensitivity of anadromous cutthroat trout populations to exclusion by coho and steelhead at the juvenile life history stage.
- Saltwater movements and ecology, and also freshwater overwintering requirements of both juveniles and prespawning adults.
- More complete information from across the subspecies historic range about the level and distribution of genetic diversity.

References

Behnke, R. J. 1979. The native trouts of the genus *Salmo* of western North America. Report to U. S. Fish and Wildlife Service, Denver. Colorado .

Behnke, R. J. 1988. Phylogeny and classification of cutthroat trout. *American Fisheries Society Symposium* 4: 17.

Bisson, P. A. and J. R. Sedell. 1984. Salmonid populations in streams in clear-cut vs. oldgrowth forests of western Washington, in: "Fish and wildlife relationships in old growth forests," W. R. Meehan, T. R. Merrell, Jr. and T. A. Hanley, eds., pp. 121-129, American Institute of Fisheries Research Biologists, Narragansett.

Bisson, P. A., K. Sullivan and J. L. Nielsen. 1988. Channel hydraulics, habitat use, and body form of juvenile coho salmon, steelhead, and cutthroat trout in streams. *Transactions of the American Fisheries Society* 117: 262-273.

Bustard, D. R. and D. W. Narver. 1975. Preferences of juvenile coho salmon (*Oncorhynchus kisutch*) and cutthroat trout (*Salmo clarki*) relative to simulated alteration of winter habitat. *Journal of the Fisheries Research Board of Canada* 32: 681-687.

Campton, D. E., Jr. 1981. Genetic structure of searun cutthroat trout (*Salmo clarki clarki*) populations in the Puget Sound area. Master's thesis. University of Washington, Seattle.

Campton, D. E, Jr. and F. M. Utter. 1987. Genetic structure of anadromous cutthroat trout (*Salmo clarki clarki*) populations in the Puget Sound area: evidence for restricted gene flow. *Canadian Journal of Fisheries and Aquatic Sciences* 44: 573-582.

Glova, G. J. 1984. Management implications of the distribution and diet of sympatric populations of juvenile coho salmon and coastal cutthroat trout in small streams in British Columbia, Canada. *Progressive FishCulturist* 46: 269-277.

Glova, G. J. 1986. Interaction for food and space between experimental populations of juvenile coho salmon (*Oncorhynchus kisutch*) and coastal cutthroat trout (*Salmo clarki*) in a laboratory stream. *Hydrobiologia* 131: 155-168.

Glova, G J. 1987. Comparison of allopatric cutthroat trout stocks with those sympatric with coho salmon and sculpins in small streams. *Environmental Biology of Fishes* 20: 275-284.

Hall, J. D., G. W. Brown and R. L. Lantz. 1987. The Alsea watershed study: a retrospective, in: "Streamside Management: Forestry and Fisheries Interactions," E. O. Salo and T. W. Cundy, eds., Contribution Number 57, pp. 399-416, Institute of Forest Resources, University of Washington, Seattle, Washington USA.

Hartman, G. F. and T. G. Brown. 1987. Use of small, temporary, floodplain tributaries by juvenile salmonids in a west coast rainforest drainage basin, Carnation Creek, British Columbia. *Canadian Journal of Fisheries and Aquatic Sciences* 44: 262-270.

Hartman, G. F. and C. A. Gill. 1968. Distribution of juvenile steelhead and cutthroat trout (*Salmo gairdneri* and *S. clarki clarki*) within streams in southwestern British Columbia. *Journal of the Fisheries Research Board of Canada* 25: 33-48.

Hicks, B. J., J. D. Hall, P. A. Bisson and J. R. Sedell. 1991. Response of salmonids to habitat changes, in: "Influences of Forest and Rangeland Management on Salmonid Fishes and Their Habitat," W. R. Meehan, ed., pp. 483-518, American Fisheries Society Special Publication 19, Bethesda.

Johnston, J. M. 1981. Life history of anadromous cutthroat trout with emphasis on migratory behavior, in: "Proceedings of the Salmon and Trout Migratory Behavior Symposium," E. L. Brannon and E. O. Salo, eds., pp. 123-127, University of Washington, School of Fisheries, Seattle.

June, J. 1981. Life history and habitat utilization of cutthroat trout (*Salmo clarki*) in a headwater stream on the Olympic Peninsula, Washington. Master's Thesis, University of Washington, Seattle, USA.

Lestelle, L. C. and C. J. Cederholm. 1984. Short-term effects of organic debris removal on resident cutthroat trout, pp. 131-140 in W R. Meehan, T. R. Merrell, Jr. and T. A. Hanley, eds., Fish and Wildlife Relationships in Oldgrowth Forests. American Institute of Fisheries Research Biologists, Narragansett.

Moore, K. M. S. and S. V. Gregory. 1988. Summer habitat utilization and ecology of cutthroat trout fry (*Salmo clarki*) in Cascade Mountain streams. *Canadian Journal of Fisheries and Aquatic Sciences* 45: 1921-1930

Moore, K. M. S. and S. V. Gregory. 1988. Response of young-of-the-year cutthroat trout to manipulation of habitat structure in a small stream. *Transactions of the American Fisheries Society* 117: 162-170

Murphy, J. L and J. D. Hall. 1981. Varied effects of clearcut logging on predators and their habitat in small streams of the Cascade Mountains, Oregon. Canadian *Journal of Fisheries and Aquatic Sciences* 38. 137-145.

Murphy, M. L., C. P. Hawkins and N. H. Anderson. 1981. Effects of canopy modification and accumulated sediment on stream communities. *Transactions of the American Fisheries Society* 110: 469-478.

Nehlsen, W., J E. Williams, and J. A. Lichatowich. 1991. Pacific salmon at the crossroads: stocks at risk from California, Oregon, Idaho, and Washington. *Fisheries* 16, no. 2: 421.

Pearcy, W. G., R. D. Brodeur, and J. P. Fisher. 1990. Distribution and biology of juvenile cutthroat trout *Oncorhynchus clarki clarki* and steelhead *O. mykiss* in coastal waters off Oregon and Washington. *Fishery Bulletin, U. S.* 88: 697-711.

Pauley, G. B., K. Oshima, K. L. Bowers, and G. L. Thomas. 1989. Species profiles: life histories and environmental requirements of coastal fishes and invertebrates (Pacific Northwest)sea-run cutthroat trout. U. S. Fish and Wildlife Service Biological Report 8: (11.86). U. S. Army Corps of Engineers TR EL824. 21 pp.

Sedell, J. R., J. E. Yuska, and R. W. Speaker. 1984. Habitats and salmonid distribution in pristine, sediment-rich valley systems: S. Fork Hoh and Queets River, Olympic National Park, in: "Fish and Wildlife Relationships in Old Growth Forests," W. R. Meehan, T. R. Merrell, Jr., and T. A. Hanley, eds., pp.33-46, American Institute of Fisheries Research Biologists, Narragansett.

Tipping, J. 1982. Cowlitz River searun cutthroat 1979-1981. Olympia, Washington, Washington State Game Department, Fisheries Management Division Report 829.

Trotter, P. C. 1989. Coastal cutthroat trout: a life history compendium. *Transactions of the American Fisheries Society* 118: 463-473.

Waring, R. H. and J. F. Franklin. 1979. Evergreen coniferous forests of the Pacific Northwest. *Science* 204: 1380-1386.

Status of Genetic Conservation in Salmonid Populations from Asturian Rivers (North of Spain)

P. MORAN, E. GARCIA-VAZQUEZ, A.M. PENDAS, J.I. IZQUIERDO, J.A. MARTIN VENTURA* and P. FERNANDEZ-RUEDA*

1. Introduction

Salmo salar L. and *S. trutta* L. (migratory and not migratory) are the two native species of salmonids from Asturias, north of Spain. *S. trutta* is the dominant species in Asturian rivers, without apparent recession. However, *S. salar* populations are changing in two ways: catches have decreased significantly from 1953 to 1989 in three rivers, Navia, Sella, and Cares-Deva, and the proportion of grilse (one-sea-winter salmon) increased in all rivers with the increases being significant in two rivers, Narcea and Cares-Deva (Garcia de Leániz and Martinez, 1988; Nicieza et al., 1990). *S. salar* populations show signs of being disturbed.

From March to July both species are exploited by legally regulated angling (rod and line). Spawning time is from late November to early January for both species. Populations are managed by means of ban periods and restocking with both autochthonous and foreign stocks (Tables 1 and 2). All the Asturian rivers and tributaries have abundant natural spawning of *S. trutta*, except in areas isolated by hydroelectric powerstations as in the Middle-Nalón. However, *S. salar* natural spawning is restricted to some areas because of obstacles in the channel of the river and destruction of banks. Redd inventory has not been carried out in Asturian rivers. Eo, Esva, Narcea, Sella, and Cares conserve some spawning areas. Navia River has lost the most native salmon production and its spawning areas are very reduced.

The present status of genetic conservation in Asturian rivers of both species, *S. salar* and *S. trutta* is the subject of this paper.

2. Materials and Methods

To detect genetic consequences of over-exploitation and restocking, genetic studies are carried out in *S. salar* populations from the rivers Navia (low spawning), Esva (high natural spawning), Narcea (middle natural spawning) and its tributary Pigüeña (middle natural spawning); and in *S. trutta* populations from the rivers Trubia, Middle-Nalón, Aller, Pigüeña, Esva, Ollorin, and Navia. Middle-Nalón lost the native *S. trutta* population ten years ago and was intensively restocked. Samples of trout were taken by electrofishing from May to November, 1990.

Departamento de Biología Funcional (Genética), Universidad de Oviedo, 33071-Oviedo, Spain.
* Consejería de Medio Ambiente, 33007-Oviedo, Spain.

Genetic Conservation of Salmonid Fishes, Edited by J.G. Cloud and G.H. Thorgaard, Plenum Press, New York, 1993

Native salmon parr were captured before stocking in the Esva and in the Pigüeña rivers (native samples in Table 4); salmon parr were also captured five months after stocking in the Esva, Narcea, and Navia rivers and seven months after stocking in the Pigüeña river (after-stocking samples in Table 4). In addition, Table 3 shows data of Scottish stocked salmon (stocked samples) from Youngson et al. (1989).

Table 1. Stocking effort for *S. salar* in the last ten years. Thousands of individuals released in each river, origin of stocked fishes and year of stocking.

		Asturian Rivers Restocked[1]							
Year	Origin[2]	Cares	Bedon	Sella	Narcea	Esva	Navia	Porcia	Eo
1981	Scotland	3.0	-	1.5	1.5	-	-	-	-
1982	Iceland	60.0	-	60.0	45.0	-	10.0	-	-
	Norway	-	-	30.0	25.0	20.0	-	-	-
	Scotland	30.0	-	-	30.0	20.0	-	-	-
1983	Iceland	-	-	-	21.0	25.0	-	-	-
	Norway	45.0	-	45.0	-	-	5.0	-	-
1984	Iceland	3.5	-	3.5	7.1	-	1.2	-	-
1985	Norway	4.0	-	12.5	15.0	13.0	4.0	-	10.0
1986	Autocht.	-	-	-	5.0	1.2	-	-	-
	Ireland	47.2	-	42.6	63.5	25	1.5	15.0	25.0
	Norway	-	-	-	-	-	5.0	-	-
1987	Autocht.	-	-	-	15.0	-	-	-	85.0
	Ireland	-	20.0	100.0	150.0	100.0	6.4	-	-
	Scotland	100.0	-	-	-	-	1.5	30.0	-
1988	Autocht.	-	-	-	5.0	-	-	-	-
	Ireland	25.0	-	70.0	25.0	25.0	-	-	25.0
	Scotland-C	50.0	30.0	50.0	-	-	-	-	50.0
	Scotland	25.0	-	25.0	75.0	100.0	41.0	30.0	25.0
1989	Scotland	90.0	21.0	85.0	97.0	85.0	55.0	27.0	80.0
1990	Scotland	93.0	-	102.0	131.0	83.0	17.5	40.0	91.0

1 Thousands of individuals released in each river
2 Scotland = River Shin; Scotland-C = River Connor

The three hatchery stocks of brown trout employed in repopulation (Cuenca, Huesca, and Infiesto) were analyzed in this study. Huesca belongs to a German population while the origin of Cuenca and Infiesto is unknown.

The enzymatic loci analyzed were: Pgm-1*, Pgm-2*, Gpi-1,2*, Gpi-3*, Ck-1*, Ck-2*, Agp-1,2*, Mdh-1*, Mdh-3,4*, Aat-1,2*, Aat-3*, Me-2*, Idh-4*, Sod*, and Ldh-4* in *S. salar*; and Aat-1*, Aat-2*, Aat-4*, Agp-2, Mdh-2*, Mdh-3,4*, Me-1,2,3*, Pgm-1*, Pgi-1,2*, Pgi-3*, Ck-1*, Ck-2*, Ldh-1*, Ldh-4*, Ldh-5*, Idh-3*, Idh-4*, and Sod* in *S. trutta*.

Starch gel electrophoresis methods followed Verspoor and Cole (1989) for salmon and Taggart et al. (1981) for trout. Interspecific hybrids were identified by the enzymatic patterns of Pgi* and Pgm* (Vourinen and Piironen, 1984). Nomenclature of loci protein variation follows Shaklee et al. (1990).

Table 2. Stocking effort for *S. trutta*. Thousands of individuals released in each drainage system.

Drainage	Origin of stocked trouts[1]							
				I and H				C and I
	1983	1984	1985	1986	1987	1988	1989	1990
Cares	50	30	50	25	20	-	-	-
Sella	110	83	50	40	20	55	-	-
Piloña	137	120	35	60	300	70	34	24.5
Villavic.	-	17	-	20	60	50	20	1.5
Nalon	60	70	30	100	110	100	12	23
Nora	19	21	-	15	-	25	15	3
Caudal	126	85	98	320	140	158	25	14.5
Trubia	71	27	90	-	20	35	18	12
H.Narcea	60	40	111	75	70	26	-	-
L.Narcea	42	70	94	45	86	65	35	10
Esva	21	13	45	40	-	-	8	10
Negro	6	5	17.5	-	-	-	-	-
Navia	20	-	75	-	-	25	-	3
Ibias	15	-	-	-	-	-	19	-
Porcia	7	10	-	-	-	5	-	40
Eo	7	10	*	*	*	15	-	3

[1] C=Cuenca; H=Huesca; I-Infiesto

Table 3. Number of individuals of each genotype for *Salmo trutta* Ldh-5* locus in the ten samples analyzed. Stocking characteristics of each sample.

Samples	Genotypes			Characteristics
	100/100	100/90	90/90	
Infiesto	0	0	50	For stocking 1978-1990
Cuenca	0	0	60	For stocking in 1990
Huesca	0	0	50	For stocking 1983-1989
High-Trubia	50	0	0	Native spawning, no restocked
Aller	50	0	0	Native spawning, high stocking
Middle-Nalón	0	0	50	Reintroduced by stocking in 1981
Pigüeña	122	0	0	Native spawning, low stocking
Ollorin	35	0	0	Native spawning, high stocking
Esva	45	0	0	Native spawning, low stocking
Navia	10	0	1	Native spawning, high stocking

3. Results and Discussion

Asturian native populations of *S. trutta* have an electrophoretic genetic marker, the allele *Ldh-5*-100* (Morán et al., 1991) which is absent in stocking populations (Table 3). Heterozygotes have not been found and mixture of native and stocked populations was not revealed for any sample. The whole Middle-Nalón population is from stocking origin whereas in all other studied populations stocking has not been successful. *S. trutta* restocking was successful only in the areas without natural spawning. We can conclude that *S. trutta* Asturian populations are not genetically disturbed because of restocking.

For *S. salar*, García-Vázquez et al. (1991) have demonstrated that restocking with Scottish parr was not successful in the Esva river in 1989, probably due to an unadaptation of Scottish salmon to the extreme and uncommon drought of 1989 in Asturias. Nevertheless, this study was not repeated in 1990.

Table 4 shows isozyme results for the four loci of *S. salar* useful for comparing stocked and Asturian populations. Similarity of after-stocking samples from the stocked population (river Shin, Scotland) and their divergence with the natives is evident. But the four after-stocking Asturian samples have an excess of homozygotes in all the loci analyzed, G reaching statistical significance (significant deviation from the Castle-Hardy-Weinberg equilibrium) in five cases (Table 5). This finding indicates significant mixture of native and stocked populations (Walhund effect, Verspoor and Cole, 1989). In this sense, a higher mixture was found in the Esva River, with high natural spawning.

Table 4. Frequencies of *100 allele in each locus.

	Stocked	Natives		After-stocking samples			
Locus	Shin	Esva	Pigüeña	Navia	Narcea	Pigüeña	Esva
Me-2*	0.44	0.92	0.90	0.41	0.41	0.52	0.48
Mdh-3*	1.00	0.99	0.99	1.00	1.00	1.00	0.93
Pgm-1*	-	0.98	1.00	1.00	1.00	0.98	1.00
Aat-3*	0.73	0.75	0.86	0.73	0.88	0.85	0.80
N	50	64	44	40	29	65	84

N = number of individuals in each sample.

Stocked fishes are able to adapt and survive in these Spanish rivers, at least to male parr maturity. Even if the rates of return of stocked fishes were low (Bams, 1976; Reisenbichler, 1988; García de Leániz et al., 1989), mixture of genomes is possible in those rivers because of the high rate of maturity in male parr. Mature males represented the following percentages of the total of captured individuals: 13.8 in the Narcea, 16.8 in the Esva, 17.5 in the Navia and 21.5 in the Pigüeña. These percentages increased at spawning time. Since parr fertility was 5% in spawnings of adult returning females (Hutching and Myers, 1988), this is the way for a year-by-year slow introduction of foreign genomes in Asturian *S. salar* populations. On the basis of the stocking story in Asturias and the ability of stocked fishes to reach parr maturity, we can conclude that Asturian *S. salar* populations are genetically disturbed with foreign genomes.

Both species, *S. salar* and *S. trutta*, are able to cross in the wild and a genetic introgression is possible in Spanish rivers. *S. salar* x *S. trutta* hybrids have been found in the Narcea and in the Esva rivers. The rate of interspecific hybridization, measured as the percent of hybrids against the total number of individuals captured, was 1.42% in the Esva sample and from 0.96 to 4.0% in different sites in the Narcea. One individual was a mature male parr. Similar hybridization rates have been reported in other rivers from northern Spain and could be considered higher than those from other European populations (Garcíade Leániz et al., 1989). Increasing hybridization rates can be interpreted as a consequence of restocking with foreign populations (Vuorinen and Piironen, 1984) and/or of abundance of mature parr

(Crozier, 1984). Anyway, it may be assumed as an indication of genetic disturbation in Asturian salmonid populations.

As a consequence of genetic studies related to both species in Asturian rivers, the "Consejería del Medio Ambiente" is compiling a guide for the management of salmonids in Asturias. All of the rivers will be conserved without Atlantic salmon restocking and consequences will be studied. Their future evolution will be taken into account for managing all the other Asturian salmonid populations in order to preserve native genetic pools.

Table 5. Number of individuals observed and expected (in brackets) of each genotype, in the four *Salmo salar* Asturian samples. G values to test Castle-Hardy-Weinberg equilibrium.

		After-stocking samples			
Locus	Genotype	Navia	Narcea	Pigüeña	Esva
Me-2*	100/100	7 (6.8)	6 (4.5)	21 (17.8)	24 (19.4)
	100/125	19 (19.4)	10 (13.1)	26 (32.4)	33 (41.9)
	125/125	14 (13.8)	11 (9.4)	18 (14.8)	27 (22.7)
	G	0.167(ns)	1.465(ns)	4.012[1]	3.875[1]
Mdh-3*	100/100	40	29	65	76 (72.7)
	100/80	0	0	0	5 (10.9)
	80/80	0	0	0	3 (0.4)
	G				10.930[2]
Pgm-1*	-100/-100	40	29	62 (62.0)	84
	-100/-80	0	0	3 (2.9)	0
	-80/-80	0	0	0 (0.1)	0
	G			0.082(ns)	
Aat-3*	100/100	29 (21.0)	24 (22.5)	48 46.5	55 53.8
	100/50	11 (15.9)	3 (6.1)	14 16.9	24 25.5
	100/25	0	0	0	1 1.3
	50/50	0 (3.1)	2 (0.4)	3 1.5	4 3.0
	G	4.001[1]	5.176[1]	1.670(ns)	1.154(ns)

1: P<0.05; 2: P<0.04; ns = not significant

ACKNOWLEDGEMENTS. We acknowledge very much the cooperation and suggestions of Dr. F. Braña, A.G. Nicieza, and M.M. Toledo (Departamento de Biología de Organismos y Sistemas, Universidad de Oviedo). D.G. Armando González helped us to sample in the rivers. This work was supported by the FICYT.

References

Bam, R.A. Survival and propensity for homing as affected by presence or absence of locally adapted paternal genes in two transplanted populations of pink salmon (*Oncorhynchus gorbuscha*). *J. Fish. Res. Board Can.* 33:2716-2725.

Crozier, W.W. 1984. Electrophoretic identification and comparative examination of naturally occurring F. hybrids between brown trout (*Salmo trutta* L.) and Atlantic salmon (*Salmo salar* L.). *Comp. Biochem. Physiol.* 78B:785-790.

Garcia de Leaniz, C. and J.J. Martinez. 1988. The Atlantic salmon in Spain with particular reference to Cantabria, in: "Atlantic Salmon: Planning for the Future" D. Mills and D. Piggins, eds., pp. 179-209, Croom Helm, London.

Garcia de Leaniz, C., E. Vespoor, and A.D. Hawkins. 1989. Genetic determination of the contribution of stocked and wild Atlantic salmon, *Salmo salar* L., to the angling fisheries in two Spanish rivers. *J. Fish Biol.* 35(Supp. A):261-270.

Garcia-Vazquez, E., P. Moran, and A.M. Pendas. 1991. Chromosome polymorphism patterns indicate failure of Scottish stock of *Salmo salar* transplanted into a Spanish river. *Can. J. Fish. Aquat. Sci.* 48:170-172.

Hutchings, J.A. and R.A. Myers. 1988. Mating success of alternative maturation phenotypes in male Atlantic salmon, *Salmo salar*. *Oecologia* 75:169-174.

Martin Ventura, J.A. 1978. The Atlantic salmon in Asturias, Spain: analysis of catches, 1985-86. Inventory of juvenile densities, in: "Atlantic Salmon: Planning for the Future" D. Mills and D. Piggins, eds., pp. 210-227, Croom Helm, London.

Moran, P., A.M. Pendas, E. Garcia-Vazques and J.I. Izquierdo. 1991. *Ldh-5** as a genetic marker for monitoring repopulations of brown trout *Salmo trutta* in Asturias. Proceedings of International Symposium on Biochemical Genetics and Taxonomy of Fish, Queen's University of Belfast.

Nicieza, A.G., M.M. Toledo and F. Braña. 1990. Capturas de salmon atlántico (*Salmo salar* L.) en los ríos asturianos en el periodo 1953-1989. Variaciones de abundancia y estructura de edades de mar. *Rev. Biol. Univ. Oviedo* (Suplemento) IV:1-91.

Reisenbichler, R.R. 1988. Relation between distance transferred from natal stream and recovery rate for hatchery coho salmon. *N. Am. Fish. Mgmt.* 8:172-174.

Shaklee, J.B., W. Allendorf, D. Morizot and G.S. Whitt. 1990. Gene nomenclature for protein coding loci in fish. *Trans. Amer. Fish. Soc.* 119:2-15.

Taggart, J., A. Ferguson and F.M. Mason. 1981. Genetic variation in Irish populations of brown trout (*Salmo trutta* L.): electrophoretic analysis of allozymes. *Comp. Biochem. Physiol.* 69B:393-412.

Verspoor, E. and L.J. Cole. 1989. Genetically distinct sympatric populations of resident and anadromous Atlantic salmon, *Salmo salar*. *Can. J. Zool.* 67:1453-1461.

Vuorinen, J. and J. Piironen. 1984. Electrophoretic identification of Atlantic salmon (*Salmo salar*), brown trout (*Salmo trutta*), and their hybrids. *Can. J. Fish. Aquat. Sci.* 41:1834-1837.

Youngson, A.F., S.A.M. Martin, W.C. Jordan and E. Vespoor. 1989. Genetic protein variation in farmed Atlantic salmon in Scotland: comparison of farmed strains with their wild source populations. Scottish Fis. Res. Rep. 42. DAFS, Marine Laboratory, Aberdeen.

Genetic Status of Atlantic Salmon (*Salmo salar*) in Asturian Rivers (Northern Spain)

J.A. SÁNCHEZ, G. BLANCO, and E. VÁZQUEZ

1. Introduction

The Atlantic salmon (*Salmo salar*) is a species which occurs in rivers on both sides of the North Atlantic. In the last 40-50 years, man's impact has resulted in a drastic decrease of this species throughout its native distribution and in some rivers it has disappeared. In order to choose an appropriate strategy for its conservation and management, a detailed knowledge both of the population structure and the amount and distribution of genetic variation throughout the geographical distribution range of the species is necessary (Stahl, 1983, 1987; Ryman, 1983). In the last decade, several studies, using electrophoretical techniques, were carried out to provide information about the genetic diversity and structure of Atlantic salmon populations (Cross and Ward, 1980; Stahl, 1981, 1983, 1987; Vourinen, 1982; Ryman, 1983; Cross and King, 1983; Guyomard, 1987; Verspoor, 1988; Vourinen and Berg, 1989; Koljonen, 1989; Sánchez et al., 1991). However, few data on natural populations limited to southern Europe are available (Guyomard, 1987; Sánchez et al., 1991).

The Asturian rivers, in the north of Spain, constitute the southern limit of *S. salar* distribution in Europe. These rivers are short, no longer than 100 km. The natural populations of Asturian rivers produce mostly one- and two-year-old smolts which return to the river after two or three years in the sea, although in recent years greater numbers of grilse (fish returning after one year in the sea) have been observed (Mártin-Ventura, 1987, 1988). In these, the salmon are exploited only by recreational fishermen who use rod and line. In the last four years (1987 to 1989) 6,829 fish were caught ranging in weight from 10 kg spring salmon to 2 kg grilse, and in length from 75 to 85 cm. These fish constitute 60 to 80% of the total annual salmon catch in Spain (Mártin-Ventura, 1987, 1988).

The purpose of this paper is to provide a general description of the genetic structure (based on analysis of protein-coding genes) of two natural salmon populations of Asturian rivers and to examine their relationships with other European populations from drainages of the Atlantic Ocean.

2. Materials and Methods

Tissue samples of liver and muscle of mature fish collected in two rivers of Asturias were analyzed by horizontal starch gel electrophoresis. From the Esva river, 27 fish were

Universidad de Oviedo, Departamento de Biología Funcional Area de Genética, 33071 Oviedo, Asturias, Spain collected during the fishing season (March-July) in 1987 and 36 in 1988. In the river Sella, 16 mature fish were collected in 1986, 83 in 1987, and 46 in 1988.

The following 13 enzymes and 27 loci were studied: alcohol dehydrogenase (Adh); aspartate aminotransferase (Aat-1,2,3); creatine kinase (Ck-1,2); diaphorase (Dia); isocitrate dehydrogenase (Idh-1,2,3); lactate dehydrogenase (Ldh-1,2,3,4); malate dehydrogenase (Mdh-1,2,3,4); malic enzyme (Me-1,2); phosphoglucose isomerase (Pgi-1,2); 6-phosphogluconate dehydrogenase (6-Pgd); sorbitol dehydrogenase (Sdh-1,2); superoxide dismutase (Sod) and

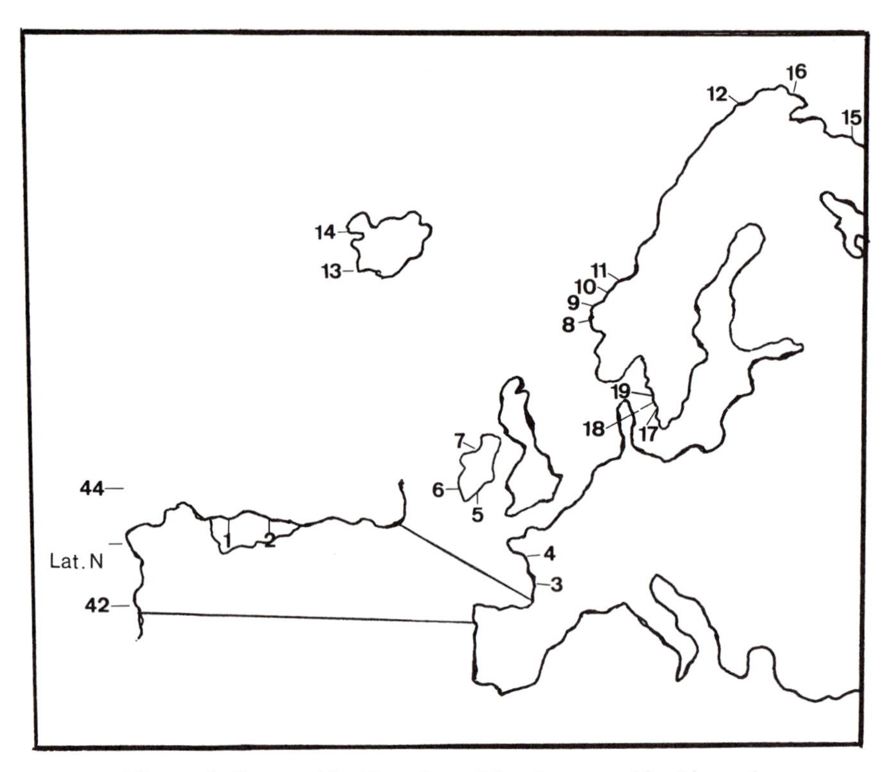

Figure 1. Geographical location of the rivers used in this study.

Asturias, Spain	1 Esva	Iceland (C)	13 Langa
	2 Sella		14 Edlida
France (A)	3 Allier	USSR (D)	15 Kola
	4 Elorn	Finland (D)	16 Tana
Ireland (B)	5 Blackwater	Sweden (C)	17 Lagan
	6 Burrishole		18 Fyllean
	7 Moy		19 Atran
Norway (C)	8 örstaelva		
	9 Bondalselva		
	10 Sokna		
	11 Mosvikelva		
	12 Alta		

A=Guyomard 1987; B=Cross and Ward 1980, Cross and King 1983; C=Stahl 1983, 1987; D=Koljonen 1989.

xantine dehydrogenase (Xdh). The electrophoretic techniques (buffer systems and staining recipes) were the same described by Sánchez et al., 1991.

The G-statistic was applied to tested heterogeneity of allele frequencies among samples (Sokal and Rohlf, 1981).

To compare with other European populations, data from 17 populations (see Fig. 1) are assembled from papers published by Cross and Ward (1980), Cross and King (1983), Guyomard (1987), Stahl (1987), and Koljonen (1989). Data for each of the two Asturian populations were pooled because there were no significant allelic frequency differences between samples. Four loci (*Aat-3*, *Idh-3*, *Mdh-3,4*, and *Sdh-1*) were selected for this analysis because they are among the five most discriminatory ones (Stahl, 1987; Blanco et al., 1992).

Overall genetic differences between samples were expressed as genetic distances (Nei, 1972) and a UPGMA dendrogram was used for visualization of these genetic distances.

To examine the relationship between genetic and geographic distances between Asturian and other European populations, a linear regression analysis was used. Geographic distance between two locations was measured by the shortest distance by water between them.

3. Results and Discussion

In the samples screened, only five loci were variable, each segregated for two alleles: *Aat-3* (100 and 74 alleles), *Idh-3* (100 and 115), *Mdh-3,4* locus (100 and 87), *Sdh-1* (100 and 72) and *Sdh-2* (100 and 28). These loci were described as variables in previous studies of Atlantic salmon populations (Cross and Ward, 1980; Stahl, 1981, 1983, 1987; Vourinen, 1982; Cross and King, 1983; Guyomard, 1987; Koljonen, 1989) and, with the ME-2 loci (not studied in these samples), account for more than 98% of the total genetic variation in Atlantic salmon (Stahl, 1987; Davidson et al., 1989)

Table 1 shows the frequencies of the common alleles (100) at each locus for the different samples. No significant allele frequency differences among samples were found for any locus and the only remarkable allele frequencies were observed at the *Idh-3* and *Mdh-3,4* loci. Both

Table 1. Frequency of the most common allele (100) and observed heterozygosity (in %) values at the five polymorphic loci.

		Aat-3	Idh-3	Mdh-3,4	Sdh-1	Sdh-2	H
Sella-86	(1)	0.937	0.968	0.875	0.968	1.000	
	(2)	12.50	6.25	25.00	6.25	——	2.08
Sella-87	(1)	0.939	0.988	0.867	0.957	0.993	
	(2)	12.04	2.40	24.09	8.43	1.20	2.00
Sella-88	(1)	0.935	0.978	0.891	0.956	0.978	
	(2)	13.04	4.34	21.73	8.69	4.34	2.17
Esva-87	(1)	0.888	0.981	0.944	0.963	0.981	
	(2)	22.22	3.70	11.11	7.40	3.70	2.50
Esva-88	(1)	0.875	1.000	0.902	0.972	0.986	
	(2)	25.00	——	19.44	5.50	2.77	2.19
	G	3.740	0.700	2.966	0.414	1.359	

H = Average heterozygosity based on total of the loci assayed (24).
(1) = frequency of the most common allele (100).
(2) = observed heterozygosity expressed in percentage.
G = values of the G-test with Williams' correction for sample size used to test heterogeneity of allele frequencies.

loci were described as variables in few European populations (Stahl, 1987; Sánchez et al., 1991) and the frequency of the 87 allele at *Mdh-3,4* locus is much higher than in other European populations and remains stable in time in both rivers.

The genetic variation at *Me-2* locus, one of the most widespread polymorphisms detected in the species, was studied only in samples of 1989 and 1990 on both rivers. According to the cline described for this locus (Verspoor and Jordan, 1989), these populations show a frequency of the 100 allele higher than North European populations. In Sella River, the frequency of the 100 allele was 0.837 for samples of 1989 and 0.833 for samples of 1990; in the Esva river, the frequency was 0.950 for 1989 and 1990 samples.

No significant departures from Hardy-Weinberg proportions were detected for the examined loci, but differences among loci between samples were observed. In Esva samples, 46% of total heterozygosity is due to heterozygosity at *Aat-3* locus, whereas in Sella samples the heterozygosity at *Mdh-3,4* locus represents 47% of the total heterozygosity. Nevertheless, the average heterozygosity (Table 1) was of the same magnitude in all samples (2.18% as mean), although the Asturian's natural populations have less variability than other Atlantic salmon stocks (Cross and Ward, 1980; Stahl, 1987; Koljonen, 1989).

On the other hand, Atlantic salmon are naturally substructured into multiple genetically differentiated and more or less reproductively isolated populations. Based on electrophoretical data, there are two major genetically distinct and geographically separate groups: North American and European populations (Stahl, 1987; Verspoor, 1988). The European populations can be subdivided into two different groups: the Baltic stocks (Baltic sea drainages) which are genetically different from the rest of the European populations, the "Atlantic stocks." In the latter group, the differentiation is relatively small and leaves open the possibility of additional genetic subdivision (Stahl, 1987; Blanco et al., 1992). For this reason, we compared Asturian populations only with European populations from drainages of the Atlantic Ocean.

Genetic distances between samples was calculated according to Nei (1972) and a UPGMA dendrogram (Figure 2) was constructed. This dendrogram shows three major clusters with a genetic distance greater than 0.065: cluster A contains the six most southern populations

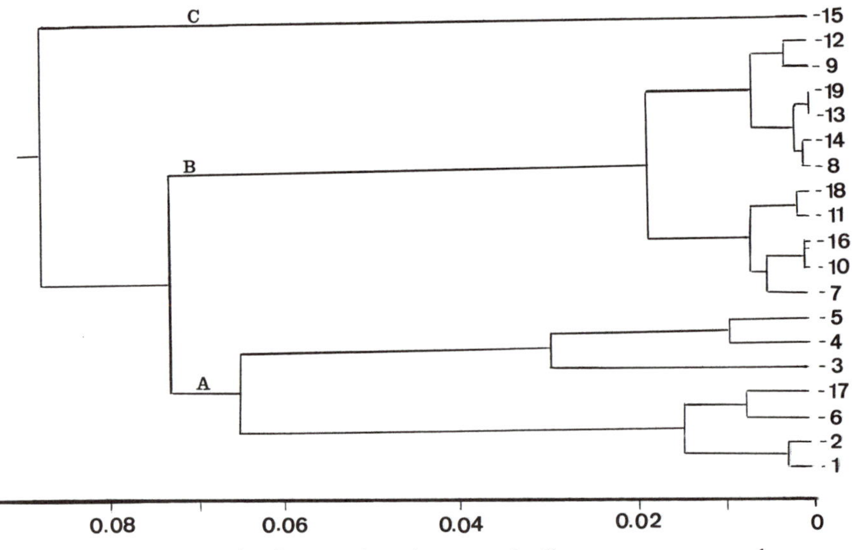

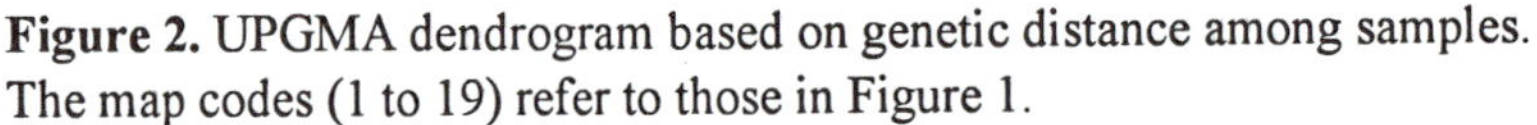

Figure 2. UPGMA dendrogram based on genetic distance among samples. The map codes (1 to 19) refer to those in Figure 1.

(including Asturian populations) plus a Swedish population; cluster B contains more northern populations; and cluster C includes the most northeasterly population.

A regression analysis (Figure 3) identified a highly significant positive relationship ($p<0.001$) between geographic and genetic distances when Asturian populations were compared with the other European samples.

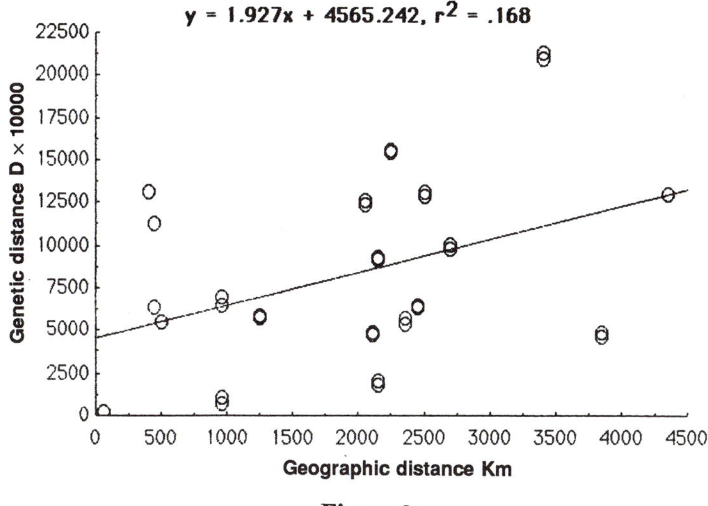

Figure 3.

In summary, the results of this study indicate that the natural salmon populations of Asturian rivers have a characteristic which distinguishes them from other European populations. These populations generally show a higher frequency of the allele 87 at the *Mdh-3,4* loci; a stability of genic frequencies was found between year classes of the same rivers. The mean heterozygosity in these populations (2, 18%) is somewhat lower than the average heterozygosity estimated in Atlantic salmon populations (around 3%, Allendorf and Utter, 1979; Cross and Ward, 1980; Stahl, 1981, 1983, 1987; Koljonen, 1989). The electrophoretic distinctiveness is in agreement with other data, both genetic (we have found differences in the distribution of basic number of chromosomes, Garciá-Vázquez et al., 1988a,b) and ecological (these natural populations produce mostly 1- and 2-year-old smolts which return to the river after two or three years in the sea whereas the populations of northern Europe consist mostly of the older smolts and many return as grilse).

All these facts about Asturian populations may be related to their geographic location (they are marginal populations in the distribution range of salmon in Europe) and may reflect an adaptation to environmental factors of this special area.

On the other hand, when we compared these populations with others from Atlantic drainages we found a positive correlation between genetic and geographic distance. In fact, Asturian natural populations are, in general, more similar to those from southern Europe (French and southern British Isles) than North Atlantic drainages.

We think that the preservation of Asturian populations must be a main objective and this distinctiveness must be taken into account in harvesting and enhancement programs.

References

Allendorf, F.W. and F.M. Utter. 1979, Population genetics, in: 'Fish Physiology," W.S. Hoar, D.J. Randall and J.R. Brett, ed., pp. 407-454., Academic Press, New York.

Blanco, G, J.A. Sánchez, E. Vázquez, J. Rubio, and F.M. Utter. 1992. Genetic differentiation among natural European population of Atlantic salmon from drainages of the Atlantic Ocean. *Animal Genetics*, 23:11-18.

Cross, T.F. and J. King. 1983. Genetic effects of hatchery rearing in Atlantic salmon. *Aquaculture* 33:33-40.

Cross, T.F. and R.D. Ward. 1980. Protein variation and duplicate loci in Atlantic salmon, *Salmo salar* L.. *Gent. Res. Camb.* 36:147-165.

Davidson, W.S., T.P. Birt, and J.M. Green. 1989. A review of genetic variation in Atlantic salmon, *Salmo salar* L., and its importance for stock indetification, enhancement programmes and aquaculture. *J. Fish Biol.* 34:547-560.

Gárcia-Vázquez, E., A.R. Linde, G. Blanco, J.A. Sánchez, E. Vázquez, and J. Rubio. 1988a. Chromosome polymorphism in far fry stocks of Atlantic salmon from Asturias. *J. Fish Biol.* 33:581-587.

Gárcia-Vázquez. E., A.M. Pendas, G. Blanco, J.A. Sánchez, E. Vázquez, and J. Rubio. 1988b. Estudio cariotípico de juveniles de *Salmo salar* en ríos Asturianos. *Bol. Cienc. Nat. IDEA* 39:129-136.

Guyomard, R. 1987. Differenciation génétique des populations de saumon atlantique: revue et interpretation des donnes electroforetiques et quantitatives, in: 'La restauration des rivieres a saumon," M. Thibaut and R. Billard, eds., pp 297-305, INRA, Paris.

Koljonen, M.L. 1989. Electrophoretically detectable genetic variation in natural and hatchery stocks of Atlantic salmon in Finland. *Hereditas* 110:23-36.

Martín-Ventura, J.A. 1987. Le saumon atlantique dans les rivieres de la province des Asturies (Espagne), in: 'La restaurations des rivieres a soumon," M. Thibaut and R. Billard, eds., pp. 133-144, INRA, Paris.

Martín-Ventura, J.A. 1988. The Atlantic salmon in Asturias, Spain: analysis of catches 1985-1986. Inventory of juvenile densities, in: "Atlantic Salmon: Planning for the Future," D. Mills and D. Piggins, eds., pp. 210-227, Crom Helm, London.

Nei, M. 1972. Analysis of gene diversity in subdivided populations. *Proc. Acad. Sci. USA*, 70:3321-3323.

Ryman, N. 1983. Patterns of distribution of bichemical genetic variation in salmonids: differences between species. *Aquaculture* 33:1-21.

Sánchez, J.A., G. Blanco, E. Vazquez, E. Garcia, and J. Rubio. 1991. Allozyme variation in natural populations of Atlantic salmon in asturias (Northern Spain). *Aquaculture* 93:291-298.

Stahl, G. 1981. Genetic differentiation among natural populations of Atlantic salmon (*Salmo salar*) in northern Sweden, in: 'Fish Gene Pools," N. Ryman, ed., Ecol. Bull. no. 34, Stockholm, pp. 95-105, Editorial Service/FRN. Sweden.

Stahl, G. 1983. Differences in the amount and distribution of genic variation between natural and hatchery stocks of Atlantic salmon. *Aquaculture* 33:23-32.

Stahl, G. 1987. Genetic population structure of Atlantic salmon, in: 'Population Genetics and Fishery Management," N. Ryman and F.M. Utter, eds., pp. 121-135, University of Washington Press, Seattle, WA.

Verspoor, E. 1988a. Reduced genetic variability in first generation hatchery populations of Atlantic salmon (*Salmo salar*). *Can. J. Fish Aquat. Sci.* 45:1686-1690.

Verspoor, E. 1988b. Identification of stocks in the Atlantic salmon, in: 'Present and Future Atlantic Salmon Management," R.H. Stood, ed., pp. 37-45, Ipswich, MA.

Verspoor, E. and W.C. Jordan. 1989. Genetic variation at the *Me-2* locus in the Atlantic salmon within and between rivers: evidence for its selective maintenance. *J. Fish Biol.* 35:205-213.

Vourinen, J. 1982. Little genetic variation in the Finnish lake salmon. *Salmo salar sebego* (Girard), *Hereditas* 97:189-192.

Vourinen, J. and O.K. Berg. 1989. Genetic divergence of anadromous and non-anadromous Atlantic salmon (*Salmo salar*) in the river Nanse, Norway, *Can. J. Fish Aquat. Sci.* 46:406-409.

Genetic Management of Natural Fish Populations

GIORA W. WOHLFARTH

1. Introduction

Serious declines in wild fish populations have been described for several aquatic systems, particularly the Great Lakes of North America. The primary causes of these declines are regarded as overexploitation of fish populations and deteriorations in aquatic environments. Fishery management, aimed at restoring wild fish populations, generally consists of controlling the fishing pressure, improving water quality, and restocking programs. No attempts are made at the genetic management of wild fish populations.

2. Overexploitation

The process of fishing involves the selective removal of the larger individuals of some target species, leaving the smaller individuals and non-target species to produce subsequent generations. The genetic impact of this negative selection became apparent by the deterioration of overexploited natural fish populations in growth and maximum size attained. This may be a genetic response to intense and repeated negative selection inherent in overexploitative fishing.

3. Rapid Environmental Changes

Rapid changes that continue to take place in the aquatic environment include pollution, infestations, topographical changes, a reduction in pH and a rise in temperature. Natural selection is not capable of changing gene frequencies in response to such rapid environmental changes. As a result many natural populations can no longer be optimally adapted to their "natural" environments.

4. Choice of Stocks

An obvious method of attempting to restore natural fisheries involves restocking programs. Wild stocks of several salmonid species have been domesticated, producing fish much easier to manage in hatchery conditions. This reduced production costs drastically, but has also reduced the adaptation of domesticated stocks to natural conditions. Wild stocks may also not be optimally adapted to natural conditions, due to negative selection and rapid environmental changes. In comparisons among wild and domestic strains of different sal-

Fish and Aquaculture Research Station, Dor, Israel

Genetic Conservation of Salmonid Fishes, Edited by J.G. Cloud
and G.H. Thorgaard, Plenum Press, New York, 1993

Table 1. Performance of domestic and wild strains and domestic x wild strain crossbreds of some salmonids in natural waters.

Species	Stock	Ratio: Recovered/ Stocked No. (%)	Weight	Reference
Cutthroat trout	Domestic	1.3	0.36	Donaldson et al, 1957
	Dom x Wild	6.5	1.30	
	Wild x Dom	6.7	1.36	
	Wild	2.5	0.87	
Brook trout	Domestic	39.1		Mason et al.1967
	Dom x Wild	46.2		
	Wild	45.3		
	Domestic	2.1	0.81	Flick and Webster. 1976
	Dom x Wild	10.7	9.90	
	Wild	1.9	2.27	
	Wild	3.1	2.90	
	Domestic	41.3	1.63	Webster and Flick, 1981
	Dom x Wild	68.1	2.95	
	Wild	52.2	2.49	
	Domestic	4.0	0.90	Fraser, 1981
	Dom x Wild	8.0	4.75	
	Domestic	6.8	0.30	Keller and Plosila, 1981
	Dom x Wild	16.9	1.89	
	Wild	10.6	1.80	
	Domestic	12.0	1.94	Lachange and Magnan. 1990
	Dom x Wild	16.4	2.35	
	Wild	16.7	2.93	
	Domestic	failed to reproduce		Fraser, 1989
	Dom x Wild	reproduced		
	Wild	successfully		
Steelhead trout	Domestic	3.6		Reisenbichler and McIntire, 1977
	Dom x Wild	3.9		
	Wild	4.5		
Pink Salmon	Purebred	0.23		Bams, 1976
	Crossbred	0.39		

monids and their first generation crossbreds, the crossbred was usually superior or similar to the wild strain in relative number and relative weight of fish recovered. Wild x domestic strain crossbreds and wild strain brook trout established self propagating populations, where earlier plantings of domestic strains had failed to reproduce. The use of crossbreds may be preferable to wild strains, due to their lower production costs.

5. Suggestions for Restoring Wild Fish Populations by Genetic Management

Genetic management is required to adapt wild fish populations to changed and changing environments. Possible strategies include:

1. Stocking samples of several genetically distinct wild fish populations from environments similar to that of the population to be rehabilitated.
2. Maintaining the variance of fitness at a level allowing the stock to perpetuate itself in the face of fishing and natural mortality. Genetic management involves maintaining a sufficiently large number of breeders and periodic introductions of fish from outside sources.
3. Stocking first generation crossbreds between domestic and wild strains. In the next generation, F2 and backcross fish will be produced, whose performance in natural conditions has not been tested. Long-term improvement may require periodic restocking and monitoring.

6. Summary

Restoring wild fishery populations requires genetic management. Genetic conservation is unlikely to succeed, since it does not attempt to adapt fish genotypes to their changed and changing environments. Genetic management requires maintaining the variance of fitness by stocking genotypes differing from the resident population. This could lead to an improvement in performance traits, particularly fitness and growth; increasing the genetic variance of these traits; and in extreme cases, increasing the effective number of breeders. The possible advantage of first generation crossbreds between wild and domestic stocks has been demonstrated. It is possible that stocking wild stocks of a strain differing from the resident one would have the same effect. It is proposed to initiate restocking programs as here suggested, and to monitor their results genetically.

References

Bams, R. A. 1976. Survival and propensity for homing as affected by presence or absence of locally adapted paternal genes in two transplanted populations of pink salmon *(Oncorhynchus gorbuscha)*. *J. Fish Res. Board Can.* 33: 2716-2725.

Donaldson, L. R., D.P. Hansler, and T.H. Buckridge. 1957. Interracial hybridization of cutthroat trout. *Salmo clarki,* and its use in fisheries management. *Trans. Am. Fish. Soc.* 86: 351-360.

Flick, W. A. and D.A. Webster. 1976. Production of wild, domestic, and interstrain hybrids of brook trout *(Salvelinus fontinalis)* in natural ponds. *J. Fish. Res. Board Can.* 33: 1525-1539.

Fraser, J.M. 1981. Comparative survival and growth of planted wild, hybrid, and domestic strains of brook trout *(Salvelinus fontinalis)* in Ontario lakes. *Can. J. Fish. Aquat. Sci.* 38: 1672-1684.

Fraser, J.M. 1989. Establishment of reproducing populations of brook trout after stocking of interstrain hybrids in Precambrian Shield lakes. *N. Am. J. Fish. Mgmnt.* 9: 352-363.

Kapuscinsky, A. and J.E. Lannan. 1984. Application of a conceptual fitness model for managing Pacific salmon fisheries. *Aquaculture* 43: 135-146.

Keller, W.T. and D.S. Plosila. 1981. Comparison of domestic, hybrid and wild strains of brook trout in a pond fishery. *N.Y. Fish Game J.* 28: 123-137.

Krueger, C.C., A.J. Gharrett, T.R. Dehring, and F.A. Allendorf. 1981. Genetic aspects of fisheries rehabilitation programs. *Can. J. Fish. Aquat. Sci.* 38:1877-1881.

Lachance, S. and P. Magnan. 1990. Performance of domestic, hybrid and wild strains of brook trout, *Salvelinus fontinalis*, after stocking: The impact of intra- and interspecific competition. *Can J. Fish. Aquat. Sci.* 47: 2278-2284.

Mason, J.W., O.M. Brynildson, and P.E. Degurse. 1967. Comparative survival of wild and domestic strains of brook trout in streams. *Trans. Am. Fish. Soc.* 96: 313-319.

Moav, R., T. Brody, and G. Hulata. 1978. Genetic improvement of wild fish populations. *Science* (Wash. D.C.) 201: 1090-1094.

Regier, H.A., J.J. Magnuson, and C.C. Coutant. 1990. Introduction to Proceedings: Symposium on effects of climate change on fish. *Trans. Am. Fish. Soc.* 119: 173-175.

Reisenbichler, R.R. and J.D. McIntyre. 1977. Genetic differences in growth and survival of juvenile hatchery and wild steelhead trout, *Salmo gairdneri*. *J. Fish. Res. Board Can.* 34: 123-128.

Webster, D.A. and W.A. Flick. 1981. Performance of indigenous, exotic and hybrid strains of brook trout *(Salvelinus fontinalis)* in waters of the Adirondack Mountains, New York. *Can. J. Fish. Aquat. Sci.* 38: 1701-1707.

Wohlfarth, G.W. 1986. Decline in natural fisheries-a genetic analysis and suggestion for recovery. *Can. J. Fish. Aquat. Sci.* 43: 1298-1306.

Genetics of Salmonids in Czechoslovakia: Current Status of Knowledge

MARTIN FLAJSHANS, PETR RAB, and LADISLAV KALAL

1. Salmonids in Czechoslovakia

The recent fish fauna of Czechoslovakia includes eight species of six genera of the fish family *Salmonidae*, subfamilies *Salmoninae, Thymallinae and Coregoninae*. The review of particular salmonid species is given in Table 1 (Lusk, 1989). It is evident that the only autochthonous species are brown trout *(Salmo trutta)*, Danube salmon or huchen *(Hucho hucho)*, European grayling *(Thymallus thymallus)* and now extinct migratory forms of whitefish *(Coregonus lavaretus oxyrhynchus*) and Atlantic salmon *(Salmo salar*). All other salmonids have been introduced.

1.1. Autochthonous Species

Two species, i.e. Atlantic salmon and sea trout, both anadromous and migrate through Labe, Vltava and Otava river systems upstream to their spawning areas, were endangered by river dam construction and have been extinct since 1930s. Moreover, the native Atlantic salmon from the Labe river system in Bohemia dissappeared at the end of the last century and therefore this river system has been heavily stocked with Atlantic salmon from Rhine River, causing hybridization, and thus, extinction of the original salmon populations before the extinction of the species itself (Andreska, 1987). The sea trout that migrate also through the Dunajec river system from the Baltic Sea (i.e., its landlocked progeny) is supposed have founded the recent ecological forms of brown trout and lake trout. Trout of the *Salmo trutta labrax* subspecies, migrating from the Black Sea through the Danube River system, (i.e., its landlocked progeny) is, according to some authors, supposed to have founded another inland sub-species, differing by a higher number of gill rakers (16-21). However, according to Balon (1967), the trout of the Danube system originated from the Rhine River and other tributaries of the northern seas, having migrated through interfluvial connections at the end of the glacial period. Egg and stockfish movements among Czech, Moravian and Slovak hatcheries on different river systems caused crossing of both these subspecies and many local forms as well, nearly eliminating the differences between them.

Research Institute of Fish Culture and Hydrobiology, Department of Fish Genetics and Breeding, CS-389 25 Vodnany, Czechoslovakia

Genetic Conservation of Salmonid Fishes, Edited by J.G. Cloud and G.H. Thorgaard, Plenum Press, New York, 1993

The Danube salmon was confined to the Danube River system but is presently stocked in other river basins as well. This large predatory salmonid species has been pushed back to the submountain zones of rivers, and its numbers have decreased considerably. The species is now threatened with extinction (Holcik, 1990b).

The houting, a migratory form of the European whitefish in the Labe River *(Coregonus lavaretus oxyrhynchus)*, which has sporadically been found even at the end of the last century, has disappeared much earlier than the Atlantic salmon population (Andreska, 1987). The European grayling *(Thymallus thymallus)* is also autochthonous in all river basins of Czechoslovakia, although it was endangered during 1950s and 1960s. Due to better water management and artificial propagation, this species is now out of danger.

1.2. Introduced Species

Since the first introduction in 1888, many imports of rainbow trout of different origins have been realized (Kalal, 1989). Table 2 reports the respective imports with the data on the

Table 1. Salmonidae in Czechoslovakia according to Lusk (1989).

1. Salmoninae	
Salmo salar (Linnaeus, 19758)	extinct
Salmo trutta trutta (Linnaeus, 1758)	extinct
Salmo trutta trutta m. fario (Linnaeus, 1758)	not threatened
Salmo trutta trutta m. lacustris (Linnaeus, 1758)	not threatened
Oncorhynchus mykiss (Walbaum, 1792)	introduced, successful acclimatization
Hucho hucho hucho (Linnaeus, 1758)	threatened
Salvelinus alpinus (Linnaeus, 1758)	introduced, acclimatization failed
Salvelinus fontinalis (Mitchill, 1815)	introduced, successful acclimatization
Salvelinus namaycush (Walbaum, 1792)	introduced
2. Coregoninae	
Coregonus albula (Linnaeus, 1758)	introduced, acclimatization failed
Coregonus autumnalis (Pallas, 1776)	introduced, acclimatization failed
Coregonus lavaretus maraena (Linnaeus, 1758)	introduced, successful acclimatization
Coregonus lavaretus oxyrhynchus (Linnaeus, 1758)	extinct
Coregonus peled (Gmelin, 1788)	introduced, successful acclimatization
3. Thymallinae	
Thymallus thymallus (Linnaeus, 1758)	out of danger
Thymallus arcticus baicalensis (Dybowski, 1876)	introduced, acclimatization failed

Table 2. The data on rainbow trout (Oncorhynchus mykiss) imports to Czechoslovakia after World War II (according to Kalal, 1989).

Year of import	Assignment	Imported amount [ex.]	Imported from	Note
1946	"local population Pd M"	eggs, 1 x 10^6	Denmark	spring spawning
1947	"local population Pd M"	eggs, 10.6 x 10^6	Denmark	dtto.
1947	"local population Pd M"	yearlings, 100 x 10^3	Denmark	dtto.
1948	"local population Pd M"	yearlings, 198 x 10^3	Denmark	dtto.
1965	Pd A	eggs, 20 x 10^3	U.S.A.	spring spawning, disappeared
1966	Pd D 66	eggs, 10 x 10^3	Denmark	kamloops form, autumn spawning
1966	Pd P 66	eggs, 12 x 10^3	Poland	disappeared
1967	Pd D 67	eggs, 20 x 10^3	Denmark	kamloops form, autumn spawning disappeared
1968	Pd D 68	eggs, 150 x 10^3	Denmark	kamloops form, spring spawning, disappeared
1975	Pd D 75	eggs, 270 x 10^3	Denmark	spring spawning
1979	Pd P 79	eggs, 10 x 10^3	Poland	autumn spawning, disappeared
1985	Pd A 85	eggs, 285 x 10^3	U.S.A.	autumn spawning, disappeared
1986	Pd F 86	eggs, 250 x 10^3	France	dtto.
1988	Pd B 88	eggs, 150 x 10^3	Bulgaria	spring spawning
1989	Pd B 89	eggs, 150 x 10^3	Bulgaria	dtto.

origin and with the assignments of the respective rainbow trout strains bred from these imports in Czechoslovakia.

Lake trout *(Salmo trutta m. lacustris)* and sea trout *(Salmo trutta m. trutta)* have been imported repeatedly from Poland. As both these trouts are the ecological forms of *Salmo trutta* only, they have changed through a few generations to brown trout form, and thus, the imports proved to be purposeless.

Brook trout *(Salvelinus fontinalis)* was introduced in 1888 and successfully acclimatized under Central European conditions.

Arctic char *(Salvelinus alpinus)* from Alpine lakes has been introduced unsuccessfully into Cerne Lake (Southern Bohemia). The same probably happened to another member of the genus *Salvelinus, S. namaycush*, originating from Norwegian and Austrian supplies and stocked into Lucina Dam Reservoir on the Mze River (Western Bohemia). The results of the latter species introduction and acclimatization have not yet been thoroughly evaluated.

Four species of the subfamily *Coregoninae*, i.e., Arctic cisco *(Coregonus autumnalis)*, vendace *(C. albula)*, European whitefish *(C. lavaretus maraena)* and northern whitefish or peled *(C.peled)*, have been introduced into Czechoslovak waters since 1882: European whitefish from Miedwie Lake in Poland, vendace from northern lakes in Poland in the 1950s, and northern whitefish from Soviet Union in the 1970s. However, the origin of peled in Europe is obscure even at present (Luczynski, pers. comm.). The acclimatization of *C. autumnalis* and *C. albula* proved to be unsuccessful, while the other two species grow and reproduce successfully in some cold-water dam lakes and highland ponds. They also are commercially cultured in some pond fish farms. Both these species, *C. lavaretus maraena* and *C. peled*, hybridize easily either under natural conditions or artificially, which causes a certain danger of rearing these crossbreds of different filial generations as pure species, thus worsening their performance.

Baical grayling *(Thymallus arcticus baicalensis*) was introduced in 1959 and stocked into Dobsina dam lake on Hnilec River. However, the acclimatization of this introduced species proved to be unsuccessful.

It should be noted that some of the previous imports and new species introductions have been evaluated, above all, from the point of view of their possible culture and, in some cases, as an enrichment of Czechoslovak *ichthyofauna*. The genetic and ecological points of view—such as avoiding genetic contamination of autochthonous or acclimatized fish species, subspecies, and forms as well as a many other problems or mistakes—have not been taken as the most important criteria.

2. Cytotaxonomy and Cytogenetics

Cytotaxonomy of the genus *Salmo* was reviewed by Flajshans and Rab (1990) and chromosome studies of *Coregoninae* have been summarized and discussed by Rab and Jankun (1991).

2.1. Karyotypes

The morphology of rainbow trout chromosomes was studied by Flajshans and Rab (1990), who analyzed the widely cultured PdD 66 Kamloops strain and tried to test its reported origin cytogenetically. The modal karyotypes, 2n = 58 with 24 pairs of bi-armed chromosomes and 5 pairs of acrocentrics, and 2n = 60 with 23 pairs of bi-armed chromosomes and 7 pairs of acrocentrics, were found in frequencies of 63.33% and 26.66%, respectively.

One crossbred individual of the PdD 68 and PdD 66 strains studied was analyzed as a spontaneous triploid with the modal 2n = 88 chromosomes, consisting of 74 metacentrics and submetacentrics, and 14 acrocentrics with the NF value = 160 (Flajshans and Rab, 1987). The possible causes of spontaneous triploidy and the breeding aspects of the occurrence of triploids in breeding stocks were discussed in this paper. The spontaneous triploidy could be a result of the egg overripening, the disposition of which could be more or less normal in different fish species relating to their reproduction characteristics or a certain genetic predisposition (Thorgaard and Gall, 1979) which could or could not be supported by full-sib mating or a result of a combination of these factors.

Brown trout karyotypes were studied in individuals of one autochthonous population as a part of a chromosome banding study dealing with localization of nucleolar organizer regions (NORs) and counterstain-enhanced fluorescence in rainbow and brown trouts (Mayr et al., 1986). This study demonstrated the correspondence between CMA3 positive fluorescent patterns and silver NORs staining in brown trout, while the CMA3 positive heterochromatin was flanking the silver NORs in rainbow trout. Later, Mayr et al. (1988) characterized the specific heterochromatins present on rainbow-, brown- and brook trout chromosomes by sequential fluorescence and established the band karyotypes.

Karyotypes of the Danube salmon from two distant populations have been studied by Rab (1980) and Rab and Liehman (1982). The diploid chromosome number was found to be 2n = 82 chromosomes in all specimens from Slovakia and Yugoslavia, while the karyotype composition differed from 13 pairs of metacentrics, 2 pairs of submetacentrics, 6 pairs of subtelocentrics and 20 pairs of acrocentrics, NF = 124 in a specimen from Slovakia to 13 pairs of metacentrics, 3 pairs of submetacentrics, 6 pairs of subtelocentrics and 19 pairs of acrocentrics, NF = 126 in a specimen from Yugoslavia. This confirmed the autochthonous origin of the Danube salmon from Slovakia. This karyotype is evidently closely related to that of Brachymystax *lenok* (2n = 90). The karyotype of the latter species is the closest to the salmonid tetraploid ancestor.

Chromosome studies in the subfamily *Coregoninae* were reviewed by Rab and Jankun (1991) and showed the wide occurence of chromosomal polymorphism in populations studied at more localities and/or those of hybrid origin. They concluded that the karyotypes of coregonids are characterized as typical salmonid sets and that the mode of their chromosomal evolution also is parallel as in other, more thoroughly studied phyletic lines of salmonids.

The Czechoslovak coregonid species, European whitefish *(C.l. maraena)* and northern whitefish *(C. peled)*, and their reciprocal hybrids were studied karyologically by Rab (in Slechtova et al., 1990). The European whitefish is polymorphic for its 2n ranging from 78 to 82, while northern whitefish has 2n = 78-80. The karyotype morphologies of both species are very similar, but their hybrids display a wider range of polymorphism than the parental species.

2.2 Genome Manipulations

The artificial polyploidization of salmonids was reviewed by Flajshans (1989), concerning on the polyploidy applications, methods of polyploidization with parameters of shock characterization and with a description of the respective methods used (i.e. cold and heat shocks, hydrostatic pressure shocks and chemical shocks).

The application of hydrostatic pressure and heat for the induction of triploidy was tested in rainbow trout strains PdD 66 and PdD 75. A special hydrostatic pressure device, using pressurized nitrogen to immediately induce the demanded pressure level, has been tested primarily for cyprinids. Its application for salmonids with the exact time optimization is still under study.

The determination of the ploidy level by means of numbers of Ag-stained NORs, as proposed by Phillips et al. (1986), was evaluated in both triploid and diploid rainbow trout fingerlings as well as in some cyprinids by Flajshans et al. (1992). They confirmed the correspondence between ploidy level and numbers of NOR sites by means of Ag-stained active NORs and CMA_3 fluorescence and analyzed statistically the minimum number of stained cells required for determination of the ploidy level. This approach was found to be one of the most simple and inexpensive methods available for analysis just after hatching or at any later time.

3. Biochemical Genetics

3.1. Transferrins

Transferrin (Tf) polymorphism has been studied in the following salmonids: brown trout (2 alleles, Kalal et al., 1971; Slechtova, 1978), rainbow trout (3 alleles, Slechtova, 1990), brook trout (2 alleles, Pavlu et al., 1971; Slechtova, 1978), Danube salmon (monomorphic, Slechtova and Slechta, 1990), European whitefish and northern whitefish (4 and 1 allele, respectively, Pavlu et al., 1971; Slechtova, 1978; Slechtova and Valenta, 1988; Slechtova et al., 1988; Slechtova et al., 1990),as well as their crossbreds (Slechtova and Valenta, 1988; Slechtova et al., 1988; Slechtova et al., 1990) and in grayling (2 alleles, Slechtova, 1978).

3.2. Lactate Dehydrogenase

The lactate dehydrogenase (LDH) isozymes in tissues have been studied by Kalal et al., 1972) in European and northern whitefish, rainbow-, brown- and brook trouts, and in American lake trout *(Salvelinus namaycush)*, where some interspecific differences in electrophoretic mobilities (except for both whitefish species) and relative isozyme intensities have been analyzed. Slechta and Slechtova (1977) have studied the same species as above including European and northern whitefish crossbreds. The extraordinary LDH systems (eye-specific isozymes) have been analyzed in rainbow-, brown- and brook trouts (Valenta et al., 1972).

Polyploidy of freshwater salmonids by gene duplication of LDH loci has been studied in both whitefish species, in rainbow-, brown-, brook- and American lake trouts, and in grayling where 5 LDH loci in all these species (duplicated A and B locus) have been found. Polymorphic LDH has been studied in both whitefish species and analyzed as identical (Slechtova, 1978) with polymorphism in LDH-C locus in European whitefish (Slechtova et al., 1990), brown trout (identical with rainbow trout except for LDH-C, Slechtova, 1978), brook trout (polymorphic with 3 alleles in duplicated LDH-B1,2 locus, Slechtova, 1978, Slechta, 1980); rainbow trout (Slechtova, 1978).

3.3. Malate Dehydrogenase

Malate dehydrogenase (MDH) isozymes in tissues of diploid and tetraploid fish have been studied for the first time by Valenta et al. (1972) in both whitefish species, brown-, rainbow-, brook- and American lake trouts, and in grayling. The existence of duplication has been proposed at least in one locus.

MDH activity and isozyme patterns in salmonid fish tissues have been studied and interspecific differencies and tissue specifity of relative isozyme amounts have been analyzed by Slechta and Slechtova (1977).

Polymorphic MDH has been studied in both coregonids (identical, Slechtova, 1978), in brown trout (polymorphism in duplicated MDH-A locus, Slechtova, 1978), rainbow trout (polymorphism in duplicated MDH-B, Slechtova, 1978; Slechta, 1980; Slechtova, 1990), and in brook trout (polymorphism in duplicated MDH-B, Slechtova, 1978; Slechta, 1980).

3.4. The Application of Biochemical Markers

Genetic traits characteristic of European and northern whitefish and their crossbreds have been studied by Slechtova and Valenta (1988), Slechtova et al. (1988) and Slechtova et al. (1990). Ten protein systems have been studied: (Tf, albumin-like protein (Alb-l), MDH, LDH, glucosophosphate isomerase (GPI), superoxid dismutase (SOD), myoglobin (Myo), glycerol-3-phosphate dehydrogenase (G3PDH), isocitrate dehydrogenase (IDH) and phosphoglucomutase (PGM). Interspecific differences have been found in Tf, Alb, GPI-2 and SOD, polymorphism has been found in Tf, IDH-l, GPI-l, PGM-2 and G3PDH (all 3 loci). In addition, Slechtova et al. (1990) has analyzed the following diagnostic loci: SOD-l, sMDH-3,4, GPI-2, Tf, Alb-l, and esterase (Est-1).

Biochemical genetic markers of selected Czechoslovak rainbow trout populations were recently analyzed by Slechtova (1990) with polymorphism found in Tf (3 alleles), Alb-l (2 alleles), SOD (3 alleles), PGM (2 alleles), sMDH-3,4 (2 alleles), LDH-1,2 (2 alleles), -3,4 (2 alleles) and sIDHP-1,2 (5 alleles). No statistically significant differences among the four populations of two strains studied (i.e., Pd M and Pd D 66) were found.

4. Breeding Work

During the last 40 years all the hatcheries and production farms producing brood or marketable fish in Czechoslovakia have been organized in the State Fisheries Enterprise and in the Czech or Slovak Anglers Unions, which besides a lot of drawbacks has brought certain advantages, as especially coordination of breeding efforts.

Although the reproduction and breeding effort in both the autochthonous and introduced salmonids have been realized since the 1890s, the unification of breeding work criteria has been made only recently—in the early 1970s. The brood fish were formerly selected according to the local criteria in each hatchery (mostly according to morphological and physiological traits, weight gain, non-specific resistance and reproductive traits). The first uniformly organized breeding work in rainbow trout was started in 1970 with instructions for practical realization of breeding work in the rainbow trout culture of the "local strain" Pd M (Kupka, 1970). At the same time, the research on heredity of growth and reproductive ability, health, resistance and feed conversion of the local Pd M strain and of the newly imported Pd D 66 and Pd D 68 strains has continued. This effort has resulted in instructions for the breeding work in rainbow trout culture in the strain Pd D 66 in the State Fisheries Enterprise (Pokorny and Kalal, 1976). The breeding work methods in all cultured rainbow trout strains are based on the following:

1. Initiating the trout selection at the age of 3 years according to the consideration of the external traits.
2. Measuring the meristic characters in positively selected fish as follows: weight, total length, body length, head length, pectoral height, anal height and comparing their characteristics with tabulated standard data.
3. In females reaching the standard size criteria, the following data were measured: egg size, the total volume of the stripped eggs and relation with the working fecundity. If these data achieve the standard, the eggs are fertilized heterospermically with sperm of at least two males.
4. Eggs are incubated separately and the water temperature, egg losses, percent eye-up, and the total duration of incubation are checked. If the losses are lower than the average from the total, the fry pass to the next culture stage.
5. During the autumn selections of yearlings, 50% of minus variants are selected out of the culture, during the second year 10% of plus variants are selected positively.
6. Young brood fish (three years) are selected as mentioned above (points 2 and 3). In older brood fish, only the negative selection according to the health and quality of sex products is done.

The breeding work efficiency was evaluated by selection effect, with results in the respective rainbow trout strains are reported in Table 3. A particular correlation analysis of these data was made by Dvorak (1985) and modified by Kalal et al. (1990), concluding that the correlation coefficient values of both strains were very close. Highly significant correlation coefficients were found between the weight and total length, body length and anal weight (r = 70). Significant correlations were found among pectoral height, weight, total length, body length, head length and anal height (r = 50-70).

As it is evident from Table 3, and as also concluded by Pokorny and Kalal (1989), the breeding work efficiency in the Pd M strain was positive, except for the egg size to which the selection should be directed; on the other hand, some worse results in selection of the Pd D 66 strain are probably caused by inbreeding, as apparent from the initial amount of eggs imported

(Table 2), environment quality changes, differing daily feeding ration qualities and quantities and by losses of some brood fish.

Breeding work in coregonids, dealing with interspecific hybridization and selection, was first carried out before Hochman and Penaz (1986) and Hochman (1987). In order to prevent undesired interspecific hybridization of both species (i.e., European and northern whitefish), it is necessary to maintain pure homozygous lines of reliable origin, using all available methods of genetics. Therefore, it is essential to select the brood fish according to the external morphological traits, degrees of performance, reliable filing, and genotype control of selected groups or parental pairs according to their offspring through diagnostic polymorphic protein variants. The uncontrolled mating of fish in various degree of heterozygosity, which can occur as a result of the former "coregonid hybridization," must be avoided. Plus variants in growth rate, fleshing, releasing gonadal products of appropriate quality and with well-expressed external morphological traits should be selected for propagation. The egg size must refer to the species-specifity. When selecting the brood fish for commercial stockfish production, only fish with poor growth and with morphological abnormalities should be selected out.

Table 3. Mean results of the rainbow trout selection of the Pd M and Pd D 66 strains, as demonstrated by the Marianske Lazne Trout Farm data, State Fishery Enterprise (P = 100%, F_3% evaluated).

	Pd M strain		Pd D 66 strain	
Trait	Males	Females	Males	Females
Weight (g)	111.9	118.9	108.0	102.5
Body length (mm)	100.8	102.8	107.7	101.0
Head length (mm)	100.2	102.4	102.5	99.0
Pectoral height (mm)	101.1	104.0	105.1	98.3
Anal height (mm)	104.3	106.7	105.4	99.1
Body condition character	105.4	109.6		
Volume of eggs (ml)		131.0		
Amount of eggs (ex.)				98.5
Egg size (mm)		99.4		102.9

Body condition character = weight (g): body length (mm), characterizing the fish nutritive status.
According to Dvorak (1985) and Kalal et al. (1990) Pd M strain and Kalal et al. (1990) Pd D 66 strain.

5. Introductions and Gene Pool Conservation

As mentioned above, certain tendencies for enhancement of breeding results, and consequently, for performance improvement of fish species under commercial culture, as well as for certain enrichment of Czechoslovak *ichthyofauna* can lead to the uncontrolled import of foreign stocks of cultured species or the uncontrolled introduction of exotic fish species. If done without thorough previous genetic, ecological and veterinary studies, this could cause a genetic and veterinary danger and more or less strong ecological competition for some autochthonous or acclimatized fish species.

In order to avoid the uncontrolled introductions of exotic fish species, Rules for Introduction Processes for Exotic Freshwater Fishes and Aquatic Invertebrates were created

(Holcik, 1990a). They require the complete proposal from the person/party interested in the introduction including the basic taxonomic and ecological data with references and aim, intended range of introduction, and the name of the scientific research organization cooperating on the introduction. The proposal must be submittted to the Introduction Committee; standpoints of the Introduction Committee and EIFAC/FAO Secretary and permissions of the Federal Ministry of Agriculture and Nature Conservation Authorities are requested. The Federal Ministry of Agriculture with the State Veterinary Authority will name the professional supervision and the quarantine conditions. After quarantine evaluation by the supervisor, veterinary authority, and Introduction Committee, the authority responsible for introductions, i.e. currently the Research Institute of Fish Culture and Hydrobiology, will determine the suitability or unsuitability of the given species introduction.

Neither any serious gene pool conservation of autochthonous and acclimatized fish species nor any complex successful breeding should be done without the exact knowledge of the genetic structure of populations. Therefore, all available data should be collected first by the appropriate methods of genetics (i.e., chromosomal analyses, electrophoretic analyses of polymorphic proteins, and by mitochondrial DNA analyses and DNA fingerprinting, as the latter two methods are not currently introduced under the conditions of Czechoslovak fish research activities).

References

Andreska, M. 1987. Rybarstvi a jeho tradice [Fisheries and its tradition]. SZN Prague, 208 pp. (in Czech).

Balon, E.K., 1967. Ryby Slovenska [Fishes of Slovakia]. Obzor, Bratislava, 420 pp. (in Slovak).

Dvorak,M. 1985. Vyhodnoceni vysledku provozni plemenarske prace pstruha duhoveho mistni formy na OZ SR Marianske Lazne [The evaluation of results of practical breeding work in rainbow trout of the local form of the State Fisheries Farm Marianske Lazne]. Post-graduate thesis, Agric. Univ. Brno, 125 pp. (in Czech).

Flajshans, M. 1989. Umela polyploidizace u lososovitych ryb (Prehled) [Artificial polyploidization in salmonid fishes (A review)]. *Bull. VURH Vodnany*, 2: 14-17 (in Czech).

Flajshans, M. and P. Rab. 1987. Nalez triploidniho pstruha duhoveho formy kamloops (Salmo gairdnerii kamloops) [A finding of the triploid rainbow trout, kamloops form (Salmo gairdnerii kamloops)]. *Zivoc. Vyr.*, 32 (3): 279-282 (in Czech with English summary).

Flajshans, M. and P. Rab. 1990. Chromosome study of *Oncorhynchus mykiss* kamloops. *Aquaculture*, 89: 1-8.

Flajshans, M. and P. Rab. 1990. Rod *Salmo* a druh pstruh duhovy *(Salmo gairdneri* Richardson, 1836) v systemu lososovitych z hlediska cytogenetiky [Genus *Salmo* and rainbow trout species *(Salmo gairdneri* Richardson, 1836) in the system of salmonid fishes from point of view of cytogenetics]. *Bull. VURH Vodnany*, 1: 23-27 (in Czech).

Flajshans, M., P. Rab, and S. Dobosz. (1992). Frequency analyses of active NORs in nuclei of artificially induced triploid fishes. *Theor. Appl. Genet.*, 85:68-72.

Hochman, L. and M. Penaz. 1986. On the problems of fish hybridization with an example on whitefish. In: Reprodukce a genetika ryb, Vodnany, Slov. zool. spol. Ichthyol. sekce: 36-40.

Hochman, L. 1987. Chov sihu [Culture of coregonid fishes]. VURH Vodnany, Methods Edition, 24, 16 p. (in Czech).

Holcik, J. 1990a. Pravidla postupu pri introdukcii exotickych druhov ryb a vodnych bezstavovcov do CSSR. [Rules for Introduction Processes for Exotic Freshwater Fishes and

Aquatic Invertebrates to CSSR]. VURH Vodnany, Methods Edition, 35, 9 p. (in Slovak).

Holcik, J. 1990b. Conservation of the huchen *Hucho hucho* (L.), *(Salmonidae)* with special reference to Slovakian rivers. *J. Fish. Biol.*, 37(A): 113-121.

Kalal, L., V. Pavlu, and M. Valenta. 1971. Polymorphismus von Transferrinen des Blutserums der Karausche *(Carassius carassius)*, der Plotze *(Rutilus rutilus)* und der Bachforelle *(Salmo trutta m. fario)* [Polymorphism of transferrins in crucian carp, roach and brown trout], in: Proc. VIIth Int. Symp. on Breeding and Genet. of the Lab. Anim., CSVTS Praha, p.35 (in German).

Kalal, L., V. Pavlu, and M. Valenta. 1972. Izoenzymy LDH v tkanich siha severniho mareny, siha pelede a nekterych dalsich druhu celedi *Salmonidae* [LDH isozymes in tissues of whitefish, peled and some other salmonid fishes]. In: Proc. Conf. on Genet. and Reprod. of Anim., Liblice. Academia Praha: 241-244 (in Czech).

Kalal, L. 1989. Prehled populaci lososovitych ryb importovanych do CSSR [A review of salmonid fish populations imported into CSSR]. In: Proc.Int. Conf.Breeding of Salmonid Fishes, Marianske Lazne, 23-25 Nov., 1988. (R. Berka, ed.) Vodnany: 75-78. (in Czech).

Kalal, L., M. Dvorak. J. Rysavy, P. Rab, J. Pokorny, O. Linhart, and M. Dvorak. 1990. The breeding of rainbow trout *(Salmo gairdneri* R)J in Czechoslovakia. Prace VURH Vodnany, 19: 14-33.

Kupka, J. 1970. Smernice pro prakticke provadeni plemenarske prace v chovu pstruha duhoveho mistni linie Pd M. Organizacni pokyny Statniho rybarstvi, op., Ceske Budejovice. [The instructions for practical realization of the breeding work in the rainbow trout culture of the local strain, Pd M. (Organizational instructions of the State Fisheries Enterprise)] (in Czech).

Lusk, S. 1989. Gene pool of fishes in Czechoslovakia: Present state and conservation efforts. Prace VURH Vodnany, 18: 15-26.

Mayr, B., P. Rab, and M. Kalat. 1986. Localizations of NORs and counterstain-enhanced fluorescence studies in *Salmo gairdneri* and *Salmo trutta* (Pisces, Salmonidae). *Theor. Appl. Genet.*, 71: 703-707.

Mayr, B., M. Kalat, and P. Rab. 1988. Heterochromatins and band karyotypes in three species of salmonids. *Theor. Appl. Genet.*, 76: 45-53.

Pavlu, V., L. Kalal, M. Valenta, and S. Cepica. 1971. Polymorfismus transferinu krevniho sera siha severniho mareny *(Coregonus lavaretus maraena)* a sivena americkeho *(Salvelinus fontinalis)* [Transferrin polymorphism in serum of whitefish and brook trout]. Zivoc. Vyr., 16: 403-407 (in Czech).

Pavlu, V., L. Kalal, M. Valenta, and S. Cepica. 1971. Serumeiweisspolymorphismus bei Grossmaraene *(Coregonu.s lavaretus maraena)* und Aesche *(Thymallus thymallus)* [Polymorphismus of serum proteins in whitefish and grayling], in: Proc. VIIth Int. Symp. on Breeding and Genet. of the Lab. Anim.,CSVTS Praha,p.34 (in German).

Phillips, R.B., K.D. Zajicek, P.E. Ihssen, and O. Johnsson. 1986. Application of silver staining to the identification of triploid fish cells. *Aquaculture*, 54: 313-319.

Pokorny, J. and L. Kalal. 1976. Pokyny pro plemenarskou praci v chovech pstruha duhoveho Pd D 66 na Statnim rybarstvi [The instructions for the breeding work in rainbow trout culture in the strain Pd D 66 in the State Fisheries Enterprise. (Organizational instructions of The State Fisheries Enterprise)]. Statni rybarstvi, op., Ceske Budejovice (in Czech).

Pokorny, J. and L. Kalal. 1989. Plemenarska prace v chovech pstruha duhoveho a sihu v CSSR [Breeding work in cultures of rainbow trout and whitefishes in CSSR]. In: Proc. Int. Conf.: Breeding of Salmonid Fishes, Marianske Lazne, 23-25 Nov., 1988 (R. Berka, ed.) Vodnany: 10-18 (in Czech).

Rab, P. 1980. Karyotypy hlavatky podunajske *(Hucho hucho)* ze Slovenska a Jugoslavie [Karyotypes of *Hucho hucho* from Slovakia and Yugoslavia] in: "Reproduction, Genetics and Hybridization of Fishes" J. Kouril, ed., Vodnany 190-194 (in Czech)

Rab, P. 1989. Genetic aspects of the conservation of fish gene resources. Prace VURH Vodnany, 18: 9-14.

Rab, P. and P. Liehman. 1982. Chromosome study of the Danube salmon *Hucho hucho (Pisces, Salmonidae)*. *Folia Zoologica*, 31,2: 181-190.

Rab, P. and Jankun, M. (in press). Chromosome studies in coregonine fishes: a review. *Can. J. Fish. Aquat. Sci.*

Slechta, V. and V. Slechtova. 1977. Isoenzymy laktatdehydrogenazy v tkanich nekterych ryb celedi *Salmonidae* [LDH isozymes in tissues of some salmonid fishes]. *Zivoc. Vyr.*, 22: 813-824 (in Czech).

Slechta, V. 1980. Studium isoenzymovych systemu laktatdehydrogenazy a malatdehydrogenazy u nekterych ryb celedi *Cyprinidae a Salmonidae* [Studies of isozyme systems of LDH and MDH in some cyprinid and salmonid fishes]. Ph. D. Thesis, Cs. Acad. Sci., Inst. Anim. Physiol. Genet., Libechov (in Czech).

Slechtova, V. 1978. Studium nekterych polymorfnich bilkovin u sladkovodnich ryb [Studies of some polymorphic proteins in freshwater fishes]. Ph. D. Thesis, Cs. Acad. 5ci., Inst. Anim. Physiol. Genet., Libechov (in Czech).

Slechtova, V. and M. Valenta. 1988. Geneticke znaky charakteristicke pro siha severniho marenu *(Coregonus lavaretus maraena)*, siha pelede *(Coregonus peled)* a jejich krizence [Genetic traits characteristic for whitefish, peled and their hybrids]. *Zivoc. Vyr.*, 33: 865-875 (in Czech).

Slechtova, V., M. Valenta, V. Slechta, and P. Rab. 1988. Vyuziti biochemickych a cytogenetickych znaku pro rozliseni puvodnich druhh sihu a jejich krizencu [Use of biochemical and cytogenetic markers for distinguishing coregonid fishes and their hybrids]. In: Proc. Int. Conf. Breeding of Salmonid Fishes, Marianske Lazne, 23-25 Nov., 1988. (R. Berka, ed.) Vodnany: 24-29 (in Czech).

Slechtova, V., V. Slechta, and M. Valenta. 1990. Biochemical markers of coregonids in Czechoslovakia. *Can. J. Fish. Aquat. Sci.* (in press).

Slechtova, P. 1990. Studium biochemickych genetickych znaku u vybranych populaci pstruha duhoveho chovaneho v Ceskoslovensku [Studies of biochemical genetic markers in selected populations of rainbow trout reared in Czechoslovakia]. M. Sc. Thesis, Agricult. Univ. Prague (in Czech).

Slechtova, V. and V. Slechta. 1990. Zlozenie polymorfnych bielkovin hlavatky, *Hucho hucho* [Polymorphic proteins in Danube salmon], in: Danube Salmon in Water Ecosystem of Slovakia. Rep. Inst. Fish. and Hydrobiol., Bratislava : 37-39 (in Slovak).

Thorgaard, G.H. and G.A.E. Gall. 1979. Adult triploids in rainbow trout family. *Genetics*, 93: 961-973.

Valenta, M., V. Pavlu, and L. Kalal. 1972. Mimoradne LDH systemy v tkanich ryb [Extraordinary LDH systems in fish tissues], in: Proc. Conf. on Genet. and Reprod. of Anim., Liblice, Academia Prague: 232-235 (in Czech).

Valenta, M., V. Pavlu, and L. Kalal. 1972. Polyploidie u sladkovodnich ryb. I.Genova duplikace LDH lokuslu [Polyploidy in freshwater fishes. I. Gene duplication of LDH loci], in: Proc. Conf. on Genet. and Reprod. of Anim., Liblice, Academia Prague: 236-240 (in Czech).

Valenta, M., V. Pavlu, and L. Kalal. 1972. Polyploidie u sladkovodnich ryb. II.Isoenzymy MDH v tkanich diploidnich a tetraploidnich ryb [Polyploidy in freshwater fishes. II.MDH isozymes in tissues of diploid and tetraploid fishes] in: Proc. Conf. on Genet. and Reprod. of Anim., Liblice, Academia Prague: 241-245 (in Czech).

Occurrence, Distribution, and Potential Future of Yugoslavian Salmonids

E. TESKEREDZIC, Z. TESKEREDZIC, E. McLEAN*, M. TOMEC, and R. COZ-RAKOVAC

1. Introduction

The borders of modern Yugoslavia (Fig. 1), a country situated in south central Europe covering 255,804 km^2, were delineated at the end of the Second World War (1945). Politically, Yugoslavia is unique in Western Europe in being surrounded by seven other countries. This fact is not inconsequential when considering the nation's freshwater flora and fauna, since plants and animals, whether endemic or introduced, like pollutants, recognize no international boundaries. Almost three-quarters of Yugoslavia is composed of mountainous areas. The country's many rivers have a total length of 118,371 km, representing an average of 0.46 km/km^2 of territory. Approximately 70% of the rivers flow into the Black Sea. The majority of the remainder empty into the Adriatic, while the Vardar and its tributaries drain into the Aegean (Fig. 1). The Danube, which forms the heart of the Black Sea drainage, together with its tributaries (Bosna, Drava, Drina, Morava, Sava), are Yugoslavia's most important rivers. The country also has many lakes, the largest and most important of which are Skadar, Ohrid and Prespa (Fig. 1).

Jurinac (1887) provided one of the first detailed lists of salmonids occurring in Croatian waters, and Heintz (1910), considered those of Bosnia and Hertzegovinia. Recently Dill (1990) reviewed the current status of Yugoslavia's freshwater fisheries. Vukovic (1977), Vukovic and Ivanovic (1971) and Bojcic et al. (1982) have presented authoritative guides to the country's ichthyofauna. However, while the general occurrence of salmonids is well established, there exists little information about their current status. Statistics for individual species catches are rare or, in most cases, nonexistent. However, it is clear that many of Yugoslavia's unique salmonid species inhabit restricted ranges, which are experiencing increasing user pressure.

2. Representative Salmonids

Of the approximately 70 species of salmonid fishes in the subfamilies Salmonidae, Thymallidae, and Coregonidae, 21 are endemic to Yugoslavian watersheds. Introductions of other species for stocking programs and aquaculture have expanded this number to 25 and,

"Ruder Boskovic" Institute, Centre for Marine Research-Zagreb, Laboratory for Aquaculture, Bijenicka 54, 41000 Zagreb, Croatia. *Department of Fisheries & Oceans, West Vancouver Laboratory, 4160 Marine Drive, West Vancouver, B.C., V7V 1N6, Canada.

Genetic Conservation of Salmonid Fishes, Edited by J.G. Cloud and G.H. Thorgaard, Plenum Press, New York, 1993

since 1980, an additional 3 species have been imported into the country, on a transient basis, for aquaculture evaluation.

The comparative abundance of salmonid representatives, together with their differential spawning seasons, permits the investigation of salmonid breeding habits year-round (Fig. 2). The following treatment provides a listing, by genus, of salmonids which occur, or have occurred in Yugoslavian waters, together with an indication of their distribution and general biology. Some typical gross morphological differences exhibited between endemic species with respect to fin structure and caecal numbers are summarized in Table 1.

Table 1. Morphological differences among representative Yugoslavian salmonid fishes.

Species	Dorsal fin rays branched/unbranched		Anal fin rays branched/unbranched		Number of caecae
Salmo trutta fario	8-11	3	8-9	2-3	40-100
S. marmoratus	11	4	9	3-4	41-71
S. farioides	10-11	4	8-9	4	36-50
S. montenegricus	10	4	8	3	46
S. macedonicus	10-11	4	8	4	32-43
S. visovacensis	10	4	12	1	N/A
S. balkanica	9	4	7-8	4	49-74
S. dentex	8-10	4	7-8	4	33-45
Salmothymus obtusirostris	9-12	4	7-9	3	48-91
Hucho hucho	8-11	3-4	8-9	3-4	200

N/A = not available.

2.1. *Salmo* spp.

Salmo trutta fario (Linnaeus, 1758) is considered widespread in cold mountainous rivers, of which Yugoslavia has an abundance. Individuals may attain ages in excess of 12 years and weights approaching 12 kg. The species matures in the 3rd-4th year of life and spawns in shallow areas where strong currents and sand-gravel beds prevail. It lays 1000-2000 eggs of 4.5-5.0 mm diameter. The diet of adults consists of copepods, insect larvae and adults, smaller fish, amphibia and mammals (Filipovic and Jankovic, 1978). Al-Sabti (1985) records a chromosome number of 80 for this species.

S. t. lacustris (Linnaeus, 1758) as implied by its name inhabits lakes, which are generally cold and oxygen rich. Individuals may live for 20 years or longer and attain weights equaling or exceeding 30 kg. The age at maturity for this species is between 4 and 7 years, and these fish spawn in rivers which drain the lakes that they inhabit. The females of this species produce up to 30,000 eggs of 5.0-5.5 mm diameter; the species is well established in Lakes Bohinj, Plitvice and Pliva among others.

S. marmoratus (Cuvier, 1817), or the marbled trout, is widespread in rivers of the Adriatic drainage (Fig. 1), and reasonable populations have been reported for the river Neretva (Aganovic and Kapetanovic, 1978). Endemic to the Dalmatian coast and northeastern Italy, this species spawns within a wide range of water temperatures (8-13°C). It is a predominantly piscivorous animal. Povz (1989) reports significant morphological variations between isolated stocks from Zadlascica brook and the Soca River. Ocvirk (1989) has recorded alarming declines of the species in the Soca and Notranjska rivers over the last 30 years, and he attributes its extinction from certain watersheds to hybridization with brown trout. Al-Sabti (1985) reports a chromosome number of 80.

S. letnica (Karaman, 1924) is only found in Lake Ohrid (Fig. 1) and has been divided into three subspecies based upon morphological peculiarities: *S. l. balcanicus*, *S. l. typicus* and *S. l. aestivalis*. The osteology of the species is recorded in detail in Dorofeeva et al. (1983). Although economically important from a fisheries viewpoint, little information is available about natural hybridizations, although this would be precluded for *aestivalis* since this morph only spawns during summer months (Fig. 2).

S. farioides (Karaman, 1937) is believed to inhabit all Adriatic drainage waters. It is readily identified by its black and red spotted coloring, with the males exhibiting denser spotting than females.

S. montenegricus (Karaman, 1933) as the name implies, is found only within the Republic of Montenegro. Populations have been reported in the Radika and Garka rivers and in the lakes Skadar and Ohrid (Fig. 1).

S. macedonicus (Karaman, 1924) and *S. pelagonicus* (Karaman, 1937) represent the only salmonids inhabiting rivers which drain into the Aegean Sea. These include the Vardar River proper and its tributaries—the Kadina for the former species and the Bijela, Zlokucanska and Dubovska for the latter.

S. peristericus (Karaman, 1937) has been recorded in the Brajcinska River, which is a tributary of Lake Prespa (Fig. 1).

S. zrmaniensis (Karaman, 1937) has only been recorded in the Zrmanja River, from which it derives its specific name.

S. visovacensis (Taler, 1951) exhibits dark body markings which take the form of crosses of "Ys." Specimens of 14 kg have been taken historically. The species is known only from Lake Visovac, where it spawns in limited areas, mainly around rapids below and above Roski slap and Skradin buk.

S. balkanica (Karaman, 1926), while generally considered lacusterine in habit, living in Lake Ohrid and Skadar (Fig. 1), is also known to occur in the Drim River. Individuals of 2 kg have been recorded.

S. taleri (Karaman, 1932) has been caught in the Danube and Sava rivers and their tributaries. This particular species has a relatively slim body which is accompanied by a narrow mouth and short dorsal fin. Body coloration always consists of some red spots and on occasion black mottling.

Historically, *S. dentex* (Heckel, 1851) is known to have inhabited the waters of the Neretva, Krka, Drim, Buna, Zeta, Moraca, Bregava and Cetina. However, the present status of this species is in doubt.

2.2. *Hucho*

The subgenus *Hucho* was established by Gunther in 1866, and a variety of authors have suggested a close relationship between it and *Salvelinus*, *Salmo*, and *Brachymystax* (reviewed in Holcik et al., 1988). However, more recent evidence, derived from mitochondrial DNA analyses, indicates that *Hucho* is more closely related to the genus *Salvelinus* (Grewe et al., 1990).

In Yugoslavian waters, and specifically in those of the Black Sea drainage (Fig. 1), the subgenus is represented by *Hucho hucho* - the huchen. Members of this population may attain 1.5 m in length and 52 kg weight. The species undertakes spawning migrations to shallow water tributaries (1.2 m depth) which offer gravelly bottoms. Schulz and Piery (1982) provide detailed commentary upon redd excavation for Sava River populations. During the first year of life, juveniles feed on a wide range of invertebrates. In the Drava river system (Fig. 1), Schulz (1985) reported that the diet of adults consisted of 44% cyprinids, 30% burbot (*Lota lota*), 9.7% pike (*Esox lucius*), and 17.3% salmonids. The occurrence of this species in Yugoslavian sections of the Drava is, however, rare due to increasing pollution. The same may also be said for the Morava, although its southernmost portions (Fig. 1) still support a population of undefined numbers. Due to high summer temperatures of the Tisa River, catches of the huchen are reported rarely, and then only between late fall and early spring.

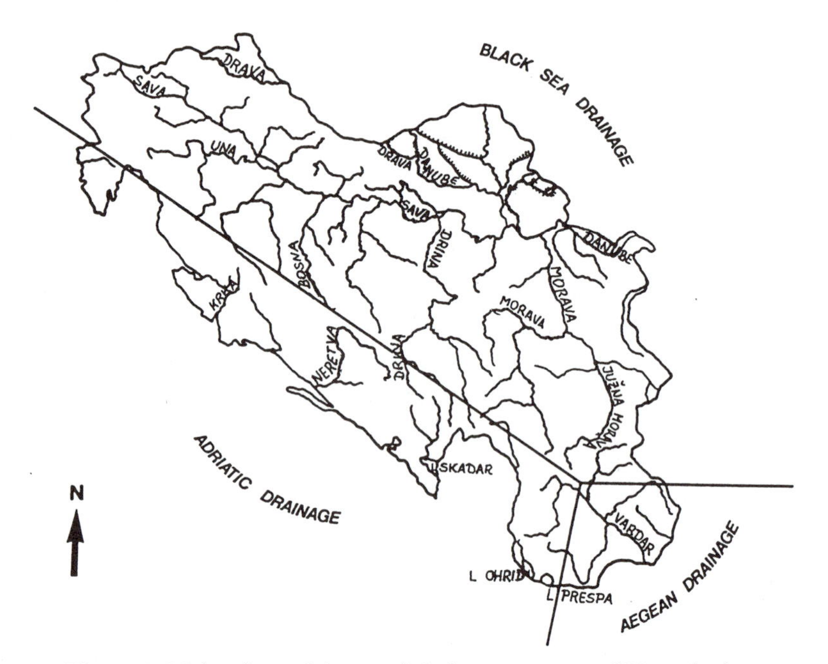

Figure 1. Major rivers, lakes, and drainage systems of Yugoslavia.

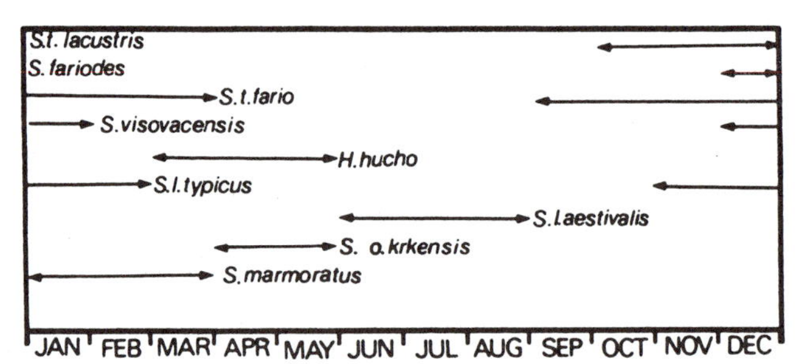

Figure 2. Example spawning periods for representative Yugoslavian salmonids.

While the Sava river (Fig. 1) has been reported to support a relatively large huchen population (Holcik et al., 1988), thermal waste from a nuclear power station 40 km north of Zagreb, together with domestic and industrial pollution from Lubljana and its environs, has reduced its range and probably its occurrence. It is likely that huchen populations only exist in good numbers in the Dobra, Korana, and Kupa rivers. Therefore, it is of note that Verce (1973) reported Slovenian catches for huchen numbered 148 in 1966 and 128 individuals in 1971. Whether the apparent decrease in numbers is representative of a general decline in numbers is however, unknown.

In the Timis and Morava rivers, the huchen is now considered extinct. The finest fishing areas remain in the Drina River, where four years ago an individual of 36 kg was taken by angling. The major reason for the health of this huchen fishery relates to the lack of pollution and habitat erosion. But, this population could be under threat, since plans for the construction of a hydroelectric power station have been drawn and are presently under consideration.

2.3. *Salmothymus* spp.

This genus is represented by two species, both of which inhabit rivers of the Adriatic drainage. It is unique to the area and is sometimes referred to as the Adriatic salmon. *Salmothymus obtusirostris* (Heckel, 1851) is colored with irregularly shaped black mottling together with a number of smaller red and orange markings. Four subspecies are recognized based upon morphological differences. *S. o. oxyrhynchus* (Steindacher, 1882) is endemic to the Neretva River system (Aganovic and Kapetanovic, 1978). The species exhibits an olive body coloration with black spots. Red and orange markings appear on the animal's tail. Males mature in their fourth or fifth year whereas the female matures between the fifth and seventh year of life. A chromosome number of 82 has been reported for this species (Berberovic et al., 1970). *S. o. salnoita* (Karaman, 1926) inhabits the Jadro River. Animals may achieve weights in excess of 2 kg. *S. o. krkensis* (Karaman, 1926) is found at the source of the Krka river and can reach weights exceeding 3 kg. *S. o. Ietnica* (Karaman, 1932) is reported for the Zeta River. Little information exists concerning the above species' current status. The other representative, *S. o. ohridanus* (Steindachner, 1892), only inhabits Lake Ohrid. This species is unique to the region and its status is uncertain.

2.4. *Thymallus*

The Thymallidae is represented in Yugoslavia by a single species, namely *Thymallus thymallus* (Linnaeus, 1758). It is widespread in waters which drain into the Black Sea and has also been transplanted into rivers of the Adriatic drainage. The species matures at an age of 45 years and specimens of 2.8 kg have been taken.

3. Introduced Species

A number of salmonids have been introduced into Yugoslav waters. Not surprisingly the first recorded transplantation was that of the rainbow trout *Oncorhynchus mykiss.* This species has been imported directly from the United States, on at least four occasions (1981, 1930, 1935, 1981). Although early imports were used for stocking programs and as game fish, subsequent transplants were employed to bolster the broodstock of Yugoslavia's trout-farming industry. While it is generally considered that self-sustaining populations are rare in Yugoslavia, it is believed that the "Kalifornijska or duzicasta pastrva" has been involved in the reduction of native salmonids. While direct evidence for such an occurrence is extremely difficult to obtain, it is known that the rainbow trout can effectively utilize and compete for niches usually inhabited by autochthonous salmonids. Perhaps more insidious than simple competitive displacements is the potential for genetic pollution by this species. While hybridization studies, undertaken at the Biological Institute of the University of Sarajevo, failed to obtain crosses between rainbow trout and *S. o. oxyrhynchus* (Anonymous, 1970), a wide variety of other studies have demonstrated the promiscuity of the rainbow trout with respect to its hybridization qualities (e.g., Schwartz, 1972,1981; Dangel et al., 1973; Chevassus, 1979).

Soon after the importation of the rainbow trout (1892), brook charr *Salvelinus fontinalis* were transplanted from "Austria", ostensibly for angling purposes. Also from Austria came stocks of the Arctic charr *S. alpinus* (1928, 1943). While populations of the latter are now believed to be self-sustaining in three lakes (Welcomme, 1988), only remnant populations of the former species are thought to exist (Welcomme, 1988). In 1977 a number of peled *Coregonus peled* were transplanted into Lake Peruca from Czechoslovakia, to examine their potential for caged aquaculture and lake stocking. At present, little information is available on the breeding status of this animal.

During the last 12 years, a number of studies have examined the potential for culturing nonindigenous salmonids on a commercial basis (see Edwards, 1984). The first species to be imported for aquaculture evaluation was the Atlantic salmon (*Salmo salar*) from Norway in 1980. Rearing temperatures at the quarantine station, Gracani, proved to be too high for this species. Later studies with coho salmon (*O. kisutch*), imported from Washington State in 1981, proved to be successful. At the present time coho are reared to the smolt stage at several locations, using waters of the Black Sea drainage; and cage culture operations currently exist in the Adriatic. In areas where temperatures may reach 23°C during summer months, coho growout is restricted from October to mid-June. However, in areas where summer temperatures are 20°C, culture operations persist throughout the year; under these conditions, some 1+ smolts have achieved 5 kg in weight in a single season (Teskeredzic *et. al.*, 1989). Although production of coho is presently only at 200 tons, plans are well under way to increase this to 3000-5000 tons by 1995. It is of note that Adriatic-based cage systems do experience escapees (McLean, 1990; personal observation); however, these individuals appear to remain within the culture facility's immediate area. Expansion of the coho industry could therefore, feasibly lead to the establishment of viable runs over the long term.

Similar to the studies with coho salmon, preliminary trials have also been initiated with amago salmon (*O. masou rhodurus*), which were imported to Yugoslavia from Japan in 1983. Results pertaining to these studies are presented in Teskeredzic and Teskeredzic (1990).

4. The Future

The general geographical distribution of salmonids in Yugoslavian waters is well characterized on an historic basis. The current status, in terms of their detailed presence and abundance in specific waters, remains in doubt. Most autochthonous species probably exhibit healthy regional populations but reduced abundance, reduced diversity, and, in some instances, extinction, are known as is adequately illustrated by the plight of the huchen.

Numerous studies have determined that Yugoslavia's major waterways (Bosna, Danube, Krka, Kupa, Morava, Neretva, Sava) have deteriorated due to inputs of substances from agriculture, mining, forestry, industrial, urban and other activities. Also, the construction of reservoirs, hydroelectric and nuclear power stations have been reported to decrease regional ichthyodiversity, although in some instances to the benefit of salmonids (Knezevic and Maric, 1989). Little information is available concerning the ability of specific Yugoslav salmonids to adapt to environmental change. This is one area of concern which requires immediate address. While protection of "significant" salmonid resources may be possible by ecosystem rehabilitation and artificial enhancement programs, the nation's present socioeconomic climate would indicate that such programs will only be considered necessary in the distant future.

Outside of national jurisdiction, but nonetheless important, is the current trend in Northern Europe to canalize major waterways to facilitate economic transportation. Canalization has, or will, ultimately link the Black Sea drainage to those of the North and Baltic Seas following construction of the Rhine-Main-Danube and Elbe canal connectors. These new waterways will inevitably expand the ranges of the more hardy ichthyofauna of the European continent, placing even more stress upon Yugoslavian salmonid populations than is already apparent. An indication of the potential dangers of opening European waters has been provided with respect to Slovenia's marbled trout populations. The introduction of the brown trout to regions inhabited by the marbled trout over the last three decades has resulted in competitive displacement and decimation of the genetic line by hybridization (Ocvirk, 1989). Since other studies have demonstrated the feasibility of crossing the brown trout with the Adriatic salmon (Anonymous, 1970) this provides further reason for concern. Associated with the expansion of the geographic ranges in fish is the increased probability of epizootics of previously unencountered diseases. This likelihood may be further increased by fish culture operations where eggs and broodstock of exotic species are transplanted. The first outbreak of enteric redmouth disease in Yugoslavia, for example, was recorded in 1988 (Ocvirk et al.). No information is available concerning its possible transmission to wild stocks.

A further hindrance to the protection and enhancement of endemic salmonid resources relates to the lack of communication between resource managers, scientists, the public and policy-makers. Some of these problems may be effectively handled with the formulation and enforcement of national policies regulating contaminant inputs, environmental degradation and modification, resource management, aquaculture and scientific priorities. As is evidenced from Yugoslavia's political locale, construction of such policies would be pointless without active international collaboration with other countries which utilize the same drainage basins. The preservation of Yugoslavia's unique salmonid fishes will inevitably require significant sacrifice. Effective resource management may represent the preferred method of stock preservation. However, to ensure the protection of all South Central European salmonid resources, the development of an International Centre for Salmonid Genetic Resources should be seriously considered.

References

Aganovic, M. and N. Kapetanovic. 1978. The age structure in the populations of some fish species from the river Neretva. *Acta Biol. Jugoslav., E: Ichthyol.* 10: 17.

Al-Sabti, K. 1985. Chromosomal studies by blood leukocyte culture technique on three salmonids from Yugoslavian waters. *J. Fish Biol.* 26: 512.

Anonymous, 1970. *FAO Fish Culture Bulletin* 2(4): 23.

Berberovic, L., M. Curcic, R. Hadziselimovic, and A. Sofradzija. 1970. Chromosome complement of *Salmothymus obstrusirostris oxyrhynchus* (Steindachner). *Acta Biol. Jugoslav. (Genet.)* 2: 55-63.

Bojcic, C., Lj. Debeljak, T. Vukovic, B. Jovanovic-Krsljanin, K. Apostolski, B. Rzanicanin, M. Turk, S. Volk, D. Drecun, D. Habekovic, D. Hristic, N. Fijan, K. Pazur, I. Bunjevac, and D. Marosevic. 1982. *"Slatkovodno ribarstvo Jugoslavije."* Jugoslavenska Medicinska Naklada, Zagreb, Yugoslavia.

Chevassus, B. 1979. Hybridization in salmonids: results and perspectives. *Aquaculture* 17: 113-128.

Dangel, J.R., P.T. Macy, and F.C. Withler. 1973. Annotated bibliography of interspecific hybridization of fishes of the subfamily Salmonidae. *NOAA Tech. Mem.*, NMFS NWFCI, 48 pp.

Dill, W.S. 1990. Inland fisheries of Europe. *EIFAC Tech. Pap.* 52.

Dorofeeva, E.A., M. Sidorovski, and N. Petrovski. 1983. Osteological peculiarities of ohrid trouts *(salmo letnica)* with reference to their taxonomy. *Zool. Zhurnal* 62: 1691-1700.

Edwards, D. 1984. Yugoslavia. Status and prospects for brackish and seawater cage culture of finfish in Yugoslavia with special emphasis on salmonid species . *FAO FI/DP/YUG/83/011/field/doc/1*, 53pp.

Filipovic, D. and D. Jankovic. 1978. Relation between the bottom fauna and fish diet of the upland streams of East Serbia. *Acta Biol. Jugoslav. E. Ichthyol.* 10: 29-40.

Grewe, P.M., N. Billington, and P.D.N. Herbert. 1990. Phylogenetic relationships among members of *Salvelinus* inferred from mitochondrial DNA divergence. *Can. J. Fish. Aquat. Sci.* 47: 984-991.

Gunther, A. 1866. *Catalogue of the fishes in the British Museum.* 6. Trustees, London.

Heintz, K., 1910. Die salmoniden Bosniens und der Hertzegowina. *Osterr. Fisch. Zeitg.* 7: 287-290.

Holcik, J., K. Hensel, J. Nieslanik, and L. Skacel. 1988. The Eurasian huchen Hucho hucho. "Largest salmon of the world." Dr. W. Junk Publ., Dordrecht. 239 pp.

Ivanovic, B. 1973. Ichthyofauna of Skadar Lake. *Rep. Biol. Station, Titograd*, 146 pp.

Jurinac, A.E. 1887. List of fishes of Croatia. *Rad. Jugoslav. Akad.* 83: 114-121.

Knezevic, B. and D. Maric. 1989. Ichthyofauna in the river Piva before and after the construction of a reservoir in Montenegro. *Ichthyos* 7: 14.

Ocvirk, A. 1989. Study of the marble trout (*Salmo marmoratus*, Cuvier 1817) population in Slovenia from 1962 to 1988. *Ichthyos* 8: 37-59.

Ocvirk, A. Janc, M., Jeremic, S. and Skalin, B., 1988. The first case of enteric redmouth disease in Yugoslavia. *Ichthyos* 6: 34-47.

Povz, M. 1989. Distribution and biometric characteristics in the marble trout (*Salmo marmoratus*, Cuvier 1817) in Slovenia. *Ichthyos* 8: 13-36.

Schulz, N. 1985. Das Wachtstum des Huchen (*Hucho hucho*, L.) in der Drau in Karnten. *Osterr. Fisch. Zeitg.* 38: 131-142.

Schulz, N. and G. Piery. 1982. Zur fortpflanzung des huchens (*Hucho hucho*, L.) - Untersuchung einer laichgrube. *Osterr. Fisch. Zeitg.* 35: 241-249.

Schwartz, F.J. 1972 World literature to fish hybrids with an analysis by family, species and hybrid. *Publ. Gulf Coast Res. Lab. Mus.* 3: 328pp.

Schwartz, F.J. 1981 World literature to fish hybrids with an analysis by family, species and hybrid: Supplement 1. *NOAA Tech. Rep.*, NMFS SSRF750, 507 pp.

Teskeredzic, E., Z. Teskeredzic, M. Tomec, and M. Hacmanjek. 1989. Culture of coho salmon (*Oncorhynchus kisutch*) and rainbow trout (*Salmo gairdneri*) in the Adriatic Sea.. *World Aquaculture* 20: 56-67.

Teskeredzic, E. and Z. Teskeredzic, 1990. A successful rearing experiment with Amago salmon (*Oncorhynchus masou rhodurus)* in floating cages in the Adriatic Sea. *Aquaculture* 86: 201-208.

Verce, F. 1973. Pogled na nase ribistvo skozi prizmo statistike. *Ribic* 32: 98-100.

Vukovic, T. 1977. "Ribe Bosne i Hercegovine (Kljuc za odredivanje), KRO "Svjetlost." OOUR Zavod za udzbenike, Sarajevo.

Vukovic, T. and B. Ivanovic. 1971. *Slatkovodne ribe Jugoslavije*. Zemaljski muzej BIH, Sarajevo, 268 p.

Welcomme, R.L. 1988. "International introductions of inland aquatic species." *FAO Tech. Pap.*, #294, 318 pp.

Genetic Differentiation and Relationship within and between Natural and Cultured Populations of *Oncorhynchus masou* Complex in Japan

MASAMICHI NAKAJIMA and YOSHIHISA FUJIO

1. Introduction

Masu salmon *(Oncorhynchus masou)* is one of the more popular fish in Japan. The distribution of masu salmon is restricted to the eastern part of the Pacific Ocean, around Japan. In Japan, many local types in each stream are separated by life history and morphological and ecological characters. This species has been called "*Oncorhynchus masou* complex." In recent studies (Kato, 1973; Yoshiyasu, 1973a; Yoshiyasu, 1973b; Kato, 1974; Honjo, 1976), this complex has been shown to consist of the following three subspecies: Sakuramasu *(Oncorhynchus masou masou)*, Amago *(O. m. macrostomus)* and Biwamasu *(O. m. rhodurus)* (Araga, 1984).

Each subspecies is clearly separated by its distribution and morphological characters. Sakuramasu has the widest distribution; it is found all around Japan except in the central part. Amago is located in the central part of Japan where Sakuramasu is not found. Biwamasu is distributed only in Lake Biwa. Morphologicaly, Amago and Biwamasu have red spots on the side of the body, but Sakuramasu does not. Two types of Sakuramasu are ecologically separated; one, called "Sakuramasu", is an anadromous type; another type, called "Yamame", is land locked. Between these two types there are no genetic differences (Nakajima et al., 1986).

Culture of these subspecies began in the middle 1960s and the subspecies are currently cultured and maintained in ponds in many parts of Japan. Since the seeds have been transferred from one cultured location to another, each cultured population is not presently located where it originated. Because the history of masu salmon culturing in Japan is relatively brief (shorter, for example, than that of rainbow trout), cultured populations may still retain many of the natural population's genetic characteristics.

In this study we aimed to determine the genetic differences and relationships among masu salmon complex populations in Japan, and to estimate the origin of cultured populations in comparison with natural population.

Department of Fishery Science, Faculty of Agriculture, Tohoku University, Sendai, Japan, 981.

Table 1. Data of *Oncorhynchus masou* complex populations examined in this study.

Population Code Number	Origin of Each Population	Sampling Date	No. of Individuals
Oncorhynchus masou masou			
MS-1	Hokkaido(Hokkaido)	Sep. 1986	101
MS-2	Hokkaido(Hokkaido)	Sep. 1986	100
MS-3*	Hokkaido(Hokkaido)	Sep. 1986	21
MS-4	Hokkaido(Miyagi)	Jul. 1982	45
MS-5	Aomori(Aomori)	Oct. 1986	34
MS-6	Aomori(Aomori)	Oct. 1986	102
MS-7	Iwate(Iwate)	Sep. 1986	103
MS-8*	Miyagi(Miyagi)	Sep.-Oct. 1986	51
MS-9	Miyagi(Miyagi)	Sep.-Oct. 1986	61
MS-10	Fukushima(Fukushima)	Jul. 1982	47
MS-11	Niigata(Niigata)	Oct. 1986	106
MS-12	Gunma(Gunma)	Oct. 1986	100
MS-13	Akita(Nagano)	Oct. 1986	100
MS-14	Hokkaido(Gifu)	Oct. 1986	50
MS-15	Saitama(Gifu)	Oct. 1986	100
MS-16	Shimane(Shimane)	Oct. 1986	99
MS-17	Saitama(Miyazaki)	Jun. 1987	100
MS-18	Miyazaki(Miyazaki)	Jun. 1987	100
O. m. macrostomus			
MC-1	Gifu(Tochigi)	Jul. 1982	25
MC-2	Gifu(Gifu)	Jan. 1989	57
MC-3	Gifu(Gifu)	Oct. 1989	37
MC-4	Nara(Wakayama)	Oct. 1989	100
O. m. rhodorus			
R	Shiga(Shiga)	Nov. 1989	89

* Indicates that the sample was collected from natural river populations, and in parenthesis the sampling location of the hatchery of each populations is indicated.

2. Materials and Methods

A total of 23 populations were examined (Table 1); 18 were Sakuramasu *(O m. masou)*, four were Amago *(O. m. macrostomus)* and one was Biwamasu *(O. m. rhodurus)*. Of the 18 Sakuramasu populations, two natural river populations were included, which were collected when they came back to the river for spawning. Since there are no genetic differences between Sakuramasu and Yamame, "Sakuramasu" is used to denote the *O. m. masou* in this study. Each cultured population has been maintained as a closed colony and these populations have been passed three or four generations under artificial conditions.

Nine enzymes were examined to estimate 26 isozymic loci (Table 2). The method of electrophoretic analysis used, as well as staining methods, were taken from Fujio (1984).

Table 2. Examined enzyme and estimated loci.

Enzyme and Enzyme Council Number	Abbreviation	Locus	Tissue*
Glycerophosphate dehydrogenase	aGPD	Gpd-1	L
(E. C. 1.1.1.8)		Gpd-2	M
		Gpd-3	M
Glucosephosphate isomerase	GPI	Gpi-1	M
(E. C. 5.3.1.9)		Gpi-2	M
		Gpi-3	M
		Gpi-4	M
Isocitrate dehydrgenase	IDH	Idh-1	M
(E. C. 1.1.1.27)		Idh-2	M
		Idh-3	L
		Idh-4	L
Lactate dehydrogenase	LDH	Ldh-l	E
(E. C. 1.1.1.27)		Ldh-2	E
		Ldh-3	E
		Ldh-4	M
		Ldh-5	M
Malate dehydroqenase	MDH	Mdh-l	M
(E, C. 1.1.1.37)		Mdh-2	M
		Mdh-3	M
		Mdh-4	M
Octanol dehydrogenase (E. C. 1.1.1.73)	ODH	Odh	L
6-Phosphgluconate dehydrogenase (E. C. 1.1.1.44)	6PGD	6Pgd	M
Phosphglucomutase	PGM	Pgm-l	L
(E. C. 2.7.5.1)		Pgm-2	M
		Pgm-3	M
Superoxide dismutase (E. C. 1.15.1.1)	SOD	Sod	L

* M means the muscle, L means the liver, and E means the eye.

3. Results

3.1 Genetic Variation

Of the 26 loci examined, genetic variations were observed at 11; these loci were Gpi-3, Idh-3, Idh-4, Ldh-l, Ldh-4, Mdh-l, Mdh-4, Odh, Pgm-l, Pgm-3 and Sod. Alternatively, Gpd-l, Gpd-2, Gpd-3, Gpi-l, Gpi-2, Gpi-4, Idh-1, Idh-2, Ldh-2, Ldh-3, Ldh-5, Mdh-2, Mdh-3, 6Pgd and Pgm-2 were monomorphic. The gene frequencies are listed in Tables 3 and 4.

Average heterozygosity, which indicates the genetic variability, of Sakuramasu varied from 0.023 to 0.074 with a mean of 0.050. In the Pacific Ocean side, the mean average heterozygosity for this population was 0.044 with a range of 0.023 to 0.062; on the Japan Sea side, the mean average was 0.059 with a range of 0.049 to 0.074. The genetic variability in the population originating in the Japan Sea was higher than that of Pacific Ocean side origin. Within cultured populations, the average heterozygosity was 0.048, and this value is very similar to the value of cultured populations reported by Nakajima et al. (1986). Among Amago, average heterozygosity varied from O to 0.049 with a mean of 0.026. Among Biwamasu, the mean was 0.017.

Table 3. Gene frequency of *Oncorhynchus masou* complex populations (Pacific Ocean side origin).

Locus/ Allele		MS-2	MS-5	MS-6	MS-7	MS-8	MS-9	MS-10	MS-12	MS-15	MS-17	MS-18	MC-1	MC-2	MC-3	MC-4	R
Gpi-3	A	0	0	0	0	0	0	0	0	0.005	0	0	0	0.280	0.176	0.170	0
	B	1.000	1.000	1.000	1.000	1.000	1.000	1.000	1.000	0.995	1.000	1.000	1.000	0.720	0.824	0.830	1.000
Idh-3	A	0.006	0.113	0	0.015	0.064	0.058	0.138	0.030	0	0	0	0	0	0	0.029	0
	B	0.912	0.807	0.897	0.987	0.883	0.925	0.755	0.935	1.000	0.940	0.985	1.000	1.000	1.000	0.770	0
	C	0	0	0	0	0	0	0	0	0	0	0	0	0	0	0	1.000
	D	0.011	0.048	0.076	0.088	0.053	0.017	0.107	0.005	0	0.060	0.015	0	0	0	0.155	0
	E	0.011	0.032	0.027	0	0	0	0	0.030	0	0	0	0	0	0	0.046	0
Idh-4	A	0	0	0	0	0	0	0	0	0	0	0	0	0	0	0	1.000
	B	1.000	1.000	1.000	1.000	1.000	1.000	1.000	1.000	1.000	1.000	1.000	1.000	1.000	1.000	1.000	0
Ldh-1	A	1.000	1.000	1.000	1.000	1.000	1.000	1.000	1.000	1.000	0.755	0.798	1.000	1.000	1.000	1.000	1.000
	B	0	0	0	0	0	0	0	0	0	0.245	.0202	0	0	0	0	0
Ldh-4	A	0	0	0	0	0	0	0	0	0	0.025	0	0	0	0	0	0
	B	1.000	1.000	1.000	1.000	1.000	1.000	1.000	1.000	1.000	0.975	1.000	1.000	1.000	1.000	1.000	1.000
Mdh-1	A	0	0	0	0	0	0	0	0	0	0	0	0	0.061	0.068	0.055	0
	B	0.740	0.971	0.938	1.000	0.951	0.910	0.907	0.890	0.990	1.000	0.945	1.000	0.939	0.932	0.930	0.955
	C	0.260	0.029	0.047	0	0.049	0.090	0.093	0.110	0.010	0	0.055	0	0	0	0.015	0.045
	D	0	0	0.015	0	0	0	0	0	0	0	0	0	0	0	0	0
Mdh-4	A	0.015	0	0	0	0	0	0	0	0	0	0	0	0	0	0	0
	B	0.985	1.000	1.000	1.000	1.000	1.000	1.000	1.000	1.000	1.000	1.000	1.000	1.000	1.000	1.000	1.000
Odh	A	0	0.088	0	0	0	0	0	0.025	0.190	0	0	0	0	0	0.016	0
	B	1.000	0.912	1.000	1.000	1.000	1.000	1.000	0.975	0.810	1.000	1.000	1.000	1.000	1.000	0.984	1.000
Pgm-1	A	0.203	0.839	0.030	0.240	0.021	0.017	0.213	0.126	0.313	0.304	0.112	1.000	1.000	1.000	0.942	1.000
	B	0	0	0	0	0	0	0.011	0	0	0	0	0	0	0	0.005	0
	C	0.779	0.161	0.970	0.740	0.979	0.983	0.776	0.864	0.646	0.652	0.878	0	0	0	0.053	0
	D	0.018	0	0	0.020	0	0	0	0.010	0.041	0.044	0.010	0	0	0	0	0
Pgm-3	A	0.606	0.719	0.534	0.607	0.765	0.667	0.711	0.805	0.780	0.558	0.199	1.000	0.796	0.905	0.789	1.000
	B	0	0	0.059	0.058	0.010	0	0	0.020	0.005	0.196	0.051	0	0	0	0	0
	C	0.394	0.281	0.407	0.335	0.225	0.333	0.289	0.175	0.215	0.246	0.715	0	0.204	0.095	0.214	0
	D	0	0	0	0	0	0	0	0	0	0	0	0	0	0	0	0
Sod	A	1.000	1.000	1.000	0.840	1.000	1.000	1.000	0.975	1.000	1.000	1.000	1.000	1.000	1.000	1.000	0.680
	B	0	0	0	0.160	0	0	0	0.025	0	0	0	0	0	0	0	0.320
	He	0.054	0.047	0.035	0.053	0.023	0.030	0.051	0.038	0.045	0.062	0.041	0	0.062	0.023	0.049	0.017

Table 4. Gene frequency of *Oncorhynchus masou* complex populations (Japan sea side origin).

Locus	Allele	MS-1	MS-3	MS-4	MS-11	MS-13	MS-14	MS-16
Gpi-1	A	0	0	0	0	0	0	0
	B	1.000	1.000	1.000	1.000	1.000	1.000	1.000
Idh-3	A	0.161	0.228	0.172	0.048	0.030	0	0.084
	B	0.661	0.663	0.813	0.952	0.745	0.793	0.675
	C	0	0	0	0	0	0	0
	D	0.116	0.065	0.015	0	0.225	0.159	0.234
	E	0.062	0.044	0	0	0	0.048	0.007
Idh-4	A	1.000	1.000	1.000	1.000	1.000	1.000	1.000
	B	0	0	0	0	0	0	0
Ldh-1	A	1.000	1.000	1.000	1.000	1.000	1.000	1.000
	B	0	0	0	0	0	0	0
Ldh-4	A	0	0	0	0	0	0	0
	B	1.000	1.000	1.000	1.000	1.000	1.000	1.000
Mdh-1	A	0	0	0	0	0	0	0
	B	0.767	0.860	0.822	0.929	0.915	0.788	0.995
	C	0.218	0.140	0.178	0.071	0.085	0.212	0.005
	D	0.015	0	0	0	0	0	0
Mdh-4	A	0	0	0	0	0	0	0
	B	1.000	1.000	1.000	1.000	1.000	1.000	1.000
Odh	A	0	0	0	0	0	0.201	0.158
	B	1.000	1.000	1.000	1.000	1.000	0.799	0.842
Pgm-1	A	0.226	0.094	0.167	0.167	0.245	0.267	0.500
	B	0	0.021	0.011	0	0	0	0
	C	0.774	0.864	0.822	0.833	0.755	0.723	0.464
	D	0	0.021	0	0	0	0.010	0.036
Pgm-3	A	0.710	0.583	0.600	0.571	0.702	0.782	0.594
	B	0.025	0	0.056	0.119	0	0.056	0
	C	0.265	0.417	0.244	0.310	0.177	0.156	0.224
	D	0	0	0	0	0.121	0.006	0.182
Sod	A	1.000	1.000	1.000	1.000	1.000	0.981	0.970
	B	0	0	0	0	0	0.019	0.030
He		0.064	0.057	0.049	0.049	0.053	0.068	0.074

3.2. Genetic Differentiation

A difference in allelic distribution was observed among the three subspecies (Tables 3 and 4). As shown in Figure 1, Idh-3 and Idh-4 loci of Biwamasu were divergent from Sakuramasu and Amago. At the Idh-3, Sakuramasu and Amago had A, B, D and E alleles, but only the C allele was observed in Biwamasu. At the Idh-4, Sakuramasu and Amago showed only the B allele while Biwamasu showed only the A allele.

Additionally, a difference of allelic distribution at Pgm-1 was observed between Sakuramasu and the other two subspecies, as shown in Figure 1; the C allele was dominant in Sakuramasu, and the A allele was dominant in Amago and Biwamasu.

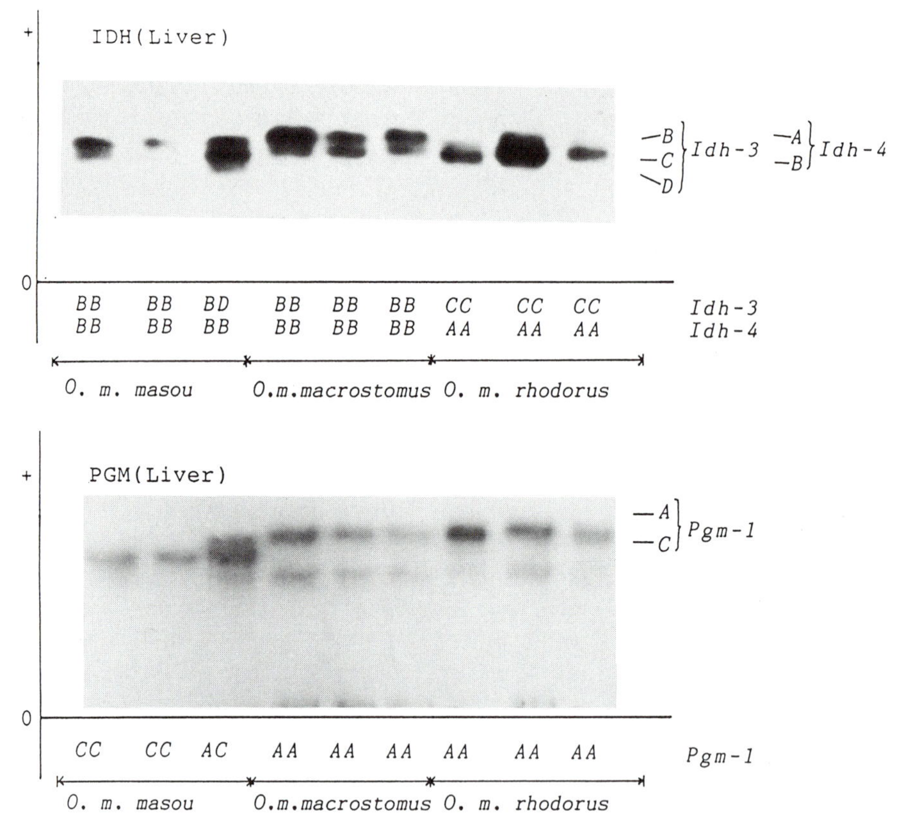

Figure 1

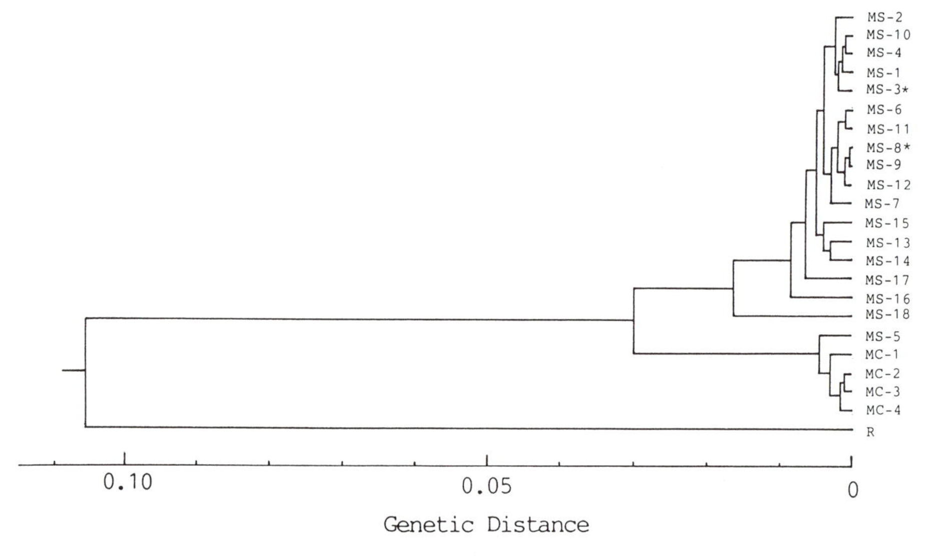

Figure 2

Nei's genetic distance (1972) was calculated using 26 estimated loci, and the resultant dendrogram shows the relationship between each population of the *Oncorhynchus masou* complex (Fig. 2). Generally speaking, 23 populations were separated into three groups. One was the group of Sakuramasu except MS-5; the second group was Amago, which includes MS-5, and the third group was Biwamasu. The level of genetic diversity, which was high between Biwamasu and the other two groups, and was low between Sakuramasu and Amago. Because the A allele frequency at Pgm-1 was high in MS-5 but low in the other populations of Sakuramasu, MS-5 was included in the Amago type. The MS-5 population, collected from the Ohata River, is called "Suginoko" and its distribution is restricted to the Ohata.

The origin of MS-1, MS-2 and MS-4 is Hokkaido, and these populations were similar to MS-3, a natural river population of Hokkaido. The same case was observed between MS-8 and MS-9. The origin of MS-9 is the Kitakami River and, although more than three generations have passed since the culture was started, it was very similar to the Kitakami's natural population, MS-8. No significant difference of gene frequency was observed between MS-8 and MS-9.

Based on an examination of isozymes, Okazaki (1986) concluded that natural populations of masu salmon (Sakuramasu) were separated into the Pacific Ocean side and Japan Sea side, including Hokkaido. To determine whether such diversity exists in cultured populations, populations were compared. Ten cultured populations which used the comparison of genetic character of natural and cultured populations by Nakajima et al. (1986) were added to this comparison. MS-5, which was included in the Amago type, and MS-3 and MS-8, which are natural river populations, were excluded in this analysis. Thus, a total of 25 populations were used.

The original location of each population was separated into three regions: Hokkaido, the Pacific Ocean side of Honshu, and the Japan Sea side of Honshu. The genetic distance within and between each region is shown in Table 5.

Table 5. Genetic distance between and within the original area of the grouped populations.

Within Hokkaido (n=6)*	0.0046 ± 0.0020
Within Honshu (n=210)	0.0110 ± 0.0078
Within Pacific Ocean side (n=91)	0.0099 ± 0.0094
Within Japan Sea side (n=21)	0.0069 ± 0.0031
Between Pacific Ocean and Japan Sea side (n=98)	0.0127 ± 0.0056
Between Hokkaido and Honshu(n=84)	0.0103 ± 0.0065

* The parentheses represents the number of combinations

Within Hokkaido, the mean genetic distance was 0.0046 with a standard deviation of 0.0020. Within Honshu, mean genetic distance was 0.0110 ± 0.0078. In Honshu, mean genetic distance was 0.0099 with an SD of 0.0094 for the Pacific Ocean side, and the mean genetic distance was 0.0069 with 0.0031 SD for the Japan Sea side. Between the Pacific Ocean side and Japan Sea side of Honshu, the mean genetic distance was 0.0127 with an SD of 0.0056. Between Hokkaido and Honshu, mean genetic distance was 0.0103 with an SD of 0.0065. Likewise, with the cultured populations, the largest genetic diversity was observed between the Pacific Ocean side and the Japan Sea side.

4. Discussion

Nei (1975) reported that the range of genetic distance between local races is about 0.01, between subspecies about 0.10, and between species more than 1.0. According to these values, the extent of genetic diversity between Biwamasu and the other two populations is at the subspecies level, and between Sakuramasu and Amago, at the local race level. One of the conditions for a certain deme to be classified into subspecies by isozymic loci is the existence of fixed alleles, which requires that reproductive isolation has extended over a long period of time. Biwamasu satisfied this condition. There are no divergent genes between Sakuramasu and Amago, however, the extent of genetic diversity is at the local race level. Included among Sakuramasu populations is MS-5 which displays the genetic character of Amago. Called "Suginoko", MS-5's distribution is at the very restricted upstream area of the Ohata River. Nakamura and Kawai (1966) reported that there were no differences between Suginoko and cultured Sakuramasu. On the other hand, Kubo (1979) reported that the parr-mark of Suginoko is more similar to the parr-mark of Amago than that of Sakuramasu.

Amago was distinguished from Sakuramasu by red spots on their body side, but there are Amago variants which have no red spots in cultured populations. So the existence of red spots is not a fixed genetic character of Amago. For this reason, Amago can be regarded as a local race of Sakuramasu. This situation invites a more detailed investigation.

The combined observations that cultured populations are similar to the natural population of their original region and that the genetic distance between populations from the Pacific Ocean and Japan Sea side is large, suggest that the genetic character observed in the natural population still exists in cultured populations.

In the process of cultivation, a decrease of genetic variability (Allendorf and Phelps, 1980, Taniguchi et al., 1983) and increase of genetic distance between cultured populations (Nakajima et al., 1986, Nakajima and Fujio, 1988, Fujio and Nakajima, 1989) has been observed. The reason for such change of genetic character has been explained by founder and/or bottleneck effects. Thus, it is expected that the genetic character of cultured populations can become completely different from natural populations; cultivation results in the subdivision of population and the resultant fixation of gene. Nakajima et al. (1986) reported that the genetic variability calculated from the total average of gene frequency of natural and cultured populations show the same values. Since the genetic variability calculated from the average of each population decreases in cultured populations, a number of populations from the same origin need to be cultivated in order for the genetic character of all cultured populations combined to show the same characteristics found in the original population.

These results suggest that culture is useful as a method to conserve the natural gene stocks. Cultivation for the purpose of gene pool conservation is becoming increasingly important, and more in-depth genetic studies are needed to develop adequate management programs.

5. Summary

Masu salmon (*Oncorhynchus masou*) is one of the more popular fish in Japan. In this species, there are many local types and these types have been called "*Oncorhynchus masou* complex." In this study, our objective was to determine the genetic differences and relationships among cultured *Oncorhynchus masou* complex populations in Japan, and to estimate the origin of cultured populations in comparison with the natural population.

The degree of genetic differentiation between Sakuramasu and Amago was determined to be at the local race level and these populations were subspecies to Biwamasu. The cultured populations of Sakuramasu in Japan were separated into two groups based on genetic distance. One group corresponded to a population which originated from the rivers on the Pacific Ocean side, and the other from the rivers of the Japan Sea side. This suggests that cultured populations have preserved the genetic characteristics of their origins.

ACKNOWLEDGMENTS: For supplying specimens, we wish to express our thanks to Freshwater Fish Research station in Hokkaido, Aomori, Iwate, Miyagi, Fukushima, Niigata, Gunma, Tochigi, Nagano, Gifu, Wakayama, Shimane and Miyazaki.

References

Allendorf, F.W. and S.R. Phelps. 1980. Loss of genetic variation in a hatchery stock of cutthroat trout. *Trans. Am. Fish. Soc.* 109:537-543.

Araga, C. 1984. "The Fishes of the Japanese Archipelago," Tokai University Press, pp. 38.

Fujio, Y. 1984. "Study of the genetic charactristics of fish and shellfish in isozymic analysis". Nosuisho Tokubetu Shiken Houkokusho, pp. 1-65 (in Japanese).

Fujio, Y. and M. Nakajima. 1989. Genetic monitoring of released population in black rockfish *(Sebastes schlegeli). Tohoku J. Agr. Res.* 40:19-64.

Honjo, T. 1976. Experiments on amphidromonus migration of "Amago trout". *The Freshwater Fishes* 2:27-35, (in Japanese).

Kato, F. 1973. Ecological study on the sea-run form of *Oncorhynchus rhodurus*, found in Ise Bay, Japan. *Japan. J. Ichthyol.* 20:225-234, (in Japanese).

Kato, F. 1973. On the destribution of a sea-run form of the salmonid fish, *Oncorhynchus rhodurus*, found in southwestern Japan. *Japan. J. Ichthyol.* 21:191-197, (in Japanese).

Kubo, T. 1979. "Suginoko oyobi sakuramasu-rui no rikufu ni tuite". *The Freshwater Fishes* 3:1()0-107, (in Japanese).

Nakajima, M., A. Kita, and Y. Fujio. 1986. Genetic features of natural and cultured populations in masu salmon (*Oncorhynchus masou*). *Tohoku J. Agr. Res.* 37:31-42.

Nakajima M., and Y. Fujio. 1988. Genetic defferentiation in cultured populations of rainbow trout *(Salmo gairdneri). Tohoku J. Agr. Res.* 38:35-48.

Nakamura, M., and Y. Kawai. 1966. "Ohatagawa jyouryu no tansuigyo suginoko ni tuite". Shigen kagaku kenkyujyo shuho 46-47:10:3-107 (in Japanese).

Nei, M. 1972. Genetic distance between populations. *Am. Natur.* 106:283-295.

Nei, M. 1975. "Molecular population genetics and evolution," North-Holland, Amsterdam, pp.288.

Okazaki, T. 1984. Genetic variation and population structure in masu salmon, *Oncorhynchus masou* of Japan. *Bull. Japan. Soc. Sci. Fish.* 52:1365-1376.

Taniguchi, N., K. Sumantadinata, and S. Iyama. 1983. Genetic change in the first and second generations of hatchery stock of black seabream. *Aquaculture* 35:309-320.

Yoshiyasu, K. 1973a. Starch-gel electropholesis of hemoglobins of freshwater salmonid fish in southwestern Japan-II. *Bull. Japan. Soc. Sci.* Fish. 39: 97-114.

Yoshiyasu, K. 1973b. Starch-gel electropholesis of hemogrobins of freshwater salmonid fishes in northeast Japan. *Bull. Japan. Soc. Sci. Fish.* 39:449-459.

Data Base for Trout Brood Stocks

HAROLD L. KINCAID

1. Introduction

Trout culture in the United States began in the 1870s with the development of rainbow trout hatcheries. Little is known of the sources or performance characteristics of the original brood stocks. The McCloud River strain of rainbow trout was widely distributed and was probably ancestral to most current brood stocks. From 1870 to 1970, new brood stocks were established for management agencies in various ways: interhatchery transfers, introductions from natural populations, crosses between brood stocks, and selection programs that enhanced specific traits. Most brood stocks used in fishery management today have been developed within the last 25 years.

Fisheries personnel have long sought to manage by matching specific strains to the production objectives of individual fisheries, but until recently, this goal was little more than a dream. In the 1970s a number of workers began work to characterize trout strains based on a variety of performance traits (Kinunen and Moring, 1978; Gall and Cross, 1978; Busack and Gall, 1980; Ihssen et al., 1981; Kincaid, 1981). Most studies were designed to characterize two to four strains for hatchery growth traits (Kincaid, 1981) or performance trials in specific fishery situations (Haskell, 1952; Ihssen and Tait, 1974; Plosila, 1977; Keller, 1979; Brauhn and Kincaid, 1982; Dwyer and Piper, 1984; Krueger et al., 1988). Smith et al. (1987) compared carcass composition of six rainbow trout strains reared concurrently. While this approach identified superior strains among those tested in a specific situation, the absence of common control lines among the trials prevents extrapolation of the information to other situations (Kincaid, 1979). The result was that data from different studies could not be pooled systematically to allow comparisons among tested strains and situations. In addition, standard protocols were not used nor were the same traits measured in each study. For managers to identify superior strains that may be stocked into a specific management situation, a standard set of information must be measured for each strain, and all strains must be evaluated in the same environment. During the past decade several federal and state agencies have published lists of the strains and brood stocks used in their management programs (Crawford, 1979; Kincaid, 1981; Claggett and Dehring, 1984; Kincaid and Berry, 1986; Kincaid and Stanley, 1989).

Regardless of the fishery program planned, information on the basic performance characteristics (i.e., growth rate, disease resistance, stress tolerance) of potential strains can

National Fishery Research and Development Laboratory, U.S. Fish and Wildlife Service, Wellsboro, Pennsylvania, U.S.A.

Genetic Conservation of Salmonid Fishes, Edited by J.G. Cloud and G.H. Thorgaard, Plenum Press, New York, 1993

help to identify specific strains that will best meet management requirements. The initial effort to develop a trout strain registry began in 1980 when state and federal fishery resource agencies throughout the United States were surveyed to identify brood stocks used in management programs. Information was gathered covering brood stock management (e.g., population size, breeding methods, and management applications), hatchery performance (growth, stress tolerance, and disease resistance) and field performance of each brood stock used in fishery management programs. A second survey in 1983-84 assembled information on changing patterns of brood stocks managed and additional field performance information (Kincaid and Berry, 1986). Results of these surveys (Tables 1, 2, and 3) are the foundation for the permanent trout strain data base being developed.

The trout strain data base will meet several objectives of fisheries management. Specific objectives are:

1. To identify trout strains that are available to fishery managers,
2. To provide standardized information on each trout strain that can be used by managers as a basis for management decisions,
3. To establish a centralized data base for trout strain information,
4. To provide an identification system for trout strains and brood stocks, and
5. To develop a distribution system for transferring trout strain information to user groups.

2. Methods

The Trout Brood Stock Data Base and Registry includes information on the six species: rainbow trout (*Oncorhynchus mykiss*), brook trout (*Salvelinus fontinalis*), brown trout (*Salmo trutta*), cutthroat trout (*Salmo clarki*), lake trout (*Salvelinus namaycush*), and Atlantic salmon (*Salmo salar*). Development of the data base was divided into five steps:

STEP 1. A trout strain advisory committee was established to include representation of federal, state, and commercial aquaculturalists interested in development and maintenance of trout strains (brood stocks). Prominent individuals were chosen to serve on this advisory group from the U.S. Fish and Wildlife Service, State Fish and Game Agencies, and commercial aquaculture. The advisory committee assisted with all aspects of the trout strain data base and registry effort. Committee meetings were held annually to review progress and develop strategies for each step of the work. The advisory committee was responsible to:

a. Develop guidelines on specific traits included in the data base and registry,
b. Develop brood stock surveys for distribution to federal, state and commercial brood stock managers,
c. Served as liaison to brood stock managers in the federal, state, and commercial sectors, and
d. Served as a review board to edit publications developed from the trout strain data base.

STEP 2. An extensive review of literature on fish strain studies was undertaken. Searches of published literature and federal aid reports were conducted. Unpublished reports from state agencies were also reviewed to identify studies that compared fish strain performance in field situations.

STEP 3. A national survey of federal, state, and commercial organizations was carried out targeting the six species. The questionnaire requested a standard set of data on each brood stock. Types of information requested were: brood

Table 1. Change in Number of Trout Brood Stocks Managed by Federal and State Agencies in 1980 and 1983 Surveys.

Species		1980	1983	Change number	% change
Rainbow	State	83	125	+42	+51
	Federal	44	18	-26	-59
	Total	127	143	+16	+13
Brown	State	40	36	- 4	-10
	Federal	8	6	- 2	-25
	Total	48	42	- 6	-13
Brook	State	35	39	+ 4	+11
	Federal	4	4	0	0
	Total	39	43	+ 4	+10
Cutthroat	State	35	34	- 1	- 3
	Federal	6	5	- 1	-17
	Total	41	39	- 2	- 5
Lake	State	8	19	+11	+138
	Federal	6	7	+ 1	+17
	Total	14	26	+12	+86

Table 2. Number of Wild and Domestic Brood Stocks Used by Fisheries Management Agencies by Species in 1983.

Species	Brood stocks			% Wild brood stocks
	Wild	Domestic	Total	
Rainbow	47	96	143	33
Brown	5	37	42	12
Brook	5	38	43	12
Cutthroat	18	21	39	46
Lake	8	18	26	31
Total	83	210	293	28

Table 3. Mean Values for Four Important Hatchery Production Characteristics of Brood Stock of Five Trout Species in 1983 (Sample Sizes in Parentheses).

Species	Characteristic			
	Food conversion[a/]	Eye %	Fry survival %	Growth[b/] (g)
Rainbow	1.5(40)	79.0(43)	52.0(42)	79.7(37)
Brown	1.5(13)	83.0(17)	75.0(17)	47.8(17)
Brook	1.3(15)	82.0(22)	73.0(22)	69.9(19)
Cutthroat	1.7(18)	79.0(26)	78.0(21)	37.8(16)
Lake	1.9(7)	72.0(7)	81.0(8)	25.21(9)

[a/] Food conversion expressed in units of feed per unit of fish weight gain.
[b/] Attained weight at one year expressed in grams.

stock history, hatchery performance traits, field performance traits, handling characteristics, etc. Each organization was also asked to identify individuals to serve as contacts for future updates of brood stock information as new data becomes available.

STEP 4. The Trout Strain Data Base and transfer information was constructed from returned survey forms (Step 3) and information extracted from published and unpublished literature (Step 2). Follow-up contacts with brood stock managers were used to clarify and expand the survey information. Data for each brood stock was compiled and summarized separately. Individual summary sheets were then prepared for each brood stock and strain.

STEP 5. A trout strain registry was prepared for distribution to user groups. Components of the registry include:

a. Review of fish strain literature,
b. Description of data collection and analysis procedures,
c. Description of the registry and guidance for its use,
d. Tables for comparison of hatchery performance traits among strains of each species,
e. Tables for comparison of field performance traits among strains of each species, and
f. Summary data sheets for individual strains and brood stocks.

3. Management Applications

Fisheries managers today are asked to manage a wide diversity of fisheries (natural, restoration, enhancement, and put and take) in streams, lakes and reservoirs. The past practice of planting fish from a single source into every situation has proven to be inefficient. In fisheries that have naturally reproducing populations, planting fish of the same species from a different gene pool can be destructive. If the natural population hybridizes with fish planted from another source, genetic introgression can occur resulting in deterioration and possibly the ultimate loss of the adapted natural population. Cataloging available information on performance characteristics for each strain will allow managers to identify those strains that are adapted to their particular fishery and management situation. Managers can use information on fish performance, health status, cost and availability to choose the specific strain that will be planted in each fishery. Commercial aquaculturists can choose the strains that more closely match the markets for their products whether the fish are destined for stocking programs by management agencies or private pond owners, or for consumption by restaurants and food retailers. Standardized information on trout strains that are available to the aquaculture industry will greatly increase the managers ability to match strain performance characteristics with the fishery being managed. The trout strain data base being developed offers the opportunity to achieve these objectives.

References

Brauhn, J. L. and H.L. Kincaid. 1982. Survival, growth, and catchability of rainbow trout of four strains. *North American Journal of Fisheries Management* 2:1-10.

Busack, C. A. and G. Gall. 1980. Ancestry of artificially propagated California rainbow trout strains. *California Fish and Game Journal* 66:17-24.

Claggett, L. E. and T.T. Dehring. 1984. *Wisconsin salmonid strain catalogue*, Administrative Report Bureau of Fisheries Management, Wisconsin Department of Natural Resources, Madison, Wisconsin.

Crawford, B. A. 1979. *The origin and history of the trout broodstocks of the Washington Department of Game*, Washington State Game Department, Fish Resources.

Dwyer, W. and R. Piper. 1984. Three-year hatchery and field evaluation of four strains of rainbow trout. *North American Journal of Fisheries Management* 4:216-221.

Gall, G. and S. Cross. 1978. A genetic analysis of the performance of three rainbow trout broodstocks. *Aquaculture* 15:113-127.

Haskell, D. C. 1952. Comparison of the growth of lake trout fingerlings from eggs taken in Seneca, Saranac and Raquette Lakes. *Progressive Fish-Culturist* 14:15-18.

Ihssen, P. and J. Tait. 1974. Genetic differences in retention of swim bladder gas between two populations of lake trout. *Journal of the Fisheries Research Board of Canada* 31:1351-1354.

Ihssen, P., H. Booke, J. Casselman, J. McGlade, N. Payne, and F. Utter. 1981. Stock identification: materials and methods. *Canadian Journal of Fisheries and Aquatic Sciences* 38:1838-1855.

Keller, W. T. 1979. *Management of wild and hybrid brook trout in New York lakes, ponds, and coastal streams.* New York Department of Environmental Conservation, Division of Fish and Wildlife, Albany, New York FWPI48(11/79).

Kincaid, H. L. 1979. Development of standard reference lines of rainbow trout. *Transactions of the American Fisheries Society* 108:457-461.

Kincaid, H. L. 1981. *Trout strain registry.* FWS/NFC-L/81-1, National Fisheries Center - Leetown, U.S. Fish and Wildlife Service, Kearneysville, WV.

Kincaid, H. L. and C.R. Berry. 1986. Trout broodstocks used in management of National Fisheries, in: "Fish Culture in Fisheries Management," R. H. Stroud, ed., pp. 211-222, American Fisheries Society, Bethesda, Maryland.

Kincaid, H. L. and J.G. Stanley. (eds.) 1989. *Atlantic salmon brood stock management and breeding handbook.* U.S. Fish and Wildlife Service, Biological Report 89(12).

Kinunen, W. and J. Moring. 1978. Origin and use of Oregon rainbow trout broodstocks. *Progressive Fish-Culturist* 40:87-89.

Krueger, C. C., J.E. Marsden, H.L. Kincaid, and B. May. 1988. Genetic differentiation among lake trout strains stocked in Lake Ontario. *Transactions of the American Fisheries Society* 118:317-330.

Plosila, D. 1977. Relationship of strain and size at stocking to survival of lake trout in Adirondack Lakes. *New York Fish and Game* 24:1-24.

Smith, R. R., H.L. Kincaid, J.M. Regenstein, and G.L.Rumsey. 1987. Growth, carcass composition, and taste of rainbow trout of different strains fed diets containing primarily plant or animal protein. *Aquaculture* 70:309-321.

The Use of Supplementation to Aid in Natural Stock Restoration

MICHAEL L. CUENCO, THOMAS W.H. BACKMAN, and PHILLIP R. MUNDY

1. Introduction

Supplementation is one of the strategies that may be used for restoring natural production of anadromous salmonid populations in the Columbia River Basin. Depending on the particular circumstances, supplementation may be used by itself or in conjunction with other management strategies for restoring natural production such as habitat restoration and maintenance, improvement of tributary and mainstem river passage survival, improvement of estuarine and ocean survival, and harvest management by escapement objectives to allow the population to optimally seed available habitat. Above mainstem dams, all measures may need to be employed simultaneously to achieve success.

1.1. Ecological Complexity and Degradation

Efforts to restore naturally reproducing salmonid fish populations in the Columbia River Basin must begin with an understanding of the diverse and complex biology and life history of these populations. Not only are many species and races involved, but these fishes use diverse freshwater, estuarine, and marine habitats during the different stages of their life cycles (Davidson and Hutchinson, 1938; Northcote, 1969; Ricker, 1972; Howell et al., 1985).

Superimposed upon this natural complexity is man's intervention in the form of timber harvest and removal of riparian vegetation (Chamberlain, 1982), forest roads (Yee and Roelofs, 1980), water transportation of logs (Sedell and Duval, 1985), agriculture and irrigation withdrawals (NPPC, 1986), livestock grazing on riparian areas (Platts, 1981), mining (Martin and Platts, 1981), urban development (NPPC, 1986), fishing (NPPC, 1986), and hydroelectric development (Raymond, 1979; NPPC, 1986). Numerous studies have been conducted to quantify the detrimental effects of these activities on anadromous populations and their habitat.

1.2. Systems Approach to Restoration

Because the Columbia River Basin is a complex system with many interacting components, an improvement in one component will not necessarily result in improvement of the system as a whole. Thus, the function of any component and its manipulation can only be

Columbia River Inter-Tribal Fish Commission, 729 N.E. Oregon, Suite 200, Portland, Oregon 97232, USA

Genetic Conservation of Salmonid Fishes, Edited by J.G. Cloud and G.H. Thorgaard, Plenum Press, New York, 1993

properly assessed in relation to the system of which it is a part. In the Columbia Basin, this approach has been called "gravel-to-gravel" management.

A first step in restoration of natural fish populations is an assessment of population "health," the stock's biological characteristics, the difference between the quality and quantity of present and former habitat, analysis of the factors limiting abundance, and modes of interaction of these factors (RASP, 1992). Understanding the physical and biological requirements for each life history stage of all stocks of concern, as well as the ways in which they use the habitat, is key. Although there will be common factors affecting many fish populations, detailed restoration plans should be based on a case-by-case analysis.

Supplementation efforts described in this paper are based on restoring anadromous salmonid populations to their historical localities and levels of production. Natural fish stock rehabilitation activities facilitated by the use of the hatchery system are commonly known in the Columbia Basin as supplementation, although the specific practices envisioned have varied among the proponents. A holistic rehabilitation plan, given the constraints imposed upon the Columbia River Basin fish production system by human activities, requires the effective use of supplementation in conjunction with improved habitat, water quality and flow, and fishery management. Wherever possible, the plan envisions actions to make it possible for anadromous salmonid stocks to be returned to their ancestral habitats through a variety of actions, including supplementation.

2. Definition of Supplementation

For the purpose of this document, *supplementation is defined as the stocking of fish into the natural habitat to increase the abundance of naturally reproducing fish populations*. Maintaining the long-term genetic fitness of the target population, while keeping the ecological and genetic impacts on nontarget populations within acceptable limits, is inherent in this working definition. This definition is consistent with efforts by other groups, such as the Regional Assessment of Supplementation Programs, to define elements of supplementation.

Supplementation includes activities where fish are stocked into barren habitats (currently unoccupied by the species). This activity is commonly referred to in the literature as transplantation or introduction (Withler, 1982; Fedorenko and Shepherd, 1986). In the Columbia Basin, supplementation activities will, in most cases, involve stocking fish into habitats that contain depressed, but existing natural fish populations.

Although artificial propagation has a central role in most supplementation activities, the definition of supplementation used here does not preclude the use of fish that have not been reared in a hatchery or other man-made propagation facility. The choice of a wild or natural stock for direct transfer should follow the genetically and ecologically sound guidelines presented as "similarity criteria" in Kapuscinski et al. (1992). However, unlike many "traditional" hatchery programs, the objective of supplementation is to increase the abundance of a naturally reproducing fish population and therefore, is oriented toward maintaining the natural biological characteristics of the population and reliance on the rearing capabilities of the natural habitat. In contrast, many traditional hatchery programs were designed to augment harvest by the development of hatchery fish populations that rely entirely on artificial spawning and rearing in the hatchery. These hatchery fish populations were not intended to contribute, and most likely have not contributed, to the abundance of naturally spawning fish populations. Typically, the juvenile fish are released into streams adjacent to the hatchery and the returning adults are guided back to the hatchery through the use of barriers and fishways.

The increase in the abundance of a naturally reproducing fish population may be self-sustaining after an initial but finite period of supplementation or it may be sustained through continual assistance from supplementation depending on the specific spawner-recruit relationship under given environmental conditions and management objectives. Supplementation measures will not obviate the need to concurrently pursue other necessary actions such as improvement of mainstem passage, habitat protection, and harvest management to rebuild stocks.

3. Uses of Supplementation

Supplementation is considered a tool for rebuilding natural fish populations, not a panacea. It can be used to assist in rebuilding natural stocks, to replace extirpated stocks, or to introduce and establish a stock in a barren habitat (Withler, 1982; Fedorenko and Shepherd, 1986)

3.1. Seeding Barren Habitat

For barren habitats that historically produced salmon or steelhead but are currently unoccupied, it is necessary to stock fish into the habitat to re-establish a desired fish population. The fish should be stocked at densities that do not exceed the carrying capacity of the habitat for the limiting life history stage of the fish being stocked. The carrying capacity of a damaged habitat will need to be re-evaluated as rehabilitation is undertaken. In evaluating carrying capacity, the potential for interspecific interactions and risks to non-candidate species must be addressed. In currently unoccupied habitat, it is essential to choose a fish population that has adaptive traits that are as similar as possible to those of the extirpated population. It is also desirable to match their genetic lineages if such information is available.

3.2. Provide Survival Advantage for Depressed Stocks

For "sparsely populated" habitats where there is an existing damaged salmonid population, the objective is to boost the population density above a certain minimum viable population size as quickly as possible (Thomas, 1990). For example, the minimum viable spawning escapement size for each stock may be calculated from the minimum effective breeding number by a transfer function, whose elements include the amount of spawning and rearing habitat available and the average total mortality. The concept is to employ a supplementation program to a level that minimizes risk of extirpation.

The primary role of supplementation in this case is to increase the survival rate of the population during its early life history (egg through smolt) relative to its survival rate under natural conditions. It is anticipated that this effort will result in increased adult returns to seed sparsely populated habitats.

For depressed stocks, the question of how many and what proportion of the natural stock to intercept for hatchery broodstock must address the need to maintain an effective breeding number in the natural and hatchery broodstocks. In practice such questions can only be resolved by carefully evaluating the impact of programs that initially take moderate fractions of the depressed population for broodstock.

3.3. Speed Rebuilding to Carrying Capacity

For a rebuilding, lightly damaged stock in healthy habitats, a potential but unresolved role of supplementation is to increase the rate at which the population rebuilds. Supplementation may be unnecessary in the long term if other factors limiting populations in the basin are corrected.

4. When to Use Supplementation

4.1. Life Cycle Analysis of Limiting Factors

A sound analysis of the population status (such as depressed or healthy), population trend (such as decreasing, stable or increasing), and the factors limiting population abundance are necessary to address the policy question of whether it is appropriate to use supplementation as a tool in increasing natural production. When a stock is considered to be at some high level of risk (nonviable status or declining numbers) and a policy decision is reached on the need to reverse the slide toward extirpation and implement restorative measures, and the physical and biological constraints on the natural stock (such as habitat conditions, passage and water quality) make its restoration feasible, supplementation should be considered as a chief form of biological support for the natural population.

Other support must aim at reducing or eliminating the original causes of decline. All available conservation actions such as reducing passage mortalities, reducing harvest mortalities, and rehabilitating spawning and rearing habitat need to be identified and prioritized to be used singly or in various combinations in concert with a supplementation program. Supplementation measures do not obviate the need to correct other factors limiting stock productivity.

4.2. Prerequisites for Supplementation

For supplementation to be part of the solution to increase the abundance of a natural fish population, the following prerequisites must also be met. In reading the text that follows, care must be taken in interpreting terms such as "carrying capacity" and "maximum escapement level." These terms are defined in the context of an undamaged environment in which the various species of anadromous salmonids complete their life cycles.

4.3. Decisions Regarding the Use of Supplementation

The potential need for and efficacy of use of supplementation depend both upon stock status and management objectives for a particular stock as well as potential impacts on other stocks and species. The decision to initiate a new supplementation program or to modify an ongoing program must be determined on a case-by-case basis. The following are eight criteria that the fishery agencies and tribes consider in determining whether to initiate or revise a supplementation program:

I. Extirpated Stock: Average spawning escapement is effectively zero. When data permit, the average spawning escapement is to be based on a period of years equal to three times the age class that represents the largest proportion of the run.
   A. Stock in the most effective manner with the most similar available genetic, phenotypic and ecotypic stock.
   B. Emphasize the use of returning adults for broodstock while allowing for escapement to the original spawing grounds.
   C. Cease stocking when average return of spawners exceeds the lesser of the minimum viable population size, MVP, (see Thomas, 1990) or 85% of carrying capacity for three generations. If successful after one or two generations, stocking can be reduced.

II. Damaged Stock:
   A. Badly damaged; average spawning escapements fall far below MVP and are be-

low the number of adults needed to produce 50% of the carrying capacity of the freshwater environment for the limiting life history stage.

1. Decreasing; the trend in average spawning escapements is declining or indeterminate for three life cycles.
    a. Take as few spawners as necessary to cross with the most similar available genetic, phenotypic and ecotypic stock (Kapuscinski et al., 1992, for some guidelines).
    b. Preserve unique characteristics of damaged stock; the ratio of donor to damaged stock should be approximately 1:1 in parental generations, leaving at least 50% F1 to spawn naturally, if survivals permit. Use the other returnees for broodstock in a 1:1 ratio with the damaged stock if survivals permit it to continue supplementation. The specific breeding protocol must be worked out with the advice of a professional geneticist on a case-by-case basis. Return (supplement) all production to the habitat from which the damaged stock was taken at an appropriate life history stage.
2. Stable or increasing; the trend in average spawning escapements is stable or increasing for three life cycles.
    a. Various proportions of native stock up to 50% may be taken for broodstock in the breeding program designed with the advice of a professional geneticist. All progeny will be returned to the stream from which broodstock was taken.
    b. As supplementation continues, use varying proportions of hatchery-reared and natural returns for broodstock for both artificial rearing and natural spawning in a professionally designed breeding program.

B. Moderately damaged; average spawning escapements fall between MVP and the number of adults needed to produce 50% of the carrying capacity of the freshwater environment for the limiting life history stage.
1. Decreasing; the trend in average spawning escapements is declining or indeterminate for three life cycles.
    a. As a first approximation, annually take no more than 25% of natural spawners over a period of two life cycles to produce offspring that are reared in isolation and returned to the spawning habitat of their parents.
    b. After evaluating one life cycle of returns, if the average rate of returns of artificially reared fish is equal to or better than returns from natural spawners, increase the percentage of natural spawners taken to no more than 50%, taking no artificially reared fish for broodstock. If the rate of returns is not better than returns from natural spawners continue at the 25% level.
2. Stable or increasing; the trend in average spawning escapements is stable or increasing for three life cycles.
    a. Supplementation is a lower priority than cases indicated above.
    b. Monitor survival and age-sex structured escapements and if population begins decreasing over one life cycle proceed as in II.B.1.

C. Lightly Damaged; average spawning escapement levels fall between the number of adults needed to produce 50% and 85% of the carrying capacity of the freshwater environment for the limiting freshwater life history stage.
1. Decreasing; the trend in average spawning escapements is declining or indeterminate for three life cycles.

a. Monitor survival and age-sex structured escapements while determining the trend in escapements, and/or the proximate cause of decline.

b. If stock is actually decreasing after one life cycle;

i. other remedies being available, apply them until the escapements increase for one life cycle.

ii. if the escapements continue to decline after two life cycles, if the average population levels fall below 50% of the carrying capacity, or if other remedies are not available, then evaluate the need for supplementation and make a decision whether to proceed as in II.B.

2. Stable or increasing; the trend in average spawning escapements is stable or increasing for three life cycles.

No supplementation action is necessary, but careful monitoring of escapements and survivals is essential.

III. Undamaged Stock: No supplementation is necessary to achieve basic conservation purposes when average escapement deficits are within O% to 15% of the maximum in undamaged habitat. Such population levels still require monitoring to evaluate survival and to obtain age and sex structured escapements.

4.2.1. Sufficient Natural Habitat Exists

The present or rehabilitated habitat should be judged capable of supporting a viable, self-perpetuating population in the face of natural stochastic events (such as floods, droughts, earthquakes) demographic factors (ability to find a mate, sex ratio, age structure) and genetic considerations (such as maintenance of an effective population size to prevent serious loss of genetic variation). This is important to ensure that the carrying capacity of the habitat does not become a limiting factor in the population's rebound. The actual numbers of fish that constitute a) minimum viable population size, b) proportions withdrawn for broodstock, and c) carrying capacity of the environment would have to be determincd on a case by case basis.

Salmon need different types of habitat during various stages of their life cycle (Reiser and Bjornn, 1979). Sufficient pre-spawning habitats (deep, cool, calm pools) should be available for spring chinook, summer chinook, and summer steelhead adults that have to hold several months before proceeding to their spawning habitat. Adequate spawning habitat should be presently or potentially available. Anadromous salmonid habitat requirements for spawning include sufficient gravel of the right size, suitable water temperatures, flow conditions, and water quality.

Adequate rearing habitat for feeding and growth should be available to support the juveniles produced during the season (spring, summer, fall) that they occupy the habitat. For juveniles that have to overwinter in fresh water, adequate habitat for this purpose must be present. In streams having high juvenile production followed by a large downstream displacement in fall, the lack of sufficient overwinter rearing habitat in downstream areas would negate production increases in headwaters. Thus, an important consideration is a periodic evaluation of habitat suitability and sufficiency on a seasonal and basinwide scale.

4.2.2. A Suitable Stock Is Available for Supplementation

The biological requirements of the stock to be used for supplementation must be carefully matched to the proposed habitat such that survival, growth and reproduction are successful. Thus, it is important to obtain knowledge of the biology, life histories, habitat use, genetic lineages and genetic diversity of candidate stocks insofar as this knowledge is obtainable. Supplementation can involve stocking of fish that descended from the natural population being supplemented or can involve the use of fish that have varying degrees of genetic distance from the supplemented population. In many cases, action on the basis of the best available information will be necessary. To maximize the chances for success, the indigenous stock should serve as the broodstock in its own enhancement program. If this is not possible, a stock with the greatest likelihood of being closely related to the potential recipient stock, or fish from environments that closely resemble the proposed recipient site, should be evaluated for their potential effectiveness as alternative broodstocks.

Among important biological traits, spawning, incubation and emergence times (temperature-dependent) must be synchronized with favorable environmental conditions such as suitable flows, temperatures, and food availability. To ensure migration success, imprinting of juveniles, migration to the ocean, and subsequent homing as adults to their natal stream must occur (Hasler and Wisby, 1951; Jensen and Duncan, 1971; Madison et al., 1973; Scholz et al., 1975; Cooper et al., 1976; Cooper and Scholz, 1976; Nordeng, 1977; Cooper and Hirsch, 1982; Hasler and Scholz, 1983; Brannon et al., 1984; Slatick et al., 1988). Because adults do not feed during their reproductive migration, they must possess sufficient energy reserves and physical stamina to travel back to their natal stream. Thus, fish from a lower Columbia Basin watershed may successfully imprint and attempt to home to an upper basin watershed, but the migrant may not have the physical stamina and energy to complete its journey.

4.2.3. Appropriate Technology

For supplementation to be successful, the artificial propagation technology must be adequate to rear and release fish that are biologically, genetically and ecologically suited to their receiving environments. The technology must provide a survival advantage sufficient to bring the spawner-recruit relation of the combined natural and hatchery-reared segments of the fish population above the replacement level. A life cycle analysis of the components of fecundity and mortality of the fish population both with and without supplementation should be conducted to compare the results under given environmental and management conditions.

5. Approach to a Supplementation Program

5.1. Phased Implementation and Adaptive Management

Although the fishery agencies and Indian tribes of the Columbia River Basin consider supplementation to have potential as a tool for increasing natural fish production, there is not yet a detailed understanding of which techniques work best under which circumstances. These uncertainties will necessitate undertaking a program of phased, experimentally designed supplementation studies as part of ongoing implementation of the supplementation program. Supplementation should proceed cautiously so that productivity of supplemented stocks can be tested. Past achievements have left concerns about meeting productivity. Procedures and techniques identified in this chapter are intended to improve supplementation practices. These procedures apply the concept of adaptive management, which relies heavily upon monitoring and evaluation.

Representative pilot sites will be identified for initial supplementation by the process described above. As knowledge and confidence are gained and natural production is increased, supplementation technology will be improved and more sites will be phased in for supplementation.

Within a given project site, the level of supplementation will also be phased in, commensurate with the number of spawners available for broodstock. For example, for a site with an estimated capacity of 200,000 juveniles, the level of effort could be increased in quarters: Phase 1, 25% of target capacity; Phase 2, 50%; Phase 3, 75%; Phase 4, 100%. As experience is gained, succeeding phases will be adjusted or the entire project aborted as warranted.

5.2. Monitoring and Evaluation

Because supplementation technology is nascent and uncertainties exist about its effectiveness and safety, it is important to incorporate a monitoring and evaluation program to assess performance of each supplementation project. Knowledge and experience gained should be incorporated into the design and operation of future supplementation projects. Kapuscinski and Lannan (1986) describe a conceptual phenotypic model to ensure the long-term reproductive fitness of stocks. Essential elements of monitoring include escapement data by sex and age, estimates of survival at each life stage, and the ability to distinguish supplemented from natural spawners visually, or by some other rapid method that does not harm the animal.

Two levels of monitoring and evaluation are envisioned. The first level is to determine the degree of success of the supplementation project. The second level is to try to answer why the project was successful or not successful and provide ways to adjust the program and to apply the results to guide other proposed supplementation projects. The procedure for monitoring and evaluation should include the following elements.

1. Clearly define supplementation objectives.
2. Identify performance measure(s) for each supplementation objective.
3. Develop experimental and sampling design.
4. Collect and analyze data.
5. Interpret results.
6. Adjust or correct the parts of the supplementation plan that are ineffective or inefficient in meeting the objective(s). Alternatively, if objective(s) are unclear, too general, conflicting or too ambitious, modify them so that the existing plan can achieve them.
7. Review the adequacy of the monitoring and evaluation plan and modify accordingly.

6. Supplementation Technology

Previous sections presented a working definition of supplementation, discussed its potential role in a comprehensive effort to address the physical and biological constraints on natural fish production, and provided general guidelines of when supplementation is appropriate. The following sections discuss some key considerations and general guidelines to provide a logical starting point in crafting the specifics of how to conduct the supplementation program. Each of the following considerations needs to be evaluated and tailored to the specific supplementation program.

6.1. Level of Technology

Since supplementation in the Columbia River Basin will most likely involve some type of artificial propagation, one of the considerations is to determine an appropriate level of technology to a specific situation. Artificial propagation encompasses a wide range of technology, from small-scale facilities (such as streamside incubators) located at tributaries to large-scale, centralized hatchery facilities. Thus, it may be useful to consider the availability, ecological and economic advantages and disadvantages of small-scale facilities and large-scale facilities.

6.1.1. Large-Scale Facility

A total of 85 hatcheries and satellite facilities in the Columbia River Basin rear an average of 7.7 million pounds of anadromous fish per year (Delarm and Smith, 1990). One approach is to consider whether some of these existing hatchery facilities, with some modifications, would be appropriate to use in a supplementation program since these facilities are already in place and operational.

In general, some key changes would have to be made for some of these facilities to be used in supplementation. It is envisioned that incubation and rearing will continue to be done in the central hatchery. However, if two or more stocks will be reared in the hatchery, provisions must be made to keep them separate. Moreover, to ensure genetic compatibility, hatchery broodstock should be collected from the natural population targeted for supplementation if possible. Thus, provisions must be made to collect adults at or near their home stream. Also, juvenile acclimation facilities should be considered for the purpose of allowing the fish to imprint and adjust gradually to the natural environment before their eventual release into their home stream. Care would be taken that the selection of rearing water not interfere with the ability of stocked fish to return home to the point where they were released.

An example that applies this approach is the East Bank central hatchery facility located near Rocky Reach Dam in the Columbia River (Rock Island Project Settlement Agreement). The central hatchery was designed with incubation and juvenile rearing facilities, but no facilities for adult collection and release of juvenile fish. Instead, satellite facilities located near the streams targeted for supplementation were designed for broodstock collection and juvenile acclimation and release.

The potential advantages of using the large-scale approach must be weighed against some potential disadvantages. Large-scale hatcheries literally put all one's eggs in one basket with all that is implied about risk taking (Larkin, 1981). Should there be a failure, human error or accident in the hatchery, large numbers of fish may be lost. The fish will not be reared in the same water into which they will be released except for the final period before release when acclimation ponds are used. Thus, imprinting may not be as complete and unequivocal. This approach may also entail more handling and transporting of adult fish from the supplemented streams to the hatchery and of juvenile fish from the hatchery to the supplemented streams. Thus, there is greater potential for stress, health impairment, fish mortality, and straying.

6.1.2. Small-Scale Facility

Another approach envisions the use of small-scale facilities that are located alongside the streams targeted for supplementation. The size of these facilities are relatively small (600 to 10,000 pounds of juveniles) and would depend on the capacity of the streams targeted for supplementation. Adult collection and juvenile release facilities would be located on site, thus eliminating or greatly reducing fish handling and transportation which may lead to stress, impaired health and mortality. This approach would include incubation and rearing facilities located on site and would use the stream water for its water supply. Because the fish would be reared using the same water where they would be released throughout their residence in the hatchery, acclimation facilities would not be necessary and imprinting of juveniles should be more complete and unequivocal. This, in turn, should improve adult homing back to the stream and reduce straying.

The use of small-scale facilities allows for considerable flexibility in managing many smaller units, so that when deemed appropriate, a project can be abandoned with limited potential ecological damage and loss in investment. Smaller releases of juveniles commensurate with the capacity of the stream should reduce potential effects from "ecological swamping." This approach is readily adaptable to individual drainages, enabling the conservation of gene pools. Because the fish would be reared in artificial conditions more similar to their natural environment, domestication selection should be reduced.

The disadvantages with the small-scale hatchery approach include potentially greater cost in constructing and operating many small facilities located in the tributary streams. Logistics for many scattered facilities may also prove difficult. Moreover, some of these potential sites may not be readily accessible (no roads).

Most of the sites in the Columbia River Basin containing large quantities of water required by large hatcheries (100,000 pounds of fish or more capacity) have already been developed (Senn et al., 1984). However, there are many potential sites still available for developing small-scale hatcheries to produce smaller quantities of fish.

An example of this approach is the streamside chinook salmon spawning and rearing facility of the U.S. Forest Service at Horse Linto Creek, a tributary to the Trinity River in northern California. The facility consists of an adult migrant trap, two hatch boxes, a filter system, two fiberglass raceways, and an earthen rearing pond. The adult migrant trap and the juvenile release facility are located on site and adjacent to the stream. This arrangement minimizes fish handling and transportation. Juvenile releases from the facility started with 5,000 fish in 1984 and have increased to 57,000 in 1989. Before the restoration project began, less than 10 spawning pairs were counted (1979-1981) in a 2.5 mile index. By 1988 and 1989, the number of spawning pairs counted had increased to 50 and 55, respectively. Forest Service biologists are confident that the project can rebuild the chinook population to the estimated stream capacity.

Another example is a Swedish program to preserve native runs of Atlantic salmon in tributaries to the Baltic Sea after they were blocked by dams (Behnke, 1986). Instead of constructing a few large centralized hatcheries, Sweden opted for constructing a smaller hatchery in each major river (17 in all). This approach was chosen to propagate the native runs of each river and to preserve the original genetic diversity. The smolt-to-adult survival rates have typically ranged from 10% to 20%.

6.2. Spawning Protocol

The goal is to conserve genetic resources and maintain the ability of the stock to survive and reproduce in the natural environment. Relevant broodstock management principles and spawning guidelines (Hershberger and Iwamoto, 1981; Kreuger et al., 1981; Kincaid, 1983; Seidel, 1983; Tave, 1986, Kapuscinski et al.,1992) should be carefully considered. Special considerations will be necessary to supplement remnant (endangered or threatened) wild stocks (Meffe, 1986; Kapuscinski and Phillip, 1988). The main points of these guidelines include:

1. Maintenance of a large effective breeding size (Ne) for each generation to minimize inbreeding and genetic drift. To protect against inbreeding depression, the following information is required: the level of inbreeding at which inbreeding depression occurs and the number of generations you wish to incorporate in a breeding program before inbreeding reaches the critical value. For example, an Ne of 250 would keep the level of inbreeding below 10% through 50 generations (Tave,

1986). To guard against genetic drift, the following information is required: the value of the rare alleles (how rare an allele would you try to save), and the probability level of saving rare alleles through the course of a given number of generations. For example, an Ne of 424 would provide a 99% probability of saving a rare allele with a frequency of 0.01 through 50 generations.

2. Insurance that the broodstock selected is representative of the natural population targeted for supplementation. To achieve this objective, a large sample size should be selected at random from the entire spectrum of the fish population (over all age groups and sizes and over the entire spawning season).
3. Implementation of a "no selection" protocol (Tave, 1986). For example, fish with poor secondary sexual characteristics, slow growth, etc. will not be culled out. This is designed to conserve the gene pool and ensure survival and reproduction in the wild.
4. Use of equal numbers of males and females as much as possible or at least keep the sex ratio within the bounds 60:40 to 40:60. This is designed to maintain a high Ne.
5. Monitoring of the wild and hatchery-reared fish for genetic and phenotypic information.
6. Consideration of the ratio of wild/natural to hatchery spawners in the natural habitat that minimizes potential genetic risks to the wild/natural spawners.
7. Consideration of the proportion of wild/natural fish used as hatchery broodstock to maintain the genetic integrity of the wild/natural stock and minimize adaptation to the hatchery.

6.3. Rearing Protocol

The basic approach is to provide more natural rearing conditions to promote the success of the fish after release to the natural environment. The objective is to mimic important natural rearing conditions (such as temperature) as much as possible but while providing a more abundant food supply and eliminating predation. Thus, the use of rearing units (earthen ponds and raceways) that provide more natural rearing conditions should be considered.

Stocking densities should more closely mimic densities in the stream at full seeding. Crowding should be reduced to help prevent stress, disease outbreaks, and disruption of territorial and other behaviors that are adaptive in the natural environment. Should territorial behavior be disrupted, it may be possible to restore it by behavioral conditioning (NFA, 1989) of the fish for two to four weeks before release (Shustov et al., 1980).

Hatchery-reared fish can exhibit diminished behavior to avoid predation in the natural environment (Bams, 1967; Mead and Woodall, 1968) and consequently result in increased mortality (MacCrimmon, 1954; Piggins, 1959; Kanid'yev, 1966; Larsson, 1985). To improve the chances of escaping predators after the fish are released, the use of predator avoidance conditioning should be considered (Thompson, 1966; Kanayama, 1968; Olla and Davis, 1989).

6.4. Disease Prevention

Disease prevention must be emphasized. For many diseases, the causative agent is almost always present in the fish's environment. Despite the presence of the pathogen, as long as the environment is not stressful to the fish, disease outbreaks are unlikely (Wood, 1974; Meyer et al., 1983). Disease prevention is based on the proper understanding and management of the interactions between the pathogen, the host, and the environment (Sniezko, 1973,1974; Wedemeyer et al., 1977; Meyer et al., 1983; Rohovec, 1988,1990). Primary attention must be given to the role of the environment in increasing the disease resistance of the fish and decreasing the virulence of the pathogen (minimizing crowding, handling, and stress; providing proper nutrition and water quality; and providing proper hygiene and sanitation).

6.5. Release Strategy

6.5.1. Level of Seeding

Sufficient numbers of fish should be stocked and matched to the biological productivity of the habitat to ensure an adequate, but not excessive level of seeding with respect to carrying capacity of the suite of natural environments encountered by the fish. As the natural stock rebuilds to higher seeding levels, higher egg-to-smolt mortality is expected due to density dependence (Major and Mighell, 1969; Bjornn, 1978; Jonasson and Lindsay, 1983; Knox et al., 1984). However, the total number of smolts produced should increase at higher seeding levels up to the carrying capacity of the habitat.

6.5.2. Life Stage to Stock

There are at least two considerations that would affect the choice of life stage to outplant. First, we want to ensure that successful imprinting to the distinctive chemical cues in the habitat and subsequent homing occur. The existence of a "sensitive" period for olfactory imprinting (SPOI) in early ontogeny in Atlantic salmon has been demonstrated (Morin et al., 1989). In salmonid species that undergo smoltification (Hoar, 1988), the SPOI appears to correspond to the smoltification period (Cooper and Hirsch, 1982; Hasler and Scholz, 1983; Smith, 1985; Hara, 1986). The SPOI was evident between three to four weeks after the onset of smoltification (the total smoltification period was eight weeks) in Atlantic salmon (Morin et al., 1989). During SPOI, the fish's capacity to store information in memory is optimal, implying that some of this capacity may persist beyond the sensitive period.

Second, we want to increase the survival rates for the hatchery-reared fish (during the time period from the egg stage through the smolt stage) compared to the corresponding survival rates typical of these life stages in the natural environment.

Stocking adults or eggs should provide better imprinting to the stream compared to stocking fry and smolts. However, stocking adults or eggs would not provide a survival advantage. Most of the mortality in the time period from the egg stage through the smolt stage occurs soon after the fish emerge from the gravel. For example, fry-to-smolt survival rates are on the order of 20% for spring chinook and 1.5% to 3.8% for steelhead in Big Springs Creek, Idaho; 2.2% to 6.7% for steelhead in Snow Creek, Washington; 5.7% for coho in Speelyai Creek, Washington; and 7.7% for coho in White Salmon River, Washington (Smith et al., 1985). In comparison, egg-to-emergent fry survival rates using streamside incubators in Oregon were on the order of 73.5% to 88.5% for spring chinook, 79.3% to 89.4% for fall chinook, 85.6% to 93% for summer steelhead, about 89% for winter steelhead, and 78% to 83.1% for coho (Smith et al., 1985). Rearing the fish in the same water as the stream where they would be released (the small-scale approach) would allow the flexibility of releasing the fish at any life stage while providing for imprinting.

6.5.3. Size and Age of Fish at Release

It is recommended that fish be released at a size and age that is compatible with those of the natural fish being supplemented to minimize potential adverse ecological interactions (such as predation between the hatchery fish and the natural fish, unequal competitive advantage) and alteration of the age composition of returning adults (increased jack returns with release of larger juveniles). To accomplish this objective, the two primary factors affecting growth (temperature, and ration levels) should reference natural rearing conditions. Since reduced rations or starvation have adverse consequences on an actively growing fish (Ivlev, 1961; Rondorf et al., 1985), primary attention should be directed at mimicking the natural water temperature in the stream.

6.5.4. Acclimation for Stress Reduction

Acclimation is a technique used to prepare fish for release in the natural habitat. It is used to provide the fish time to adjust gradually to the natural stream conditions and reduce transportation-induced stress. This is important when fish are not reared in the same kind of water in which they will be released. In contrast, acclimation would not be necessary when the small-scale approach is used because the fish would be reared in the same water as the receiving stream. Care must be exercised to minimize stress from physical handling, confinement of large numbers of fish in small containers, and sudden changes in water quality parameters (such as temperature) when the fish are transferred from one water to another. Such stress can lead to mortality and can also impair a fish's ability to learn for up to several weeks (Sandoval, 1979; Olla and Davis, 1989). This could block imprinting processes needed for subsequent adult homing.

6.5.5. Timing of Release

Timing of release of juveniles into the natural habitat from hatcheries is a major determinant of survival success. This timing involves the coincidence of various biologic factors (fish size, readiness of fish to migrate and adapt to ocean conditions, outmigration of other hatchery and natural stocks, and estuarine and marine conditions such as food availability, predator abundance, competition for food from other fish stocks) and physical factors (migration flows, operation of mainstem dams, mainstem and tributary water temperature patterns and estuarine and marine conditions such as temperature and upwelling). Volitional release (allowing the fish to exit the rearing facility when they want to) is favored over forced release. Releasing fish at the proper time of day can also reduce initial predation losses and facilitate the adaptation of the fish to a new environment.

6.5.6. Dispersal

Past supplementation programs commonly released the fish at a single location in the stream (Steward and Bjornn, 1990). This practice may lead to limited dispersal and poor utilization of the habitat. The relative effectiveness of scattered and point releases should be considered in a supplementation program.

7. Risk Analysis

This section will address only the potential risks of the kind of supplementation as proposed in this chapter, not any other kind of supplementation or hatchery program.

7.1. Risk of Extirpation or Reduction of Natural Stock

7.1.1. No Supplementation Action

The risk of contributing to the further decline of a fish population through the use of supplementation must be weighed against the risk of continued decline and eventual extirpation when no further action, of any kind, is taken to restore the population.

7.1.2. Partial or Total Failure in the Hatchery

Possible loss of a significant portion of the natural stock through partial or complete failure in the hatchery (loss of electric power, pump failure, loss of rearing water through leakage, human error and accidents) is a risk that must be minimized. Efforts to reduce this risk include building fail-safe features in critical hatchery components, reducing the proportion of natural fish that are brought into the hatchery, rearing the fish in two or more facilities to avoid the risk of failure at one facility, and the use of small-scale facilities.

7.1.3. Predation

Increased or decreased predation on wild fish may occur due to large point-source stocking of hatchery-reared fish in the stream (Steward and Bjornn, 1990). To minimize the impact of predators on young salmon, it would be necessary to understand which predators are present and their capability to consume salmon prey. Different predators respond differently to increased prey abundance. Birds have been shown to have a nonlinear and depensatory functional response (Mace, 1983), whereas predator fish can show a compensatory response at low prey densities, but depensatory at higher densities (Peterman, 1987). This risk can be minimized by avoiding large point-source releases. Instead, the stocked fish should be dispersed throughout the target stream area. In addition, because we are attempting to restore the natural population to historic levels, an increase in predation should be no more than what the population sustained when it was at abundant levels.

Another concern is potential predation between stocked fish and the supplemented natural fish. If there are significant size differences, predation between hatchery-reared fish and the supplemented fish cannot be ruled out. Thus, this risk can be avoided by stocking fish at a size consistent with that of the natural fish.

7.1.4. Competition

Competition for food and space between the natural fish being supplemented and the stocked fish is influenced by the capacity of available rearing habitat. There is a paucity of information on the potential competition between the supplemented salmon population and other fish populations inhabiting the target stream (Steward and Bjornn, 1990). Since the goal is to restore the natural population to historic levels commensurate with the carrying capacity of the habitat, the adverse effects of competition should be no more than those experienced by the population when it was at higher abundance levels. The supplementation strategies described in this chapter seek to minimize or eliminate any differences between the stocked fish and the wild fish so that they are a single population.

7.1.5. Disease

Not much is known about the role of disease in natural fish populations. There is little evidence that hatchery-reared fish cause widespread transmission of disease to natural fish (Steward and Bjornn, 1990). Fish can carry pathogens and not show any outward signs of the disease. As a consequence, subclinically infected fish are probably released into natural waters more often than is realized (Marnell, 1986). In any case, it is desirable to avoid introduction of pathogens and disease to the supplemented stock. This precaution includes introduction of exotic pathogens and also endemic ones that present a threat to the healthy naturally spawning population.

7.2. Loss of Genetic Variability between Populations

Hybridization between different populations typically increases gene diversity (heterozygosity) within the hybridizing populations at the expense of a loss of gene diversity between populations. The concern is that a variety of locally adapted populations will be replaced with a smaller number of relatively homogeneous populations (Allendorf and Leary, 1988). This consolidation will tend to limit the potential of the species to adapt to new environmental conditions and reduce its capacity to buffer total productivity of the resource against periodic or unpredictable changes (Riggs, 1990).

7.2.1. Outbreeding Depression

Outbreeding is the mating of unrelated or distantly related individuals. The potential for outbreeding depression, specifically when hatchery fish mate with wild fish, is a concern related to supplementation in the Columbia Basin. Depending on the specific mating and on the genetic distance between the hatchery and wild fish, the hybrids may display increased fitness (heterosis or hybrid vigor) or decreased fitness (outbreeding depression) (Waples, 1991). Heterosis is more likely when the hybridizing gene pools are inbred and not too different genetically (Waples, 1991). As genetic distance between the parental stocks increases, however, genetic incompatibilities become more likely and the fitness of the hybrids may decline (outbreeding depression). Current genetic theory on hybridization indicates that the potential effects of hybridization result from genetic variance due to the interactions among alleles (Tave, 1986). Because this form of genetic variance depends on interactions, it is disrupted during meiosis and cannot be transmitted from parent to offspring. This genetic variance is created anew and, in different combinations each generation, its effects are basically those based on chance.

A series of studies carried out during the last 30 years showed that crossbreeds between wild and domestic stocks are superior to domestic stocks and may eqaul or even surpass wild stocks in performance in the wild (Wohlfarth, 1991, this symposium). A few studies have indicated the potential for outbreeding depression when hatchery fish are mated with wild fish. Those studies did not look at outbreeding depression per se, but rather compared the performance of hatchery fish with wild fish in the natural environment. A study in the Deschutes River in Oregon compared the progeny from hatchery, wild, and hatchery-wild parents in the natural environment (Reisenbicher and McIntyre, 1977). The authors concluded that wild eggs survived better than hatchery-wild eggs and hatchery eggs. Juvenile fish differed in size among the treatments. Hybrid juveniles were larger than the non-hybrid juveniles. This suggests that there were differences between the hatchery fish and the wild stocks. This study may not support the conclusion that outbreeding depression occurs when wild and hatchery stocks are mated. It is unclear whether or not the hatchery stock originated from a stock different than that of the wild stock. If genetic lineages of the hatchery and wild stocks were different, the likelihood of outbreeding depression or heterosis would increase independently of potential hatchery effects on genetic makeup and performance (Kapuscinski and Miller, 1992). In a study in Washington, the reproductive success of a Skamania Hatchery steelhead stock was compared with the reproductive success of wild Kalama River steelhead stock in the natural environment of the Kalama River (Chilcote et al., 1986). The success of hatchery fish in producing smolt offspring was only 28% of that for wild fish. The failure of the hatchery fish to produce as many offspring as the wild fish can be attributed to the use of a hatchery stock that was genetically and ecologically poorly matched to the natural environment of the Kalama River. The Skamania Hatchery stock originated from wild stock indigenous to the Klickitat and Washougal rivers (a different drainage) and has been subjected to artificial selection for hatchery production traits for many years (Leider et al., 1990). The hatchery fish were subjected to more unfavorable flooding conditions in the Kalama River because they spawned earlier than the indigenous wild fish.

The effectiveness of using hatchery fish to rebuild wild populations was evaluated in 15 supplemented and 15 control streams in the Oregon coast (Nickelson et al., 1986; Solazzi et al., 1990). Although the summer density of hatchery and wild juveniles increased in the supplemented streams, the density of only the wild juveniles was reduced. Adult returns to the supplemented streams were not significantly different from returns to the control streams and the hatchery fish produced juveniles at a lower rate than wild fish. The failure of this program can be attributed to the use of a hatchery stock that was incompatible with the wild population and genetically and ecologically poorly matched to the natural environment. The hatchery fish were subjected to more unfavorable conditions because they spawned earlier than the wild fish. Also, the hatchery fish outcompeted the wild fish because they were released at a much larger size than the wild fish.

All these studies indicate the fundamental importance of selecting a stock that is compatible with the target stock and releasing fish at life stages and with biological features that are adaptive in the target stream environment. Since the level of outbreeding that causes outbreeding depression is not known, it is impossible to predict whether a particular hatchery and wild cross will result in outbreeding depression. In addition, the variable (hatchery fish) being tested is quite undefined and imprecise, which results in a variety of effects given the same variable. Hatchery fish are spawned, incubated, and reared in many different hatchery environments using many different hatchery practices. A comparison between hatchery fish and wild fish lumps too many complicating factors that cannot be separated from each other. Thus, it is not known exactly what is being tested.

There is no clear evidence that a well-managed supplementation program, as described here, would pose a serious genetic risk to the natural population through outbreeding depression. The supplementation program described here attempts to eliminate or minimize any serious divergence between the hatchery broodstock and the target stock by ensuring that the hatchery broodstock is representative of the natural stock and by minimizing divergent natural selection by minimizing important differences between the hatchery and wild environments.

7.3. Loss of Genetic Variability within Populations

7.3.1. Inbreeding Depression

Inbreeding is the mating of related individuals (Tave, 1986). Genetically, all inbreeding does is increase the homozygotes at the expense of the heterozygotes. Because almost every organism carries deleterious recessive alleles that are hidden in the heterozygous state, increasing homozygosity increases the likelihood that deleterious recessive alleles will be paired and expressed. The pairing of detrimental recessive alleles produces a general trend toward lowered viability, survival, growth, egg production and increased abnormalities (Ryman and Stahl, 1980; Allendorf and Phelps, 1980; Gall, 1983; Allendorf and Ryman, 1987). This phenomenon is called inbreeding depression.

To protect against inbreeding depression, a basic approach is to maintain a large, effective population size (Ne). To calculate the Ne that is needed to prevent inbreeding from reaching levels that decrease productivity, two pieces of information are required: the critical level of inbreeding at which inbreeding depression occurs, and the number of generations to incorporate in a breeding program before inbreeding reaches the critical level (Tave, 1986). For example, an Ne of 250 fish would keep the level of inbreeding below 10% through 50 generations.

Thus, an integrated approach, to guard against both outbreeding and inbreeding depressions, must travel the middle road. If we move too far to one side (mating closely related individuals), we risk falling into the inbreeding depression ditch. If we move too far to the opposite side (mating distantly related individuals), we risk falling into the outbreeding depression ditch.

7.3.2. Genetic Drift

Genetic drift refers to random changes in gene frequency caused by sampling error between generations (Tave, 1986). The effect of genetic drift is the loss of some alleles and the fixation of others (inbreeding). Rare alleles are easily lost, but more common alleles can also be lost via genetic drift. The loss of alleles (reduction in genetic variability) will limit the potential of a population to adapt to changes in environmental conditions and compromise its ability to exploit new environments.

A narrow genetic variability in a fish population would not necessarily result in low productivity or fitness to a particular environment. It would depend on the degree of adaptation of the population to the given environment and the magnitude and rate of change in the environment. Introduced chinook, coho, and pink salmon in the Great Lakes are examples of fish populations that were initiated from very small founding populations. Despite this narrow genetic variability, these fish have been thriving well for the past 20 to 40 years (Tanner, 1988).

To guard against genetic drift, the basic approach is to maintain a large Ne. The Ne that is needed depends on the following pieces of information: how rare an allele you would try to save and the desired probability level of saving rare alleles through the course of a given number of generations. For example, an Ne of 424 fish would provide a 99% probability of saving a rare allele with a frequency of 0.01 through 50 generations.

7.3.3. Selection

Anadromous salmonids are reared in a hatchery environment for only a portion of their life cycle. For the majority of their life cycle, hatchery fish are exposed to the same natural environment and subjected to the same natural selection process as wild fish. However, this does not mean that the selection (artificial and natural) that may occur in the hatchery is not important as far as the abilily of the fish to survive and reproduce successfully in the natural environment.

Artificial selection in the hatchery might select for hatchery production traits that are not adaptive in the natural environment. For example, broodstock might be selected from only the early part of the run because the egg-take quota has been filled or to produce fish that are larger at release or can be released earlier. Only large spawners or spawners that are ripe when the hatchery manager wants to spawn fish might be spawned. Also, throughout the rearing period slow-growing fish may be culled out.

Moreover, the natural selection process that occurs in the hatchery is most likely different from that which occurs in the wild by virtue of the difference between the hatchery and wild environments. Egg-to-smolt survival rates are typically 5% to 15% in the natural environment while the corresponding rates under artificial propagation are about 60% to 80% (Howell et al., 1985). This difference represents a reduced intensity of selection than occurs in the wild (i.e., most of the fish that would have died in the wild survive in the hatchery). Conversely, a higher percentage of wild smolts may typically survive to return as adults than will hatchery smolts. The high post-release mortality for hatchery fish allows ample opportunity for selection against traits that are adaptive for hatchery conditions, but not for the wild, thus counteracting, to some extent, adaptation to the hatchery environment. Thus, the degree to which the hatchery fish would diverge from the wild fish will depend on the degree to which the wild smolts are genetically representative of the hatchery smolts.

The supplementation scheme described in this chapter seeks to eliminate any divergence between the hatchery fish and the wild fish by using representative samples of the wild population as hatchery broodstock, by avoiding any artificial selection, and by minimizing the difference between the natural and the hatchery environments. Although one can hypothesize that exposure to the hatchery environment, for even a small portion of the fish's life cycle, allows some genetic divergence from the natural genome, the degree and consequences of the change remain unknown.

7.3.4. Hatchery versus Natural Environment

The kind of supplementation described in this chapter is based on the underlying principle that the fish population must be adapted to its environment to thrive. Since genetic fitness is partially a function of the environment, it is important to evaluate the hatchery environment vis-à-vis the natural environment with respect to those parameters that are related to performance and genetic fitness traits. A complicating factor that must be taken into account is that the natural environment and, to a lesser extent, the hatchery environment change daily, seasonally, and from year to year. In general, the quantity and diversity of salmonid habitats in the Columbia Basin have been greatly reduced (NPPC, 1986). For example, over one-third of the spawning and rearing habitat has been eliminated by impassable dams. Thus, any program of supplementation that emphasizes restoration of natural stocks and maintenance of their genetic diversity must be accompanied by an equal emphasis on restoration of habitat quality and quantity to which these stocks have adapted.

7.4. Genetic Risk of Other Activities

The genetic concern associated with supplementation must not preclude needed attention on other equally important genetic risks associated with habitat loss and degradation, alteration of the migrational environment, and harvest. Habitat loss and degradation can affect the genetics of wild populations by depressing the Ne of the population and by natural selection for increased fitness in the new environment, which may decrease the value of the resource (Kapuscinski and Jacobson, 1987). Harvest management can affect the genetics of wild fish in at least two ways (Kapuscinski and Jacobson, 1987). First, high exploitation rates can reduce the Ne of a stock so that rates of genetic drift and inbreeding are increased. Second, fishing methods and regulations can act as artificial selection programs that can cause genetic change in the stock over time (Handford et al., 1977; Favro et al., 1979; Ricker, 1981). The migrational environment of Columbia Basin anadromous salmonids has been drastically transformed from fast flowing streams to slow-moving reservoirs. Flow allocations and spill at dams favor juvenile fish whose outmigration timing coincides with the flow.

8. Research Needs

1. How different is a hatchery-reared fish population from its wild counterparts in terms of important performance traits such as survival, growth, reproduction, migration? How are these traits affected by the hatchery environment and by the natural environment?
2. In a natural fish population, which genotypes and gene frequencies comprise the typical 5% to 15% of the fish eggs that survive to smolt stage in the natural environment? Do the survivors represent a random sample of the total eggs deposited or are they a result of natural selection in the wild?
3. In a hatchery-reared fish population, which genotypes and gene frequencies comprise the typical 60% to 80% of the fish eggs that survive to smolt stage in the hatchery environment? Do the survivors represent a random sample of the total eggs spawned or are they a result of natural selection in the hatchery?
4. Assuming that the fish spawned in a hatchery were derived from and are genetically representative of a given wild fish population, are the hatchery smolts that survive in the hatchery environment genetically similar to the wild smolts from the same wild fish population that survive in the natural environment?
5. What is the level of inbreeding that causes inbreeding depression? How do we measure inbreeding? How do we measure inbreeding depression? Is there a qualitative aspect to inbreeding (i.e., one population with the same level of inbreeding as another population exhibits inbreeding depression where the other will not)?
6. What is the level of outbreeding that causes outbreeding depression? How do we measure outbreeding? How do we measure outbreeding depression? Is there a qualitative aspect to level of outbreeding (i.e., will one population with the same level of outbreeding as another population exhibit outbreeding depression where the other will not)?
7. Does natural selection reduce genetic variability within a population; between populations?
8. What are the most effective means of ensuring no artificial selection in hatcheries?
9. What are we doing in the hatchery that renders fish less fit when released into the natural environment? Does the hatchery environment influence the expression of traits or behaviors that are not adaptive in the wild environment? For example, hatchery reared fish are not exposed to predators nor provided natural food. Hence, many fish that would have died in the wild due to predation or starvation survive in the hatchery. In the wild, the juveniles learn to avoid predators and learn to capture prey, but in the process incur mortalities through predation and starvation.
10. Evaluate food availability during different life stages in terms of survival, growth, and reproduction.
11. Compare levels and variability in important parameters (such as temperature) in natural stream environments with those in hatchery environments. Look at temporal variability (diel, seasonal, interannual) and spatial variability (location of

hatchery or stream, locations within hatchery or stream - pools, riffles, different reaches). Is there more seasonal variability in a given location than between locations (streams) for a given season?

ACKNOWLEDGEMENTS. This presentation was based on the supplementation guidelines of the Integrated System Plan For Salmon and Steelhead Production in the Columbia River Basin prepared by fishery agencies and Indian Tribes of the Columbia River Basin Fish and Wildlife Authority. The manuscript was thoughtfully reviewed and improved by Dr. Anne Kapuscinski, L. Miller, Dr. Dale McCollough, and Dr. Douglas Tave.

References

Allendorf, F.W. and R.F. Leary. 1988. Conservation and distribution of genetic variation in a polytypic species, the cutthroat trout. *Conserv. Biol.* 2:170-184.

Allendorf, F.W. and S.R. Phelps. 1980. Loss of genetic variation in a hatchery stock of cutthroat trout. *Trans. Am. Fish. Soc.* 109:537-543.

Allendorf, F.W. and N. Ryman. 1987. Genetic management of hatchery stocks, in: "Population Genetics and Fishery Management," F. Utter, ed., pp. 141-159, University of Washington Press, Seattle.

Bams, R.A. 1967. Differences in performance of naturally and artifically propagated sockeye salmon migrant fry, as measured with swimming and predator tests. *J. Fish. Res. Board Can. 24:1117-1153.*

Behnke, R.J. 1986. Atlantic salmon, *Salmo salar.* Trout Unlimited, Inc. 3:4217.

Bjornn, T.C. 1978. Survival, production, and yield of trout and chinook salmon in the Lemhi River, Idaho Bulletin 27, College of Forestry, Wildlife, and Range Sciences, University of Idaho, Moscow, Idaho.

Brannon, E.L., R.P. Whitman, and T.P. Quinn. 1984. Responses of returning adult coho salmon to home water and population specific odors. *Trans. Amer. Fish. Soc.* 113:374-377.

Chamberlain, T.W. 1982. Influence of forest and rangeland management on anadromous fish habitat in western North America - No. 3. Timber harvest. General Technical Report, PNW-136. U.S. Dept. of Agriculture, Forest Service, Pacific Northwest Forest and Range Experiment Station, Portland, Oregon, pp. 130.

Chilcote, M.W., S.A. Leider, and J.J. Loch. 1986. Differential reproductive success of hatchery and wild summer-run steelhead under natural conditions. *Trans. Amer. Fish. Soc.* 115:726-735.

Cooper, J.C. and P.J. Hirsch. 1982. The role of chemoreception in salmonid homing, in: "Chemoreception in Fishes," T.J. Hirsch, ed., pp. 343-362, Elsevier, Amsterdam.

Cooper, J.C. and A.T. Scholz. 1976. Homing of artificially imprinted steelhead trout. *J. Fish. Res. Board Can.* 33:826-829.

Cooper, J.C, A.T. Scholz, R.M. Horrall, A.D. Hasler, and D.M. Madison. 1976. Experimental confirmation of the olfactory hypothesis with artificially imprinted homing coho salmon, *Oncorhynchus kisutch. J. Fish. Res. Board Can.* 33:703-710.

Davidson, F.A. and S.J. Hutchinson. 1938. The geographic distribution and environmental limitations of the Pacific salmon (genus *Oncorhynchus*). Bureau of Fisheries, U.S. Dept. of the Interior, Bulletin No. 26, vol. 48, pp. 667-692.

Delarm, M.R. and R.Z Smith. 1990. Assessment of present anadromous fish production facilities in the Columbia River Basin, Volumes 1 - 5. Project No. 89-045, Bonneville Power Administration Portland, Oregon.

Favro, L.D., P.K Kuo, and J.F. McDonald. 1979. Population genetic study of the effects of selective fishing on the growth rate of trout. *J. Fish. Res. Board Can.* 36:552-561.

Fedorenko, A.Y. and B.G. Shepherd. 1986. Review of salmon transplant procedures and suggested transplant guidelines. *Canadian Technical Report of Fisheries and Aquatic Sciences No. 1479*, 144 pp.

Gall, G.A. 1983. Genetics of fish: a summary of discussion. *Aquaculture* 33:383-394.

Handford, P., G. Bell, and T. Reimchen. 1977. A gillnet fishery considered as an experiment in artificial selection. *J. Fish. Res. Board Can.* 34:954-961.

Hara, T.J. 1971. Role of olfaction in fish behavior, in: "The Behavior Teleost Fishes," T.J. Pitcher, ed., pp. 152-176, Croom Helm, London.

Hasler, A.D. and A.T. Scholz. 1983. "Olfactory Imprinting and Homing in Salmon," Springer-Verlag, New York, pp. 134.

Hasler, A.D. and W.J. Wisby. 1951. Discrimination of stream odors by fishes and relation to parent stream behavior. *Amer. Nat.* 85:223-238.

Hershberger, W.K and R.N. Iwamoto 1981. "Genetics Manual and Guidelines for The Pacific Salmon Hatcheries of Washington," University of Washington, Seattle, Washington, 83 pp.

Hoar, W.S. 1988. The physiology of smolting salmonids, in: "Fish Physiology, Volume 11, The Physiology of Developing Fish, Part B, Viviparity and Posthatching Juveniles," W.S. Hoar and D.J. Randall, eds., pp. 275-343, Academic Press, San Diego.

Howell, P., K. Jones, D. Scarnecchia, L. LaVoy, W. Kendra, D. Ortmann, C. Neff, C. Petrosky, and R. Thurow. 1985. Stock Assessment of Columbia River Anadromous Salmonids. Vol. I: Chinook, coho, chum, and sockeye salmon stock summaries. Vol. II: Steelhead stock summaries, stock transfer guidelines, information needs. Bonneville Power Administration, Portland, Oregon.

Ivlev, V.S. 1961. "Experimental Ecology of the Feeding of Fishes," Translation by Douglas Scott, Yale University Press, New Haven, Connecticut.

Jensen, A. and R. Duncan. 1971. Homing in transplanted coho salmon. *Prog. Fish.-Cult.* 33:216-218.

Jonasson, B. and R. Lindsay. 1983. An ecological and fish cultural study of Deschutes River salmonids. Fish. Res. Proj. F-88-R-13. Annual Progress Report, Oregon Dept. of Fish and Wildlife, Portland, Oregon.

Kanayama, Y. 1968. Studies of the conditioned reflex in lower vertebrates. X. Defensive conditioned reflex of chum salmon fry in group. *Mar. Biol.* 2:77-87.

Kanid'yev, A N. 1966. Tolerance of hatchery-reared juvenile chum (*Oncorhynchus keta*) to rate of flow and to predaceous fishers. *Tr Murmansk Morsk Biology Institute* 12:101-111.

Kapuscinski, A.R. and L.D. Jacobson. 1987. "Genetic Guidelines for Fisheries Management," Minnesota Sea Grant, University of Minnesota, Duluth, Minnesota. 66 pp.

Kapuscinski, A.R. and J.E. Lannan. 1986. A conceptual genetic fitness model for fisheries management. *Can. J. Fish. Aquat. Sci.* 43:1606-1616.

Kapuscinski, A.R. and L.M. Miller. 1992. A Review of: The use of supplementation to aid in natural stock restoration. Northwest Power Planning Council (Agreement 92-031).

Kapuscinski A.R. and D.R. Philipp. 1988. Fisheries genetics: issues and priorities for research and policy development. *Fisheries* 13:4-10.

Kapuscinski A.R., C.R. Steward, M.L. Goodman, C.C. Krueger, J.H. Williamson, E. Bowles, and R. Carmichael. 1992. Genetic conservation guidelines for salmon and steelhead supplementation. Proceedings of the Sustainability Workshop, Cascade Lodge. Northwest Power Planning Council, Portland, OR. (in review).

Kincaid, H.L. 1983. Inbreeding in fish populations used for aquaculture. *Aquaculture* 33:215-227.

Knox, W., M. Flesher, B. Lindsay and L. Lutz 1984. Spring chinook studies in the John Day River. Annual Progress Report, Fish Res. Proj. DE-AC79-84BP39796. Oregon Dept. of Fish and Wildlife, Portland, Oregon.

Krueger, C.C., A.J. Gharrett, T.R. Dehring, and F.W. Allendorf 1981. Genetic aspects of fisheries rehabilitation programs. *Can. J. Fish. Aquat. Sci.* 38:1877-1881.

Lande, R. and G.F. Barrowclough. 1987. Effective population size, genetic variation, and their use in population management, in: "Viable Populations for Conservation," M.E. Soule, ed., pp. 87-123, Cambridge University Press, Cambridge.

Larkin, P.A. 1981. A perspective on population genetics and salmon management. *Can. J. Fish. Aquat. Sci.* 38:1469-1475.

Larsson, P.O. 1985. Predation on migrating smolt as a regulating factor in Baltic salmon, *Salmo salar* L., populations. *J. Fish Biol.* 26:391-397.

Leider, S.A., P.L. Hulett, J.J. Loch, and M. Chilcote. 1990. Electrophoretic comparison of the reproductive success of naturally spawning transplanted and wild steelhead trout through the returning adult stage. *Aquaculture* 88:239-2S2.

MacCrimmon, H.R. 1984. Stream studies on planted Atlantic salmon. *J. Fish. Res. Board Can.* 11:362-403.

Mace, P.M. 1983. Bird predation on juvenile salmonids in the Big Qualicum River, Vancouver Island, Canada. *Canadian Technical Report of Fisheries and Aquatic Sciences 1176.*

Madison, D.M., A.T. Scholz, J.C. Cooper, and A.D. Hasler. 1973. I. Olfactory hypothesis and salmon migration: a synopsis of recent findings. Fish. Res. Bd. Can. Tech. Rept. No. 414. 37 pp.

Major, R.L and J.L. Mighell. 1969. Egg-to-migrant survival of spring chinook salmon (*Oncorhynchus tshawytscha*) on the Yakima River. Washington Fishery Bulletin 67:347-359.

Marnell, L.F. 1986. Impacts of hatchery stocks on wild fish populations, in: "Fish Culture in Fisheries Management," R.H Stroud, ed., pp.339-347, American Fisheries Society, Bethesda, MA.

Martin, S.B. and Platts, W.S. 1981. Influence of Forest and Rangeland Management on Anadromous Fish Habitat in Western North America - No. 8. Effects of mining. General Technical Report, PNW-119. U.S. Dept. of Agriculture, Forest Service, Pacific Northwest Forest and Range Experiment Station, Portland, Oregon. 15 pp.

Mead, R.W. and W. Woodall. 1968. Comparison of sockeye salmon fry produced by hatcheries, artificial channels, and natural spawning areas. International Pacific Salmon Fisheries Commission, Progress Report No. 20.

Meffe, G.K 1986. Conservation genetics and the management of endangered fishes. *Fisheries* 11:14-23.

Meyer, F.P., J.W. Warren and T.G. Carey (eds.) 1983. A guide to integrated fish health management in the Great Lakes basin, Great Lakes Fishery Commission, Ann Arbor, Michigan. Spc. Pub. 82-3. 272 pp.

Morin, P.P., J.J. Dodson and F.Y. Dore. 1989. Cardiac responses to a natural odorant as evidence of a sensitive period for olfactory imprinting in young Atlantic salmon, *Salmo salar*. *Can. J. Fish. Aquat. Sci.* 46:122-130.

NFA (Norsk Forening For Akvakulturforskining). 1989. Ethology in Aquaculture. Norwegian Society of Aquaculture Research, Bergen, Norway. 84 pp.

Nickelson, T.E., M.F. Solazzi, and S.L. Johnson. 1986. Use of hatchery coho salmon (*Oncorhynchus kisutch*) presmolts to rebuild wild populations in Oregon coastal streams. *Can. J. Fish. Aquat. Sci.* 43:2443-2449.

Nordeng, H. 1977. A pheromone hypothesis for homeward migration in anadromous salmonids. *Oikos* 28:155-159.

Northcote, T.G. 1969. "Symposium on Salmon and Trout in Streams." A Symposium Held at the University of British Columbia, February 22 to 24, 1968. H.R. MacMillan Lectures in Fisheries. Institute of Fisheries, University of British Columbia, Vancouver, B.C.

NPPC (Northwest Power Planning Council). 1986. "Compilation of information on salmon and steelhead losses in the Columbia River Basin." March 1986. NPPC, Portland, Oregon. 252 pp.

Olla, B.L. and M.W. Davis 1989. The role of learning and stress in predator avoidance of hatchery-reared coho salmon (*Oncorhvnchus kisutch)* juveniles. *Aquaculture* 76:209-214.

Peterman, R.M. 1987. Review of the components of recruitment of Pacific salmon, in: "Common Strategies of Anodromous and Catadromous Fishes, American Fisheries Society Symposium 1," M.J., Klauda, R.J. Moffit, C.M. Saunders, R.L. Rulifson, R.A. Cooper, J.E. Dadswell eds., pp. 417-429, American Fisheries Society, Bethsda, MA.

Piggins, D.J. 1959. "Investigation on predators of salmon smolts and parr." Salmon Research Trust Ireland 5, Appendix 1.

Platt, W.S. 1981. Influence of forest and rangeland management on anadromous fish habitat in western North Arnerica - No. 7. Effects of livestock grazing. General Technical Report, PNW-124. U.S. Dept. of Agriculture, Forest Service, Pacific Northwest Forest and Range Experiment Station, Portland, Oregon. 25 pp.

RASP (Regional Assessment of Supplementation Programs). 1992. Supplementation in the Columbia River Basin, Parts 1 and 3, (Project Number 85-12). Bonneville Power Administration, Portland, OR.

Raymond, H.L. 1979. Effects of dams and impoundments on migrations of juvenile chinook salmon and steelhead from the Snake River, 1966 to 1975. *Trans. Amer. Fish. Soc.* 108:505-529.

Reisenbichler, R.R. and J.D. McIntyre. 1977. Genetic differences in growth and survival of juvenile hatchery and wild steelhead trout, *Salmo gairdneri. J. Fish. Res. Board Can.* 34:123-128.

Reiser, and T.C. Bjornn. 1979. Influence of forest and rangeland management on anadromous fish habitat in western North America - No. 1. Habitat requirements of anadromous salmonids. General Technical Report, PNW-96. U.S. Dept. of Agriculture, Forest Service, Pacific Northwest Forest and Range Experiment Station, Portland, Oregon.

Ricker, W.E. 1972. Hereditary and environmental factors affecting certain salmonid populations, in: "The Stock Concept in Pacific Salmon," R.C. Simon and P.A. Larkin,eds., pp. 19-160, H.R. MacMillan Lectures in Fisheries, University of British Columbia, Vancouver.

Ricker, W.E. 1981. Changes in average size and average age of Pacific salmon. *Can. J. Fish. Aquat. Sci.* 38:1636-1656.

Rieman, B.E., R.C. Beamesderfer, S. Vigg, and R.P. Poe. 1988. Predation by resident fish on juvenile salmonids in a mainstem Columbia reservoir: Part IV. Estimated total loss and mortality of juvenile salmonids to northern squawfish, walleye and smallmouth bass, in: "Predation by resident fish in juvenile salmonids in John Day Reservoir. Vol. 1," R.P. Poe and B.E. Rieman, eds., Final Report on research.

Riggs, L.A. 1986a. "Genetic Considerations in Salmon and Steelhead Planning: Final Report," Technical paper prepared for the Northwest Power Planning Council, Portland, Oregon. May 1986.
Riggs, L.A. 1986b. "Genetic Considerations in Salmon and Steelhead Planning: I. Area Below Bonneville," Prepared for the Northwest Power Planning Council, Portland, Oregon.
Riggs, L.A. 1986c. "Genetic Considerations in Salmon and Steelhead Planning: II. Bonneville Dam to the Snake River," Prepared for the Northwest Power Planning Council, Portland, Oregon.
Riggs, L.A. 1986d. "Genetic Considerations in Salmon and Steelhead Planning: III. The Mid-Columbia to Upper-Columbia," Prepared for the Northwest Power Planning Council, Portland, Oregon.
Riggs, L.A. 1986e. "Genetic Considerations in Salmon and Steelhead Planning: IV. The Lower Snake River," Prepared for the Northwest Power Planning Council, Portland, Oregon.
Riggs, L.A. 1990. "Principles for Genetic Conservation and Production Quality: Results of a Scientific and Technical Clarification and Revision," Report on a workshop held December 15, 1989, Portland, Oregon. Prepared for the Northwest Power Planning Council. GENREC, Genetic Resource Consulting, Berkeley, California.
Rohovec, J.S. 1988. Integrated health management in salmonid aquaculture. *Food Rev. Intnl.* 6(3):389-397.
Rohovec, J.S. 1990. Infectious diseases of salmonid fish: transmission, prevention and control. Proceedings of the International Seminar, Santiago, Chile, October 19-21, 1988. pp 19-21.
Rondorf, D.W., M.S. Dutchuk, A.S. Kolok and M.L. Gross. 1985. Bioenergetics of juvenile salmon during the spring outmigration. Bonneville Power Aadministration, Portland, OR. 78 pp.
Ryman, N., and G. Stahl. 1980. Genetic changes in hatchery stocks of brown trout (*Salmo trutta*). *Can. J. Fish. Aquat. Sci.* 37:82-87.
Sandoval, W.A. 1979. Odor detection by coho salmon (*Oncorhynchus kisutch*): a laboratory bioassay and genetic basis. M.S. Thesis. Oregon State University, Corvallis, Oregon, 43 pp.
Scholz, A.T., R.M. Horral, J.C. Cooper, A D. Hasler, D.M. Madison, R.J. Poff, and R. Daly. 1975. "Artificial Imprinting of Salmon and Trout in Lake Michigan," Wisconsin Department of Natural Resources Fisheries Management, Report 80, Madison, Wisconsin, 46 pp.
Seidel, P. 1983. Spawning guidelines for Washington Department of Fisheries. March 1983. 15 pp.
Sedell, J.R. and W.S. Duval. 1985. Influence of forest and rangeland management on andromous fish habitat in western North America - No. 5 Water transportation and storage of logs. General Technical Report, PNW-186. U.S. Dept. of Agriculture, Forest Service, Pacific Northwest Forest and Range Experiment Station, Portland, Oregon, 68 pp.
Senn, H., Mack and L. Rothfus. 1984. Compendium of low cost Pacific salmon and steelhead trout production facilities and practices in the Pacific Northwest. Project No. 83-353. Bonneville Power Administration, Portland, OR. 488 pp.
Shustov, Y.A., I.L. Shchurov and Y.A. Smirnov. 1980. Adaptation times of hatchery salmon, *Salmo salar* to river conditions. *J. Ichthyology*, 20:156-159.
Slatick, E., L.G. Gilbreath, J.R. Harmon, T.C. Bjornn, R.R. Ringe, K.A. Walch, A.J. Novotny and W.S. Zaugg. 1988. Imprinting hatchery reared salmon and steelhead trout homing,

1978-1983. Vol. 1 Narrative. Final Report. Bonneville Power Administration, Portland, OR. 143 pp.

Smith, E.M., B.A. Miller, J.D. Rodgers, M.A. Buckman. 1985. Outplanting anadromous salmonids: a literature survey. Bonneville Power Administration, Portland, OR. 68 pp.

Smith, R.J.F. 1985. The Control of Fish Migration. Springer-Verlag, Berlin and New York.

Sniezko, S.F. 1973. Recent advances of scientific knowledge and developments pertaining to diseases of fishes. *Adv. Vet. Sci. Comp. Med.*, 17:291-314.

Sniezko, S.F. 1974. The effects of environmental stress on outbreaks of infectious diseases of fishes. *J. Fish. Biol.* 6:197-208.

Steward, C.R., and T.C. Bjornn. 1990. Supplementation of salmon and steelhead stocks with hatchery fish: A synthesis of published literature, in: "Analysis of Salmon and Steelhead Supplementation," W.H. Miller, ed., Technical Report 90-1, Part 2, Bonneville Power Administration, Portland, OR.

Tave, D. 1986. "Genetics for Fish Hatchery Managers." AVI Publishing Co., Inc. Westport, Connecticut. 299 pp.

Thomas, CD. 1990. What do real population dynamics tell us about minimum viable population sizes? *Conserv. Biol.* 4(3):324 - 327.

Thompson, R.B. 1966. "Effects of Predator Avoidance Conditioning on the Post-Release Survival Rate of Artificially Propagated Salmon," Ph.D. Thesis, University of Washington, Seattle, Washington.

Theeland, R.R., R.J. Wahle, and H. Arp. 1975. Homing behavior and contribution to Columbia River fisheries of marked coho salmon released at two locations. *Fish. Bull.* 73:717-725.

Waples, R.S. I990. Conservation genetics of Pacific salmon, I. Temporal changes in allele frequency. *Conserv. Biol.* 4(2):144-156.

Waples, R.S. I990b. Conservation genetics of Pacific salmon, II. Effective population size and the random loss of genetic variability. *J. Hered.* 81(4):267-276.

Waples, R.S. l990c Conservation genetics of Pacific salmon, III. Estimating effective population size. *J. of Hered.* 81(4):277-289.

Waples, R.S. 1991. Genetic interactions between hatchery and wild salmonids: lessons from the Pacific Northwest. *Can. J. Fish. Aquat. Sci.* 48(suppl. 1):124-133.

Wedemeyer, C., F. Mever and LS. Smith. 1977. Environmental stress and fish diseases. TFH Publications, Neptune City, NJ. 200 pp.

Withler, F.C 1982. Transplanting Pacific salmon. Can. Tech. Rept. Fisheries and Aquatic Sciences 1079. 27 pp.

Wohlfarth, G.W. 1991. Genetic Management of Natural Fish Populations, in: "Genetic Conservation of Salmonid Fishes," J.G. Cloud, ed., pp. 221-224, Plenum Press, New York.

Wood, J.W. 1974. Diseases of Pacific salmon, their prevention and treatment. Washington Department of Fisheries, Hatchery Division.

Yee, C.S. and T.D. Roelofs. 1980. Influence of forest and rangeland management on anadromous fish habitat in western North America - No. 4. Planning forest roads to protect salmonid habitat. General Technical Report, PNW-109. U.S. Dept. of Agriculture, Forest Service, Pacific Northwest Forest and Range Experiment Station, Portland, Oregon. 26 pp.

Status of the Genetic Resources of Pacific Rim Salmon

A. J. GHARRETT[1], B. E. RIDDELL[2], J. E. SEEB[3], and J. H. HELLE[4]

1. Introduction

Although there are numerous serious threats to the persistence of individual wild pink (*Oncorhynchus gorbuscha)*, chum (*O. keta*), sockeye (*O. nerka*), coho (*O. kisutch*), and chinook (*O. tshawytscha*) salmon populations in their native range, these species as a whole are relatively healthy compared to other salmonids reviewed in this text.

The threats to wild Pacific salmon populations can be attributed to three very general sources: (1) competition for resources, (2) over exploitation, and (3) aquaculture. Competition is most often associated with human population density and is manifested as a loss of habitat (for example, to urbanization and dam construction) or as degradation of habitat (for example, by pollution, agriculture, logging, mining, and loss of water to irrigation). The primary effects of over exploitation on salmon populations are direct and obvious (reduced population sizes and loss of less productive stocks or species). Over exploitation also imposes indirect, but significant, effects on the genetics and abundance of salmon populations from fishery management strategies (generally intended to prevent over harvest) and aquaculture programs (designed to compensate for over harvest or increase harvest). Aquaculture can effect wild populations as a result of interactions between wild fish and escaped or purposely released cultured fish. These interactions may involve the introduction of pathogens, the alteration of genetic compositions, or the imposition of increased harvest pressure on wild populations.

Generally, the problems that affect salmon populations at the southern ends of their ranges where more humans live are more extensive and varied than the problems faced by salmon in the center of their range. Here we will list some of the most important threats to salmon populations. Because the nature of these threats is often related to population density and is regional, we proceed by considering three geographic areas: the southeastern range of salmon in the eastern Pacific Ocean, the southwestern range in the western Pacific Ocean, and the northern range including the Gulf of Alaska, Bering Sea, and Okhostsk Sea.

[1]Juneau Center, School of Fisheries and Ocean Sciences, University of Alaska Fairbanks, Juneau, Alaska 99801, U.S.A. [2]Pacific Biological Station, Department of Fisheries and Oceans, Nanaimo, British Columbia V9R 5K6, Canada. [3]Fisheries Rehabilitation, Enhancement, and Development Division, Alaska Department of Fish and Game, Anchorage, Alaska 99502, U.S.A. [4]National Marine Fisheries Service, Alaska Fisheries Science Center, Auke Bay Laboratory, Auke Bay, Alaska 99821, U.S.A.

Genetic Conservation of Salmonid Fishes, Edited by J.G. Cloud and G.H. Thorgaard, Plenum Press, New York, 1993

2. Southeastern Range

We define the southeastern range of Pacific salmon as extending from California to central British Columbia (south of the Skeena River, approximately). This region includes the major human population centers of the Pacific Coast of North America. At the southern margins of their historic North American range (from the Columbia River south), pink and chum salmon have nearly disappeared; numerous populations of coho, chinook, and sockeye salmon are now extinct, and many others are considered to be at risk of extinction (Nehlsen et al., 1991). Many other stocks show significant decline in run sizes (Konkel and McIntyre, 1987; Nicholas and Hankin, 1988).

2.1. Habitat Effects

Loss of spawning and rearing habitat of salmon has occurred throughout this region. Perhaps some of the most dramatic examples of point impacts, other than dams along the Columbia River, have occured in the Fraser River drainage, British Columbia. In 1913-14, large rock slides in the Fraser Canyon impeded migration to the upper Fraser River (Thompson, 1945; Ricker, 1987). Although the block has been removed and fish passages built, re-establishment of some species, notably pink salmon, has been slow and salmon populations have not returned to previous levels (Ricker, 1989). A logging dam on the upper Adams River, built in 1907, precluded migration of summer migrating sockeye and resulted in the extinction of this stock and the loss of production from potentially two million spawners (Williams, 1987). In the lower Fraser River Valley, urbanization and agricultural development have led to significant losses of wetland and small stream habitats. For example, 80% of the Fraser River delta wetlands have been converted to other uses, primarily agriculture, and the annually flooded area reduced by about 70 % (Environment Canada, 1986).

2.2. Management Effects

The problems facing salmon populations located near human population centers involve over exploitation and loss of habitat to urbanization and pollution. Over exploitation often results when multiple users compete for a common resource. Extensive catch allocation conflicts, competitive sequential fisheries along migration routes, and a shift from biological management to political management have all contributed to frequent and persistent over exploitation.

Further, over exploitation and habitat loss have compound effects on the sustainability of salmon populations. Sustainable exploitation is less if habitat loss reduces the rate of adult production per spawner. Over harvest and habitat loss each can cause the decline or loss of stocks. However, their combined effects exacerbate the problems and increase the rates of decline and loss. Moreover, the combined problems are usually very complex, and their solutions may be difficult to reach and may arrive slowly. Examples can be observed for many salmon stocks in the Columbia River system. Following construction of numerous dams on the Columbia and Snake River drainages, nearly every upstream salmon stock has declined, many seriously enough that numerous populations of chum, sockeye, coho, and chinook are extinct, and many others are considered at high or moderate risk of extinction (Howell et al., 1985; Nehlsen et al., 1991). Although the dams have destroyed large areas of spawning and rearing habitat and have severely impeded migration, some of the decline can be attributed to allocation/management problems. For example, some chinook populations run a gauntlet of harvest starting in southeast Alaska troll and sport fisheries and continuing along the British

Columbia and Washington coasts and within the Columbia River. In addition, although Native Americans are entitled to one-half of the returning salmon and steelhead, often more than one-half have been intercepted by non-native and native fishermen before they return to the Columbia drainage.

2.3. Aquacultural Effects

Aquaculture per se need not be detrimental to conserving salmon genetic resources, but inappropriate siting or uses of hatcheries has been. Hatcheries can have two effects on nearby wild populations. The first is that genes from the hatchery stock may introgress into nearby wild populations carried by straying adults. This may lead directly to disruption of allele frequencies and the adaptedness of the wild populations. The introgression of one hatchery stock into several local wild populations may also result in the loss of interpopulational diversity, an important component of long-term survival of a species. The second problem is the apparently strong tendency for an established hatchery stock to be distributed broadly, rather than to establish new hatchery stocks from local wild fish (Simon, 1972).

This homogenization is a strong threat to interpopulational diversity but has been common, especially in the states of Oregon and Washington. Coho salmon in Oregon and Washington have been impacted by over exploitation and subsequent hatchery production and practices. For example, in the lower Columbia River, large hatchery programs have replaced wild production, and in fact, wild coho stocks appear to be being managed out of existence (Cramer, in press). Historically, fish were transferred freely from one hatchery to another. The problem has been recognized in some areas; for example, the Washington Department of Fisheries is presently attempting to rectify a problem in the Puget Sound in which hatchery stocks of chum salmon were moved to non-native areas (S. R. Phelps, Washington Department of Fisheries, Olympia, Wash., personal communication).

An indirect effect, usually by large-scale enhancement programs, has been the continued depression of wild salmon abundance and risk of genetic change or losses by genetic drift or inbreeding. Hatcheries have frequently been established to increase catch above that of the natural spawning populations. Ironically, a common effect of hatcheries has been to accelerate over exploitation of wild populations because fisheries harvest the abundant hatchery stock which is frequently mixed with the less productive wild stock. These effects have been particularly pronounced in the Puget Sound and in the Strait of Georgia.

3. Southwestern Range

In Asia, human population pressure also threatens salmonid genetic resources. In the Primoria region of the USSR, salmon stocks are severely depleted. Mining and logging operations have damaged many salmon streams; but perhaps the most notable example of the pollution problem in this region is the Amur River. The Amur is an enormous river that separates China and USSR before turning north and emptying into the Sea of Okhostsk. Historic returns of salmon to this great river were large and diverse (Sano, 1966, 1967). Human and industrial waste as well as over harvest by fishermen of several countries have greatly reduced the salmon resources in the once productive Amur River.

In Japan, there are no longer any vigorous wild populations of chum, pink, or anadromous sockeye salmon. Coho and chinook salmon were never abundant in Japan; however, viable wild populations of masou (cherry) salmon (*O. masou*) do exist throughout Japan. Japan's huge chum salmon production now results entirely from its hatchery programs. In recent decades, both hatchery practice and fishing patterns have combined to change demonstratively Japan's chum salmon stocks (Kaeriyama, 1991). In addition, the large releases may be showing density dependent effects that include increased age but decreased size at maturity (Hayashizaki and Ida, 1991).

4. Northern Range

Toward the center of the range of salmon (northern British Columbia, Alaska, Kamchatka, and the northern Okhostsk Sea), the primary threats to wild salmon are over exploitation and aquaculture.

Mixed stock fishery management is one of the major problems. Because the commercial value of salmon is higher before they mature, generally further away from the spawning grounds, there is a concerted effort to catch them before they segregate into local spawning stocks. As a result, it is difficult to ensure that each spawning population achieves an optimal escapement. The mixed stock fishery problem is especially acute in the southeast Alaskan and northern British Columbian troll fisheries where south-migrating chinook salmon from Oregon, Washington, Idaho, and southern British Columbia are intercepted in addition to fish from Alaska and northern British Columbia.

Fishery harvest strategies that attempt to assure adequate escapements can also create problems. A management procedure that is still in use for some Alaskan pink salmon fisheries allows an escapement of early returning fish and and increased harvest of the later returning fish. Because there is a strong genetic component in timing, this strategy selects for early returning fish (Smoker et al., in press; Gharrett et al., in press; Gharrett and Smoker, this volume). It is likely that pink salmon run timing has already been modified by historic fishing pressures (Alexandersdottir, 1987). Ricker suggests that gear selectivity has produced directional changes in the size of fish and that fishing pressure has selected for earlier returning, smaller fish (Ricker et al., 1978; Ricker, 1981). In contrast, some nonprofit hatcheries sell the earlier part of their returns to recover operating costs. These fish are bright and have a higher value than the fish that return later and which are used for broodstock. This strategy selects for later returning fish; either strategy can reduce genetic diversity and potential productivity of the run.

Incidental by-catch is another management problem. This can be exacerbated by hatchery production. For example, the size of the sockeye run in Prince William Sound (PWS) has declined substantially since the development of a large (600 million released fry) pink and chum salmon hatchery program in PWS. The sockeye are caught by the seine fleet incidentally with the pink and chum salmon they are targeting. It is likely that the large hatchery release of pink salmon in PWS have produced the hatchery "homogenizing" and over harvest effects on local wild stocks.

Another practice that hatcheries are introducing is the remote release of hatchery produced fish. If applied wisely, remote releases may divert fishing pressure from sensitive wild stocks. Unfortunately, outplanting is ordinarily done to make more fish available to particular fishing gear groups and may result in imposing transplanted hatchery fish on local wild populations. In such a situation, the hatchery fish have no place to return to spawn other than the steams in which the wild fish spawn which can lead to introgression.

Fishing pressure is also the primary threat to northern range Asian stocks and in some places, such as southwestern Kamchatka, has reduced populations substantially. Industrial pollution, and mining in particular, have also caused some point source impacts.

5. Conclusions

As a whole, Pacific salmon are relatively healthy and large numbers of wild populations exist over a broad geographic area only somewhat smaller than their historic range. In fact, the trend in the number of salmon escaping fisheries to spawn has actually shown an increase for many northern stocks (Konkel and McIntyre, 1987). Some recent pink and sockeye salmon runs have exceeded recorded historic levels.

In spite of the vigor of stocks at the center of the range, there is no question that many populations in the southern portions of the range are threatened. Chinook salmon, in particular, have been extensively impacted. Chinook salmon have the largest body size of these species and have been highly valued in commercial, sport, native, and subsistence fisheries. Because of their large size, less spawning habitat is available in streams and rivers for chinook than for the other species and their juveniles usually require freshwater habitat. Throughout much of their southeastern range, most of the naturally spawning populations are associated with hatchery production to mitigate habitat impacts, to increase harvest, or both. Throughout the rest of their range, they are quite sensitive to overfishing and they are targeted heavily by commercial, sport, and subsistence fisheries.

The coho salmon is another species that deserves close attention. Coho salmon juveniles generally reside in small tributaries which are among the first to be sacrificed to urbanization and other land uses. These small tributaries are also very sensitive to environmental changes, such as temperature changes, decreased water quality, and water discharge fluctuations that result in areas that have been logged. Some aquacultural practices also pose a threat to the genetic resources of North American coho populations.

Present aquacultural practices and hatchery program management also threaten many wild populations. Goodman (1990) suggests the concerns are real enough to merit federal regulation of U.S. hatcheries to alleviate some of the problems and conflicts. Also, there is a real possibility that some enhancement programs may be approaching the carrying capacity of the marine environment (Hayashizaki and Ida, in press; Kaeriyama, in press). If this is the case, questions such as 'how much enhancement is wise?' and 'who should do it?' must be addressed.

The most important step that can be taken to ensure the genetic "health" of these species is to educate fisheries scientists and managers about the importance of maintaining genetic diversity both within and among populations. These professional resource managers should in turn inform the interested public that the fundamental requirement for the long-term survival of a species is the long-term survival of the diverse populations of which the species is comprised. Further, they should inform the users that conservation of this diversity is probably not some academic or "green" desire, but the production basis and future of their resource. Also, they must communicate the value in preserving small populations; small but independent relict populations are valuable genetic reserves for restoration or preservation of the genetic diversity in a species when the relict populations are considered as a whole.

References

Alexandersdottir, M. 1987. *Life history of pink salmon (*Oncorhynchus gorbuscha*) and implications for management in southeastern Alaska*, Ph.D. thesis, University of Washington, Seattle, WA.

Cramer, S. In press. Dynamics of the decline in wild coho (*Oncorhynchus kisutch*) populations in the lower Columbia River, *Canadian Journal of Fisheries and Aquatic Science Special Publication.*

Environment Canada. 1986. Fact sheet on wetlands in Canada: a valuable resource. Lands Directorate, Environment Canada, Ottawa (Cat. No. En 73-6/86-4E).

Gharrett, A.J., S. Lane, A.J. McGregor, and S.G. Taylor. In press. Use of a genetic marker to examine genetic interaction among subpopulations of pink salmon (*Oncorhynchus gorbuscha*). *Canadian Journal of Fisheries and Aquatic Science Special Publication.*

Gharrett, A.J., and W.W. Smoker. 1992. Genetic components in life history traits contribute to population structure, in: "Genetic Conservation of Salmonid Fishes," J.G. Cloud, ed., Plenum Press, New York.

Goodman, M.L. 1990. Preserving the genetic diversity of salmonid stocks: a call for federal regulations of hatchery programs. *Environmental Law* 20:111-116.

Hayashizaki, K., and H. Ida. In press. Size decrease of chum salmon, *Oncorhynchus keta*, in Tohuku districts, Japan. *Canadian Journal of Fisheries and Aquatic Science Special Publication.*

Howell, P., K. Jones, D. Scarnecchia, L. LaVoy, W. Kendra, and D. Ortman. 1985. *Stock assessment of Columbia River anadromous salmonids*, vol. I, *Chinook, coho, chum, and sockeye stock summaries*. Final Report to U.S. Department of Energy, Bonneville Power Administration, Division of Fish and Wildlife Contract No. DE-AI79-84BP12737, project No. 83-335.

Kaeriyama, M. In press. Production trends of salmon enhancement in Japan. *Canadian Journal of Fisheries and Aquatic Science Special Publication.*

Konkel, G.W., and J.D. McIntyre. 1987. Trends in spawning populations of Pacific anadromous salmonids. *Fish and Wildlife Technical Report 9*, United States Department of the Interior Fish and Wildlife Service, Washington, D.C.

Nehlsen, W., J.E. Williams, and J.A. Lichatowich. 1991. Pacific salmon at the crossroads: stocks at risk from California, Oregon, Idaho, and Washington. *Fisheries* 16:4-21.

Nicholas, J.W., and D.G. Hankin. 1988. Chinook salmon populations in Oregon coastal river basins: descriptions of life histories and assessment of recent trends in run strengths. *Informational Reports*, Oregon Department of Fish and Wildlife, Fish Division, Number 88-1.

Ricker, W.E., H.T. Bilton, and K.V. Aro. 1978. Causes of the decrease in size of pink salmon (*Oncorhynchus gorbuscha*). *Fisheries Marine Service Technical Report* 820: 93 pp.

Ricker, W.E. 1981. Changes in the average size and average age of Pacific salmon. *Canadian Journal of Fisheries and Aquatic Sciences* 38:1636-1656.

Ricker, W.E. 1987. Effects of the fishery and obstacles to migration on the abundance of Fraser River sockeye salmon (*Oncorhynchus nerka*). *Canadian Technical Report of Fisheries and Aquatic Sciences* 1522:75 pp.

Ricker, W.E. 1989. History and present state of the odd-year pink salmon runs of the Fraser River region. *Canadian Technical Report of Fisheries and Aquatic Sciences* 1702:37 pp.

Simon, R.C. 1972. Gene frequency and the stock problem, in: "The stock problem in Pacific salmon" (R. C. Simon and P. A. Larkin, eds.), pp. 161-169, H. R. MacMillan lectures in fisheries, University of British Columbia, Institute of Fisheries, Vancouver, B.C., Canada.

Sano, S. 1966. Salmon of the North Pacific Ocean - Part III. A. review of the life history of North Pacific salmon. 3. Chum salmon in the Far East, International North Pacific Fisheries Commission. *Bulletin* 18:41-57.

Sano, S. 1967. Salmon of the North Pacific Ocean - Part IV. Spawning populations of North Pacific salmon. 3. Chum salmon in the Far East, International North Pacific Fisheries Commission. *Bulletin* 23:23-41.

Smoker, W.W., A.J. Gharrett, and M.S. Stekoll. In press. Genetic variation of timing of anadromous migration within a spawning season in a population of pink salmon. *Canadian Journal of Fisheries and Aquatic Science Special Publication.*

Thompson, W.F. 1945. Effect of the obstruction at Hell's Gate on the sockeye salmon of the Fraser River. *Bulletin of the International Pacific Salmon Fisheries Commission*, No. 1.

Williams, I.V. 1987. Attempts to re-establish sockeye salmon (*Oncorhynchus nerka*) populations in the upper Adams River, British Columbia, 1949-1984, in: "Sockeye salmon population biology and future management" H. D. Smith, L. Margolis, and C. C. Wood, eds., pp. 235-242, Canadian Special Publications in Fisheries and Aquatic Sciences 96.

Status of the Atlantic Salmon, *Salmo salar* L., Its Distribution and the Threats to Natural Populations

D. THOMPSON

1. Introduction

This short review presents the findings of a discussion group of scientists from North America and Europe which took place at the recent NATO Symposium "The Genetic Conservation of Salmonid Fishes" held at Moscow (Idaho) and Pullman (Washington) in the USA. It cannot and does not attempt to describe in detail the threats to individual populations but tries to give a brief overview of the status of Atlantic salmon stocks and the methods available for conserving the genetic integrity of natural populations.

The present geographical range of the Atlantic salmon extends from the northeast of the North American continent, where it is found as far south as New England, across the North Atlantic, including Greenland and Iceland, to Europe from northern Norway to Portugal, including the Baltic. There are hundreds, if not thousands, of breeding populations distributed in rivers throughout this range. Many populations are very small. There are also land-locked populations in Newfoundland, Labrador, Norway, Sweden, Finland and the USSR.

Within this widespread range many natural populations have already been lost due to water pollution by industrial development, the modification of rivers by hydroelectric schemes and the acidification of rivers by atmospheric pollution.

In many areas hatchery-supported mitigation programmes are being carried out to attempt to counteract declines in natural populations.

Atlantic salmon have also been introduced, with varying degrees of success, into a number of countries including New Zealand, Australia, Argentina, the United States, and the Kerguelen Islands.

A recent significant event has been the rapid development of the commercial salmon farming industry which intensively rears large numbers of fish in floating sea cages. This industry, developed in Norway, now extends to Sweden, Finland, the USSR, Iceland, the United Kingdom, Ireland, the Faroes, France, Canada, the USA, Chile and Tasmania. The current annual production is ca. 200,000 metric tonnes of fish, with the majority of this in Norway and Scotland.

Directorate of Fisheries Research, MAFF, Fisheries Laboratory, Pakefield Road, Lowestoft, Suffolk NR33 OHT, United Kingdom

Ranching, which exploits the migratory behaviour of Atlantic salmon and relies upon the release of large numbers of hatchery-reared juvenile fish, takes place in Iceland, the Faroes, Norway, Sweden, Denmark, Finland and the USSR.

Atlantic salmon stocks are exploited in three major ways:

1. *High seas interception fisheries.* These take place at western Greenland, where there is a drift net fishery catching fish from both North America and Europe and in the Faroese area, where there is a long-line fishery exploiting European fish. In 1990 the catches were 227 tonnes and 312 tonnes respectively and have declined from historic maxima of ca. 2,700 tonnes (1971) and ca. 1000 tonnes (1981) in two fisheries, respectively, partly as a result of quota controls.
2. *Coastal or home-water net fisheries.* Most of the countries around the North Atlantic which have salmon rivers have some form of coastal or estuarine net fishery. Exploitation rates vary quite considerably between fisheries but in some may be as high as 90% on some stock components.
3. *Rod and line (angling) fisheries within rivers.* Once again all the countries with salmon rivers have these fisheries. Frequently, catches from individual rivers are very low which reflect the small size of many local populations or low exploitation rates.

In an attempt to prevent overexploitation, all these forms of fishery are subjected to regulations controlling fishing: methods, effort, and, in some areas, some form of either international or local seasonal and/or quota controls.

Like other migratory salmonids, the Atlantic salmon has been severely affected by other human activities in addition to fishing. In rivers, environmental degradation has taken place due to water pollution caused by industrial, agricultural or domestic practices. Hydroelectric schemes have interfered with many watercourses, denying access to spawning grounds and affecting flow rates in rivers. Flow patterns are also affected by the abstraction of water for domestic, agricultural and industrial purposes. The reductions in flow rates not only affect the movements of adult and juvenile fish within rivers but also allow river beds to silt up, destroying suitable spawning areas. All of these changes have taken place throughout the species range.

Over the last two decades, many studies conducted throughout the geographical range have shown evidence of a detailed and highly organised population structure in Atlantic salmon. Electrophoretic studies of genetically controlled enzyme and protein polymorphisms and recent mtDNA studies have shown differences in populations: (a) between continents; (b) between river systems within continents and; (c) between tributaries within river systems. Cytological and morphological variations between populations have also been reported.

Whilst the cause, function, and importance of many of these variations is unknown, there is evidence of the genetic basis of differences in growth rate, smolting rate, maturation rate, migration timing, and temperature related processes such as food conversion efficiency and embryonic development rate. Fecundity, egg size, and resistance to disease can probably be added to this list.

Although the complex populations structure of Atlantic salmon, with detailed differences which may be of evolutionary, behavioural or adaptive significance, has been influenced by both natural and man-made environmental changes for decades, the effects of fishing and historical attempts at restocking, it is only recently that new threats have emerged which are capable of major changes in the size and genetical structure of populations. These stem from the rapid expansion of the intensive cage-rearing salmon farming industry and from hatchery-generated enhancement and reintroduction programmes.

The threats may be divided into genetical and ecological risks to Atlantic salmon populations.

2. Genetical Risks

The escape of large numbers of cage-reared farmed salmon into coastal waters may swamp local populations. Whilst this may not be perceived to be an immediate problem if these fish increase local catches, although any increase in fishing effort and exploitation could be detrimental to the native wild fish, it is the ancestry of the escaped fish that causes concern. For example, Norwegian origin broodstock are used to produce fish for rearing in cages off the coasts of Ireland and Scotland. Irrespective of origin of broodstock, deliberate or accidental selective breeding may cause a loss of genetic variation within a hatchery stock.

The escape of large numbers of non-native fish which may interbreed with local populations may interfere with the genetically controlled, local adaptations of native fish. Similar risks may be present in hatchery supported enhancement, restocking and ranching programmes.

Although programmes to enhance salmon populations by the introduction into rivers of eggs and juvenile fish (usually fry) have been taking place for over 100 years these have generally been of limited effectiveness. However, the salmon farming industry can make available large numbers of juvenile salmon which may be surplus to their production requirements. It is tempting for fishery managers to introduce these fish into rivers in attempts to increase population size given the evidence of declining catches in rod and line fisheries. The threats to the genetical structure of native populations, which may represent high levels of local adaptation, are similar to those posed by the escapes from fish farms.

Even when hatchery-based enhancement programmes use locally caught broodstock, poor husbandry practices can cause loss of genetic variation by inbreeding. Such a loss must also be avoided when hatchery-reared fish are used to reintroduce salmon into traditional salmon rivers which suffered industrial degradation but have been restored by environmental improvement.

3. Ecological Risks

Diseases and parasites may be transmitted by the movement of eggs and fish between areas. For example, large mortalities are occurring in river stocks in Norway following the introduction of *Gyrodactylus salaris,* a trematode parasite, possibly transmitted by fish moved from the Baltic; furunculosis, a disease common in intensive fish farming, has been found for the first time in salmon rivers in central Norway.

4. Conclusion

In conclusion, there are several guidelines which should be considered when formulating policies to conserve the size, distribution and structure of salmon populations.

1. Exploitation by fishing should be controlled by management regulations which take into account catch levels and rates, type and selectivity of gear, fishing seasons, and geographical limits.
2. Environmental improvements to rivers must continue, not only to improve water quality and flow patterns, but to maintain adequate spawning and nursery areas and to improve access to allow migratory movements of fish within rivers.

3. Enhancement programmes should take into account any local adaptations displayed by native fish be they behavioural, morphological, physiological, or genetical. Introductions should not destroy local population structures.
4. Restocking into "new" areas should be with fish from a similar local environment with high levels of genetic variation to provide for a fit and viable population capable of adaptation to any local selection pressures.
5. Where local populations are in danger of extinction or are threatened by disappearance, amongst a mass of escaped farmed salmon genetical conservation may be expedient. Cryopreserved sperm banks or living gene banks of broodstock may be needed in times of crisis. The long-term holding of live gene banks may be both difficult and costly and may only be considered as a short-term emergency measure. Cryopreserved gametes may be of longer-term value but are, in reality, only temporary measures whilst steps are being taken to allow the restoration of the population to its native environment. The inability to cryopreserve eggs limits the effectiveness of this technique as donor females from other populations have to be used.
6. The possibilities and probabilities of minimising the genetical threat to national populations caused by escapes from fish farms should be considered, e.g., is the farming of all-female triploid fish a practical and/or an acceptable option?

This report is not intended to be a definitive account of the status, problems and possible remedies associated with Atlantic salmon populations throughout the range of the species. Many of the ideas and policies mentioned will be the subject of more detailed presentations during the symposium. It does, however, represent a concensus of opinion from participants from eleven countries represented at the Atlantic salmon discussion group, held during the NATO Symposium on "Genetic Conservation of Salmonid Fishes."

Selective Breeding and Domestication

Harold L. Kincaid

A discussion group identified a series of hatchery practices that have the potential to induce selection and thereby to modify the gene pool of wild populations during hatchery rearing. The initial step in our discussion of selective breeding and domestication was to define these terms. Selective Breeding was defined as the intentional selection by man to modify the gene pool of a brood stock or population, and domestication was defined as the consequence of all selection pressures, both selective breeding and natural selection imposed by the environment, that modify the gene pool of a population for life in domestic environments.

1. Types of Selection in Hatchery and Natural Environments.

The hatchery environment applies two different types of selection pressure on fish populations. First, there is the pressure to remove unfit genotypes, i.e. individuals not adapted to the hatchery environment. This type of selection is seen in the mortality that occurs at hatch, first feeding, following disease epizootics, following treatment with therapeutic chemicals, and during periods of poor water quality (low oxygen). The second type of selection receives less attention but can be equally powerful to modify the gene pool. It is the lack of selection that results in the survival of genotypes in the hatchery that would have been lost in the wild environment. This lack of selection contributes to increased genetic variability which undoubtedly accelerates the domestication process. It also contributes to reduced survival of hatchery fish after stocking because of the breakup of adapted genotypes that have evolved in natural environments. Regrettably, while we know this type of selection occurs during hatchery rearing, very little is known about how to measure its effect or how to control its occurrence.

2. Hatchery Practices Can Modify the "natural" Gene Pool.

Some of the hatchery practices that contribute to gene pool changes in the domestic environment were discussed.

- Mate selection. Studies have shown that mate selection occurs in some natural populations. In contrast, the most common procedure used in hatchery operations is random pairing of mature individuals. Survival or reproductive advantages from mate selection in nature are lost in the hatchery population.

National Fishery and Development Laboratory, U.S. Fish and Wildlife Service, Wellsboro, PA 16901, U.S.A.

Genetic Conservation of Salmonid Fishes, Edited by J.G. Cloud and G.H. Thorgaard, Plenum Press, New York, 1993

- Partial spawning of total run. A common practice in hatchery spawning of natural runs is to begin the egg collection effort early in the spawning run and to terminate the operation when egg production goals are met. This results in a selection pressure toward earlier returning fish in future runs. This change in the spawning run is unlikely to be advantageous to the population in the long term.
- Exclusion of jacks, grilse, and precocious parr. Frequently, this practice is justified on the basis that it is selection for the larger multiple sea year fish preferred by fishermen. The result of this practice is exclusion of these "young males" which can change the natural spawning structure.
- Exclusion of individuals with "undesired phenotypes." Fish, which are otherwise capable of successful spawning, are sometimes culled as brood stock based on body conformation or physical deformities. Since the culturist cannot generally evaluate the individuals genotype based on the physical appearance, the practice is usually ineffective and serves only to reduce the brood stock effective population size.
- Pooling gametes (male and female). This practice can be deceptive in that differential fertility of eggs and motility of milt can significantly reduce the actual effective population number below that expected based on number of male and female parents contributing to the population.

3. Recommendations

The following recommendations were proposed to reduce the effects of selection and domestication commonly associated with production hatchery situations.

- Fisheries management agencies should develop genetics training programs for fishery biologists and hatchery managers to familiarize them with basic genetic principles and the potential long term benefits that can accrue through the application of genetic principles throughout the cultural period.
- Fishery management agencies should review all cultural practices (spawning, incubation, disease treatment, fish transport, etc.) to identify those that have a potential to apply selective pressures during production. Once identified, steps can be taken to modify those practices that exert "significant selective pressures."
- Fishery management agencies should take the lead in developing breeding guidelines and genetically sound cultural procedures for managing fish populations in the hatchery environment. This would include guidelines for handling spawning fish, loading densities, disease treatments, water quality limits, etc.
- Fishery managers and geneticists must work together to develop a permanent national system for cryopreservation of gametes to protect the genetic diversity of threatened and endangered fish populations. More research to develop procedures for cryopreservation of fish eggs and embryos must be encouraged.
- Progeny from wild fish intended for supplementary stocking of natural fisheries should be retained in the hatchery for the minimum time period required to meet production goals.
- Domestic brood stocks should not be developed for supplementary stocking of natural fisheries, if natural brood stocks are available and capable of meeting annual egg production goals.
- When a natural brood stock is reduced and inadequate to meet egg production needs, a captive brood stock should be established and used as the primary egg

source for production lots. Plans must be developed to systematically reintroduce gametes or fish from the natural brood stock to the captive brood stock on a regular basis.

- Plans for cross breeding fish populations need to be developed to increase genetic variability and restore population vitality in situations where populations are reduced in number and show evidence of inbreeding.

Index